W0227844

Brain Injury and Recovery

Theoretical and Controversial Issues

Brain Injury and Recovery

Theoretical and Controversial Issues

Edited by
STANLEY FINGER
Washington University
St. Louis, Missouri

T. E. LEVERE
North Carolina State University
Raleigh, North Carolina

C. ROBERT ALMLI
Washington University School of Medicine
St. Louis, Missouri

and
DONALD G. STEIN
Rutgers University
Newark, New Jersey

Plenum Press • New York and London

Library of Congress Cataloging in Publication Data

Brain injury and recovery.

  Includes bibliographies and index.
  1. Brain damage — Patients — Rehabilitation. I. Finger, Stanley. [DNLM: 1.
Brain Injuries — rehabilitation. WL 354 B8137]
RC387.5.B73 1988   616.8   88-12593

ISBN-13: 978-1-4612-8256-3   e-ISBN-13: 978-1-4613-0941-3
DOI: 10.1007/978-1-4613-0941-3

Cover: "Un Infirme," an etching by Jacques Callot
(1592–1635) of a person with left hemiplegia.

© 1988 Plenum Press, New York

Softcover reprint of the hardcover 1st edition 1988

A Division of Plenum Publishing Corporation
233 Spring Street, New York, N.Y. 10013

All rights reserved

No part of this book may be reproduced, stored in a retrieval system, or transmitted
in any form or by any means, electronic, mechanical, photocopying, microfilming,
recording, or otherwise, without written permission from the Publisher

Contributors

C. ROBERT ALMLI • Programs in Occupational Therapy and Neural Sciences, Departments of Anatomy and Neurobiology, Preventive Medicine, and Psychology, Washington University School of Medicine, St. Louis, Missouri 63110

PAUL BACH-Y-RITA • Department of Rehabilitation Medicine, University of Wisconsin, Madison, Wisconsin 53792

PAUL D. COLEMAN • Department of Neurobiology and Anatomy, University of Rochester Medical Center, Rochester, New York 14642

ROBERT W. DYKES • Departments of Physiology, Surgery, and Neurology and Neurosurgery, McGill University, Montreal, Quebec H3A 1A1, Canada

SVEN O. E. EBBESSON • Institute of Marine Science, University of Alaska–Fairbanks, Fairbanks, Alaska 99775-1080

STANLEY FINGER • Psychology Department and Neural Sciences Program, Washington University, St. Louis, Missouri 63130

DOROTHY G. FLOOD • Department of Neurology, University of Rochester Medical Center, Rochester, New York 14642

GABRIEL P. FROMMER • Department of Psychology, Indiana University, Bloomington, Indiana 47405

FRED H. GAGE • Department of Neurosciences, University of California, at San Diego, La Jolla, California 92093

ROBERT B. GLASSMAN • Department of Psychology, Lake Forest College, Lake Forest, Illinois 60045

SUSAN GRAY-SILVA • Behavioral Neuropsychology Laboratory, Department of Psychology, North Carolina State University, Raleigh, North Carolina 27695-7801

SAMUEL H. GREENBLATT • Department of Neurological Surgery, Medical College of Ohio, Toledo, Ohio 43699

ROBERT L. ISAACSON • Department of Psychology and Center for Neurobehavioral Science, University Center at Binghamton, Binghamton, New York 13901

ANDREW KERTESZ • Department of Clinical Neurological Sciences, Research Institute, St. Joseph's Hospital, University of Western Ontario, London, Ontario N6A 4V2, Canada

BRYAN KOLB • Department of Psychology, University of Lethbridge, Lethbridge, Alberta T1K 3M4, Canada

SCOTT LAURENCE • Behavioral Neuropsychology Laboratory, Department of Psychology, Clark University, Worcester, Massachusetts 01610

N. DAVIS LEVERE • Dorothea Dix Hospital, Raleigh, North Carolina 27611

T. E. LEVERE • The Brain Research Laboratory, Department of Psychology, North Carolina State University, Raleigh, North Carolina 27695-7801

RAJU METHERATE • Departments of Physiology, Surgery, and Neurology and Neurosurgery, McGill University, Montreal, Quebec H3A 1A1, Canada

DONALD R. MEYER • Department of Psychology, The Ohio State University, Columbus, Ohio 43210.   *Present address:* 476 Overbrook Drive, Columbus, Ohio 43214

ULF NORRSELL • Departments of Physiology and Neurology, University of Göteborg, S-400 33 Göteborg, Sweden

JERROLD S. PETROFSKY • Department of Psychiatry, University of California at Irvine, Irvine, California 92717

GEORGE P. PRIGATANO • Section of Neuropsychology, Barrow Neurological Institute, Phoenix, Arizona 85013

MARY D. SLAVIN • Behavioral Neuropsychology Laboratory, Department of Psychology, Clark University, Worcester, Massachusetts 01610

AARON SMITH • University of Michigan NHR Project, Ann Arbor, Michigan 48103

DONALD G. STEIN • Dean of the Graduate School and Associate Provost for Research, Rutgers University, Newark, New Jersey 07102

SILVIO VARON • Department of Biology, University of California at San Diego, La Jolla, California 92093

IAN Q. WHISHAW • Department of Psychology, University of Lethbridge, Lethbridge, Alberta T1K 3M4, Canada

Preface

The idea for the present volume grew from discussions that the four of us had among ourselves and with our colleagues at recent scientific meetings. All of us were impressed by the wealth of empirical data that was being generated by investigators interested in brain damage and recovery from both behavioral and biological orientations. Nevertheless, we were concerned about the relative paucity of attempts to evaluate the data provided by new technologies in more than a narrow context or to present new theories or reexamine time-honored ideas in the light of new findings.

We recognized that science is guided by new technologies, by hard data, and by theories and ideas. Yet we were forced to conclude that, although investigators were often anxious to publicize new methods and empirical findings, the same could not be said about broad hypotheses, underlying concepts, or inferences and speculations that extended beyond the empirical data. Not only were many scientists not formally discussing the broad implications of their data, but, when stimulating ideas were presented, they were more likely to be heard in the halls or over a meal than in organized sessions at scientific meetings.

There are probably many explanations for this relative lack of theory. One might be that funding in the brain sciences currently emphasizes the development of new technologies and the measurement of discrete biological events. Another factor is that as graduate training has become progressively more specialized and technique-oriented, the holistic–historical perspective, which is conducive to generating broad theories, has been displaced. Related to this, the consensus today clearly seems to be that the safest path to a successful career is the generation of large numbers of empirical, "hypothesis-free" experiments that are uncontroversial and thus guarantee quick publication. Moreover, since many journals are largely data-oriented, publishing a theoretical paper can be difficult, especially if the concepts are controversial or if the idea is not yet supported by an extensive data base.

Because of these factors, and because the study of recovery from brain damage is still in a relatively early stage of development, the four of us decided that a volume encouraging established scientists to present or evaluate theoretical issues in this field would be an interesting and informative endeavor. We wanted to have an opportunity not only to present new ideas but also to evaluate existing theories as well as the contributions of selected historical figures such as John Hughlings Jackson, Kurt Goldstein, and Margaret Kennard.

The result of this effort is the present volume, which examines both molar and molecular contemporary theories about the effects of brain injuries and processes of recovery. Selected chapters also look at the origins and current status of ideas presented by earlier theorists, and some even question how "recovery" should be defined, and why this field continues to be so controversial.

The production of this book was a very stimulating experience for all of us. It forced us to stand back and think—to go beyond the data at hand, to look at the "big picture," and to ask whether the experiments being conducted were even capable of answering some of the questions being asked. We think that the contributions to this book will likewise entice the reader to think about the issues presented here, the broad implications of his or her own specific scientific pursuits, and the direction of current research on the topic of recovery of function. We hope that this volume will stimulate and promote formal discussions of the issues that face us as we try to understand the dynamics of the nervous system and the various events that follow brain injuries.

Stanley Finger
St. Louis
T. E. LeVere
Raleigh
C. Robert Almli
St. Louis
Donald G. Stein
Newark

Contents

Chapter 5

Kurt Goldstein and Recovery of Function

GABRIEL P. FROMMER AND AARON SMITH

Chapter 6

Assumptions about the Brain and Its Recovery from Damage

ROBERT L. ISAACSON

Chapter 7

Mass Action and Equipotentiality Reconsidered

BRYAN KOLB AND IAN Q. WHISHAW

Chapter 8

Margaret Kennard and Her "Principle" in Historical Perspective

STANLEY FINGER AND C. ROBERT ALMLI

Chapter 9

Infant Brain Injury: The Benefit of Relocation and the Cost of Crowding

N. DAVIS LEVERE, SUSAN GRAY-SILVA, AND T. E. LEVERE

Chapter 10

Arguments against Redundant Brain Structures

ULF NORRSELL

Chapter 11

Another Look at Vicariation

MARY D. SLAVIN, SCOTT LAURENCE, AND DONALD G. STEIN

Chapter 12

Hughlings Jackson's Theory of Localization and Compensation

SAMUEL H. GREENBLATT

Chapter 13

The Parcellation Theory and Alterations in Brain Circuitry after Injury

SVEN O. E. EBBESSON

Chapter 14

Trophic Hypothesis of Neuronal Cell Death and Survival

FRED H. GAGE AND SILVIO VARON

Chapter 15

Sensory Cortical Reorganization following Peripheral Nerve Injury

ROBERT W. DYKES AND RAJU METHERATE

Chapter 16

Is Dendritic Proliferation of Surviving Neurons a Compensatory Response to Loss of Neighbors in the Aging Brain?

PAUL D. COLEMAN AND DOROTHY G. FLOOD

Chapter 17

Practical and Theoretical Issues in the Use of Fetal Brain Tissue Transplants to Promote Recovery from Brain Injury

DONALD G. STEIN

Chapter 18

*Functional Electrical Stimulation and Its Application for the
Rehabilitation of Neurologically Injured Individuals*

JERROLD S. PETROFSKY

Chapter 19

*Recovery of Language Disorders: Homologous Contralateral or
Connected Ipsilateral Compensation?*

ANDREW KERTESZ

Chapter 20

Sensory Substitution and Recovery from "Brain Damage"

PAUL BACH-Y-RITA

Chapter 21

Emotion and Motivation in Recovery and Adaptation after Brain Damage

GEORGE P. PRIGATANO

Chapter 22

Recovery of Function: Sources of Controversy

STANLEY FINGER, T. E. LEVERE, C. ROBERT ALMLI, AND
DONALD G. STEIN

Brain Injury and Recovery

1

Toward a Definition of Recovery of Function

C. ROBERT ALMLI and STANLEY FINGER

1. THE PROBLEM DEFINED

The clinical and research literatures on the effects of brain damage can be very confusing because, all too often, the same word or term is used to define or describe different phenomena. Descriptions of various behavioral outcomes after brain damage lack consistent and precise terminology, and no where is this more apparent than with the use of the term "recovery of function." Four case studies illustrate the problem.

Case 1. A 40-year-old construction worker was hit on the head by a piece of falling metal. The man was rendered unconscious for a few hours, and after he awakened he appeared to be confused. Testing revealed that his memory was grossly impaired for a few days after the injury. Within a week, however, marked improvements were noted, and 1 month later there was no trace of memory impairment. Attentional and cognitive processes also appeared to be normal. The construction worker soon was able to return to his old job.

Case 2. A newborn baby was found to have a tumor on the left side of the brain in the region of Broca's area. The tumor was removed, and the child's progress was followed. The child was late in developing spoken language, but by the time this young girl was of school age, she was speaking fluently. Her teachers found her indistinguishable from her classmates in language functions.

C. ROBERT ALMLI • Programs in Occupational Therapy and Neural Sciences, Departments of Anatomy and Neurobiology, Preventive Medicine, and Psychology, Washington University School of Medicine, St. Louis, Missouri 63110. STANLEY FINGER • Psychology Department and Neural Sciences Program, Washington University, St. Louis, Missouri 63130.

Case 3. Laboratory rats were given bilateral lesions of the lateral hypo-thalamus and monitored for food and water intake. Those animals not given special care suffered from dehydration and starvation and died within a week after surgery. In contrast, those who were force-fed and provided with highly palatable foods were able to maintain themselves by drinking water and eating standard laboratory chow. At this point, the latter animals were examined more carefully, and it was found that they still were below normal for body weight, that they were more finicky in dealing with foods made slightly bitter, and that they were unable to defend manipulations of body fluid and nutrient levels by altering drinking and feeding habits.

Case 4. A soldier was paralyzed on the right side of his body as a result of a gunshot wound to the head. The man, now confined to a wheelchair, was discharged from the army and given extensive rehabilitation training. Although once strongly right-handed, the soldier learned to use his left hand to eat, write, and brush his hair. When he returned to his home town, he obtained a job as a history teacher. Although he was forced to rely on his wife to drive him to and from work, this physically handicapped individual was doing well at his job and was accepted as an active and important member of the community.

These four case studies are presented because they share an important common feature. In each instance the brain-damaged subject showed notable improvement in overcoming at least some of the immediate effects of an injury to the central nervous system. Yet, on closer examination, each case is distinctly different. In the first case, there appeared to be a complete remission of symptoms over time, and in the second case, the symptoms that would have been expected in an adult never emerged in the child. In the third case, there were some residual effects of the brain injury that indicated that not everything was normal. And, in the fourth case, the salient feature was that the lost function remained lost even though the individual showed the stamina, motivation, and resourcefulness to adjust to his losses by using prosthetics and new strategies.

Are all of these cases showing "recovery of function"? Depending on the definition used and the inferences one is willing to make, one, two, or perhaps even all of the cases might be classified as "recovered," even though each case is obviously unique. The different ways in which the term "recovery" has been used have made it difficult to distinguish between it and other phenomena in the literature, phenomena such as "compensation" and "sparing," which we believe should be differentiated from recovery. In addition, the interchangeable use of descriptive words has at times resulted in such high levels of ambiguity and confusion that it has impeded progress in the field. It is in this context that we felt that a set of clear and precise definitions that could be used by both clinicians and experimentalists should be generated, presented, and defended.

In this chapter we show why recovery of function should be narrowly defined and why different definitions should be given to related phenomena. By defining a number of important terms, demonstrating the salient features of each, and showing why recovery should be viewed only as an inference and not as a fact, we hope that we can come one step closer to presenting a set of definitions that can be widely accepted by workers in this field.

2. DEFINITIONS OF RECOVERY OF FUNCTION

Although there is a large literature on "recovery of function," it is important to note that the term itself has been defined in few places and that these definitions have been relatively imprecise. Further, widely divergent meanings have been associated with the term "recovery" by individuals coming from different disciplines.

Perhaps the broadest definitions of recovery come from the fields of rehabilitation and clinical medicine. For example, "good recovery" is defined on the Glasgow Outcome Scale as a "resumption of normal life even though there may be minor neurological and psychological deficits" (Jennett and Bond, 1975, p. 483).

This definition, although possibly useful for governmental, legal, and institutional records, is a poor guide for controlled research emphasizing brain–behavior relationships. This is because a "normal life" can have multiple connotations, even if how that normal life is achieved may not be the most pressing issue from a clinical perspective. As possible outcomes, an individual may be left with less severe deficits that do not diminish further with time, or the individual may be forced to rely on prosthetic devices, or all of the symptoms may in fact disappear. Yet in each case, with a definition this broad, one could say that recovery has taken place.

In 1978, Laurence and Stein, physiological psychologists, discussed how recovery of function could be defined in either of two basic ways. The first simply asks whether the "goal" was achieved. From this perspective, if a brain-damaged laboratory rat learned a discrimination, even if it required many more trials than an unoperated cagemate, one could say that recovery of function was exhibited. One could also argue that the soldier in Case 4 who learned to use his unaffected hand to substitute for his paralyzed hand had recovered if a function such as handwriting reached a certain level of proficiency. Nevertheless, these authors point out that most scientists would prefer to see the term "recovery" used not only when the goals are reached but when they are obtained in ways, or by means, that are similar to those utilized before the injury. When defined in this second way, the term "recovery" would not be applicable to situations in which one hand is used for the other or one sensory system is substituted for another by brain-damaged subjects. Major disagreements about the concept of recovery typically do not revolve around the issue of whether goals are reached but rather are based on how those goals are achieved.

The definition of "recovery" proposed by Laurence and Stein (1978, p.

370) is "a return to normal or near-normal levels of performance following the initially disruptive effects of an injury to the nervous system."

This definition has broader applicability than the one coming from the Glasgow Outcome Scale, but it still could be "tighter," because it is unclear how near to normal performance must become before the use of the term "recovery" can be justified. For example, is any improvement in scores on a task (moving toward normal) adequate reason to assume "recovery"? And if a problem eventually can be solved, but in more days or with more trials than would be required by control subjects, would that be considered "recovery"?

Another conceptualization of recovery comes from Marshall (1985). As a neurobiologist, Marshall (p. 201) suggested that "recovery" should refer to "impairments in behavioral or physiological functions that abate as time since injury increases."

This definition is similar to that of Laurence and Stein in that it does not specify the degree to which deficits must decrease for a determination of recovery. Further, it seems to rest on the questionable assumption that recovery is a direct function of time alone.

One might contrast the Marshall definition with one suggested by Braun (1978) in a chapter examining specific conditions that might affect posttraumatic performance. From a strongly experimental orientation, Braun (p. 178) defined behavioral recovery as taking place "when persistent behavioral deficits are reduced by special training or by pharmacological, surgical, or other independent manipulations."

The Braun definition is interesting in that it downplays spontaneous recovery and excludes those lesion effects that may diminish soon after trauma, such as those that may result from shock to a system or diaschisis. Nevertheless, the idea of "persistent deficits" is also relative: Braun does not specify how long deficits must be present to be called persistent. Although Braun's operational definition still may be useful for some aspects of laboratory animal research, the required independent variable manipulations would make this type of definition too restrictive for human case studies and even for some types of experiments using animal models.

Since these four definitions of recovery of function are representative of the divergent ways the term is currently being used, one can see how problems in nomenclature can contribute to some of the confusion in this field. The Braun (1978) definition would essentially exclude much of what would ordinarily be considered recovery, and the clinical definition (Jennett and Bond, 1975) defines recovery so broadly that it is of limited research value. The remaining definitions by Laurence and Stein (1978) and Marshall (1985) are probably representative of the most common and agreed-on definitions of recovery, but these definitions do not resolve certain ambiguities in part because of their relative, open-ended nature.

These observations show that there is a need for a more uniform vocabulary, and especially for a definition of recovery that has more precision and states more clearly when the term should be used, and also when it should not be used. One step in this direction was made when LeVere (1980, p. 298) proposed that "recovery" is "the postlesion reinstatement of the specific behaviors that were disrupted by the brain injury."

This definition appears to avoid many of the major problems of the definitions discussed above by defining recovery in an absolute way (not using terms like "near normal") and by calling for the reemergence of the specific behaviors that were disrupted (not other behaviors that could achieve the same goals). We believe that these two elements are needed for a good working definition of recovery. However, before proposing our own specific definition of recovery, we must lay the groundwork by contrasting "recovery" with other terms that sometimes have been used interchangeably with "recovery" in the clinical and experimental literatures.

3. RECOVERY OR BEHAVIORAL SPARING?

It is important to distinguish those brain damage cases in which there is a specific deficit that subsides over time or with treatment from those cases in which subjects fail to show the specific deficit. The term "behavioral sparing" should be applied to the latter phenomenon, and it is proposed that this term be defined as "the absence of a specific performance deficit after a brain lesion."

Sparing should be differentiated from recovery because recovery implies that a deficit had been present at one time. The deficit might be one that could have been observed earlier in the case under study, or it could be inferred from the results obtained by testing comparable brain-damaged subjects at earlier times. Nevertheless, differentiating between sparing and recovery can be challenging because a knowledge of the absence of lesion effects throughout the postlesion course is required. This information is not always available for clinical subjects, and many experimental designs with laboratory animals involve delays of weeks or months between surgery and testing.

One problem for which there is not a simple solution is how to classify the performance of a patient with impaired consciousness after a brain injury or a laboratory animal that was given an operation that rendered it "untestable" for a number of days after surgery. Would the term "sparing" be precluded in such a case because the subject would be defective on all tests right after surgery or injury? The problem revolves around the meaningfulness of behavioral testing and measurements so soon after injury. This issue is not so pressing when describing global functions such as consciousness and orientation, on which the decision to test for specific abilities may depend. But in testing for specific abilities, it is important that the subject be "testable," and one must recognize that one can only hypothesize about the status of a specific system in the "gap" between the injury and the time when "meaningful" testing for specific abilities can take place.

The most frequently cited examples of behavioral sparing come from studies in which subjects sustain focal lesions early in life (Almli and Finger, 1984; Finger and Almli, 1984). For instance, Akert *et al.* (1960) claimed that when newborn monkeys given dorsolateral frontal cortex lesions were tested 3–4 months later, they did not exhibit the delayed response deficits that characterized animals operated on in adolescence or at maturity. The second case presented at the beginning of this chapter, involving the child with the tumor of Broca's area, also could be considered an example of sparing. In some studies of early brain damage, however, it is questionable whether the label "behavioral sparing" should apply (Almli, 1978, 1984; Johnson and Almli, 1978). In particular, this relates to situations in which subjects sustaining brain damage early in life are tested at ages at which even normal subjects would not be capable of showing the behavior in question. Under such conditions, there could be a "delayed" deficit that might only appear after the system achieved a certain maturational status (see Finger and Almli, Chapter 8, this volume).

The concept of sparing does not appear to be limited to selected cases involving the developing nervous system. The term is also applicable to some studies on fast- versus slow-growing lesions (staged lesions: Finger, 1978; McMullen and Almli, 1980). If subjects with slow-growing lesions do not exhibit symptomatology at a time when other subjects with matching lesions of rapid onset are showing functional deficits, one might proceed on the assumption that the former were showing behavioral sparing.

Thus, it does not necessarily follow that recovery of function has occurred when behavioral deficits are not found after brain damage. Nevertheless, as noted by Laurence and Stein (1978), many behavioral studies are designed in ways that make it very difficult to determine whether it is sparing, recovery, or some new strategy or action that is guiding performance. This is especially true in studies with infants and in studies with mature organisms when long time intervals intervene between trauma and testing, since the question of what happened in the period before testing took place can only be surmised.

4. RECOVERY OR COMPENSATION?

Recovery of function after brain damage is reported relatively frequently when goal achievement is the major criterion for a recovery determination. But goals might be achieved in many different ways. Normal subjects appear to rely on a hierarchy of cues to achieve ends, such that deprivation of the primary cue can result in a switch to less-preferred but still relevant cues. This general ability to switch cues and adapt to changing situations can also characterize brain-damaged individuals (LeVere, 1980). Thus, a brain-damaged subject may solve a task and superficially appear to have recovered, yet the subject may be solving that task quite differently because he really has not recovered.

The distinction between a new way or procedure to solve a problem and recovery of the previously lost function can be illustrated in a series of experiments conducted by Goldberger on monkeys (1972, 1974). This investigator showed that "recovery" from an abnormal grasp reflex after supplementary motor cortex lesions was actually caused by the use of new behavioral strategies that facilitated grasping and releasing. The goals were achieved, signifying "recovery," but the behavioral program differed considerably from that utilized by normal animals, signifying "no recovery."

New behaviors like these must be differentiated from recovery of function. The terms most frequently encountered in the journals to specify such changes are "compensation" and "behavioral substitution." We suggest using the term "compensation" to describe the switch to different receptors, effectors, and/or new strategies in an attempt to adjust to losses resulting from brain damage.

In some cases it is easy to determine if a goal is being achieved by compensation, such as with the soldier (Case 4) who, after being paralyzed on one-half of his body, learned to write and comb his hair with the previously nonpreferred hand. There are, however, many less obvious instances in which the recovery that was suggested on the basis of a cursory analysis was later recognized as a compensatory response (Finger and Stein, 1982). The split-brain literature provides some excellent examples of this, with both monkeys (Gazzaniga, 1966) and humans (Geschwind, 1974).

The ability to determine whether compensation or recovery has taken place can be exceedingly difficult under some conditions. To appreciate this, one might think of a mute, brain-damaged child who may now use visual imagery to solve simple subtraction problems that previously were handled by rote memory, or a cat with cortical damage who now explores a tactile stimulus by rubbing a bit harder than before. Even a sophisticated observer may not recognize that compensatory charges have occurred in these cases, especially if only days to criterion, errors to criterion, or percentage correct responses were measured.

5. RECOVERY AS ABSOLUTE AND INFERENTIAL

Recovery can be considered in a relative sense if one were describing a process. For example, we could say that a patient is recovering. However, when recovery is defined as a state, as it typically is in the recovery-of-function literature, one can only say that the person has or has not recovered at any point in time. For this reason, an "absolute" or "closed-ended" definition of recovery is needed.

With an absolute definition of recovery, the issue of whether a brain-damaged subject has recovered if he learns a task in 50 trials while control subjects average ten has a clear answer—"no." But what if the person with the lesion takes 11 or 12 trials? Is that recovery? Here, for an initial answer, one can rely on statistics and ask whether a score is significantly different from the control group mean at the 0.05 level of confidence. But although statistical definitions can be useful, they do not take into account individual differences. For example, although a brain-damaged subject's drawing of a bicycle may look satisfactory, the same criteria should not be applied to a truck driver, a salesman, and a professional artist. The premorbid history of the subject should be considered if at all possible.

The use of a closed-ended definition of recovery does not limit its application to any specific level of analysis. The three levels of analysis usually encountered in the recovery literature are (1) the organism overall, (2) the system, and (3) the dimension examined by a particular task. For example, consider a person who showed vision, hearing, and somatosensory impairments after a head injury and whose hearing and somatosensory functions returned to normal. On an organismic level, one would have to say that because of persisting visual deficits the subject could not be considered recovered. In contrast, on a systems level one could say that audition and somesthesis had recovered. If visual testing showed a return to premorbid levels for color vision and visual tracking but no improvement for pattern vision, one could also use the term "recovery" to refer specifically to color perception and visual tracking but not to pattern vision. The important point illustrated by this case is that when the term "recovery" is used, the investigator must be very careful to specify the level of analysis under consideration.

It is always possible to argue that one would find that so-called "recovered" laboratory animals or human patients are not really recovered if more sensitive or prolonged testing were conducted. The idea that one more test might reveal some critical behavioral deficit can never be disproved (see Chapter 22 in this volume). Thus, the term "recovery" should only be used in a theoretical or hypothetical sense: one can merely say that based on the data thus far collected, the subject appears to show recovery of function.

With all of these considerations in mind, we have chosen to define "recovery" in the following way: *Recovery is a theoretical construct that implies a complete regaining of identical functions that were lost or impaired after brain damage.*

The important features of this definition are that it (1) requires an initial deficit, (2) demands that returned behavior be identical to the one that was lost, (3) views recovery as a "yes-or-no" affair, and (4) recognizes that a determination of recovery is only a theoretical statement or an initial working hypothesis.

6. MECHANISMS OF RECOVERY

Armed with this definition of recovery, one may now ask where the best evidence for recovery can be found, and what mechanisms are most likely to account for it. A review of the literature shows clearly that the best evidence for recovery is to be found in the first few weeks after neural insult. Attesting to this, the journals are filled with examples of people losing and regaining speech or memory relatively soon after strokes or accidents. As for mechanisms that have been advanced to account for improved performance soon after insult, those most often mentioned are a lack of neuronal destruction *per se* and transient "shock" effects.

The possibility that there may not have been permanent damage but only a transient event that caused the symptoms should be considered when recovery is witnessed. For example, specific symptoms might follow ischemia, a reduction in the blood supply to a region because of mechanical obstruction. But if the obstruction can be disengaged (mechanically or pharmacologically) before too many cells being fed by the vessel are lost, the remission of symptoms following the "reversal of the lesion" could be dramatic (Goldstein *et al.*, 1970).

Another reason to expect some improvement relatively soon after insult comes from the possibility that although there may in fact be an area of permanent damage, there can also be transient effects in other areas that may disrupt their ability to function normally. This possibility does not deal with whether there should be permanent lesion effects or recovery of those functions supposedly mediated by the area directly affected. The emphasis is only on the secondary, transient effects of the lesion on other areas (e.g., edema, blood flow changes), particularly those near to the site of the injury and those related to the damaged area through interconnections. As factors like these are reversed, the expectation would be that metabolism in these distal areas would improve and that the suppressed functions would reemerge.

In a broad sense, the model might be called "neural shock," but in the hands of different investigators it has taken on more specific meanings. For example, von Monakow, who worked at the turn of the century, and whose name is most closely associated with this type of thinking, used the term "diaschisis" specifically to denote those distal effects that he believed were mediated neurally (see Finger and Stein, 1982; Glassman and Smith, Chapter 4, this volume).

Redundancy is another explanation sometimes given for relatively rapid improvements in performance after brain damage, although the viability of the "spare parts" concept has been questioned (see Glassman and Smith, Chapter 4, this volume; Norrsell, Chapter 10, this volume). But even if one were to accept the premise that a system could be diffuse and redundant for a particular function, redundancy theory by itself would predict that there should be "behavioral sparing," not "recovery of function" following a subtotal lesion. Nevertheless, redundancy could be incorporated into a two-factory theory. For example, there could be an initial loss followed by some degree of behavioral improvement if a subtotal lesion also resulted in diaschisis to disrupt temporarily the functioning of surviving elements in a redundant system.

The question of plausible underlying mechanisms is more challenging when the time period after the lesion extends beyond a few days or weeks, especially if the subject seems alert, attentive, and normal on all but one or a few tests. Although some of the potential recovery mechanisms mentioned previously might extend into this time frame, attention has centered on denervation supersensitivity and, to a greater extent, the issue of CNS reorganization.

Denervation supersensitivity can be thought of as a recovery mechanism either by itself or when combined with other theories such as redundancy or neural shock (Cannon and Rosenblueth, 1949; Finger and Stein, 1982). The idea in this case would be that following a loss of some of their inputs, there would be changes in the receptors of postsynaptic neurons that would make those neurons much more sensitive to the remaining inputs. If the inputs are homogeneous and only partially damaged, the remaining elements initially may not be able to drive the postsynaptic cells, and symptoms of brain damage would appear. But if the receptors became supersensitive and the limited inputs now became capable of driving the postsynaptic cells, these deficits could recede. Of course, if the structure had two very different types of inputs and one were lost, supersensitivity could exaggerate the lesion-induced imbalance and serve to magnify the deficit. Thus, although we might imagine that changes in sensitivity could be beneficial under some circumstances, it is also necessary to consider the opposite outcome (Finger and Almli, 1985).

As for "reorganization," one can say that the term has been used in about as many ways as the word "recovery," and even synonymously and tautologically with recovery. The word originally seemed to imply that one part of the brain could take over the functions of another part after injury, although early investigators rarely speculated on the underlying mechanisms. Today, the emphasis has shifted more toward anatomy and the possibility that new and different types of connections can help to mediate recovery of previously lost functions.

Reorganizational hypotheses were very popular in the last century and were championed by such notables as Broca (1865), with reference to language, and Munk (1877), who described perceptual memory processes in cortically damaged dogs. The liberal use of the term "reorganization" was questioned much more in the middle of this century as it became clearer that the neuronal changes that might underlie the process remained to be identified and when it was realized that the evidence for functional reorganization could be interpreted in a variety of ways. (The case usually rested on making a lesion, observing a deficit and then "recovery," and watching the deficit reappear when a second lesion was made in a previously "neutral" area.) The recognition that the brain is constantly undergoing change, and recent descriptions of neuroanatomic events after injury, served to renew interest in the possibility that neuronal reorganization can mediate recovery of function.

Three types of axonal events are currently being emphasized in the brain-lesion literature. Two, rerouting of growing axons to different locations (Schneider and Jhavari, 1974) and axons not following the program to retract at set times following damage to a system (Land and Lund, 1979), appear to be confined to cases of very early brain damage and, if anything, appear to qualify more as mechanisms for sparing than for recovery. The third type of event, the establishment of new synapses by healthy, remaining axons in vacated receptor sites, is the one most pertinent to recovery as defined here and has been called "collateral sprouting" or "reactive synaptogenesis" (Cotman and Nadler, 1978).

The idea that reactive synaptogenesis takes place primarily to protect the individual from the deleterious effects of brain damage (i.e., represents a "recovery mechanism" *per se*) has been questioned by the present authors in a recent theoretical review (Finger and Almli, 1985). Both experimental results and evolutionary considerations were thought to suggest that sprouting and related events are not present to heal damaged brains, although they may contribute to recovery under some circumstances.

From the perspective of evolution, three important but related points have been raised to challenge the notion that dynamic neuroanatomic events like these now occur because they aid in recovery of function. They are (1) that acute brain damage did not occur very often, (2) that when it did occur, chances of survival were probably quite low, and (3) even if there were survival, the notion that reproduction would follow and that the advantageous genetic material would be passed to future generations would still be suspect.

The second argument against the view that reactive synaptogenesis should be considered a recovery mechanism *per se* was based on the assumption that when these reorganizational events are present, they should be associated with recovery of function, or at least they should not be associated with maladaptive behaviors. An examination of the research literature showed that sprouting has been correlated positively with recovery in only some instances, and that even for these cases there have been problems of replication and questions raised about the meaning of the observed temporal correlations. In addition, in a number of instances sprouting was correlated with neurological dysfunction.

Taken as a whole, the research findings strongly suggested that these events could be better understood in some other context, and the case was made that they are more likely to represent basic "growth" processes (Almli, 1985) stimulated by the availability of receptor sites resulting from the lesions. It was postulated that whether these active neuronal growth events may enhance recovery, be neutral, or be maladaptive could relate to whether they could restore a state of "balance" to the system or whether they would further exaggerate the imbalance. Under natural conditions, where the characteristics of the lesion could not be controlled, one should assume that the case for axonal reorganization mediating recovery would be even weaker than it would be under highly controlled laboratory conditions.

Finger and Almli (1985) emphasize that their major point is not that events such as sprouting and supersensitivity have nothing to do with recovery or behavioral improvement but rather that from a growth perspective, recovery would be an epiphenomenon. They also emphasize that these anatomic events are taking place against a matrix of a damaged brain and that these processes are likely to be out of sequence with the program for normal development and at odds with the blueprint for normal brain organization. This could further explain why the presence of reorganizational events may not guarantee improvement in performance after damage and why axonal growth can have deleterious effects under some conditions.

7. SUMMARY AND CONCLUSIONS

The present chapter was written with the intent of trying to define the word "recovery" in a meaningful way and then examining some of the mechanisms that might account for recovery. It was argued that "recovery" lacks a widely accepted definition in the research and clinical literatures and that a good, workable definition of recovery should demand an initial deficit, that the returned behavior be identical to the one that was temporarily lost (behavioral isomorphism), and that recovery be treated as a "yes-or-no" affair rather than in a relative sense. It was also proposed that recovery can only be regarded as an inference based on the data at hand. These considerations were paramount in our defining recovery as "a theoretical construct that implies a complete regaining of the identical functions that were lost or impaired after brain injury."

Considerable attention was devoted to the fact that it is sometimes difficult to distinguish between recovery and "sparing" or "compensation," terms used in the brain-damage literature to denote the absence of specific deficits and the use of different strategies, receptors, or effectors to solve a problem, respectively. The difficulties involved in the accurate classification of events are largely related to the science of behavior and to the need for more refined observations and behavioral measures over longer time periods. In particular, the demand is greatest for more information about the time course of behavioral change and for knowledge about how goals are reached.

The best case for recovery seems to come from observations made relatively soon after damage is inflicted. In these instances, the recovery could reflect only a "transient" lesion or perhaps "shock effects" that are temporarily suppressing areas at a distance from the lesion site but whose disrupted functions are affecting scores on the tests being administered. The case for recovery, as opposed to compensation, taking place well after the more general effects of a lesion have subsided is more difficult to make. In this context, it was argued that those neuronal reorganizational events that might be thought of as recovery mechanisms *per se* seem more likely to represent basic developmental processes that are triggered by signals of synaptic availability. Although such events (e.g., reactive synaptogenesis) might account for recovery under some circumstances, it was emphasized that growth processes that would serve to exaggerate an imbalance in an already damaged system would be more likely to result in maladaptive behaviors than in recovery of function (see Finger and Almli, 1985).

It is apparent that the term "recovery" may have been used too readily in the past. Nevertheless, this should not be perceived as a reason to diminish attempts to understand recovery of function or to downplay other responses to brain lesions as uninteresting or unimportant. Recovery, sparing, and compensation should all be accepted as extremely relevant phenomena by individuals in the behavioral and neural sciences and by practitioners who must work with brain-damaged patients and their families. All would acknowledge that the most important goals are to understand the cause and nature of the deficit and to get the brain-damaged patient to do the best he can with that which is left.

REFERENCES

Akert, K., Orth, O. S., Harlow, H., and Schiltz, K. A., 1960, Learned behaviors of rhesus monkeys following neonatal bilateral prefrontal lobotomy, *Science* **132**:1944–1945.

Almli, C. R., 1978, The ontogeny of feeding and drinking: Effects of early brain damage, *Neurosci. Biobehav. Rev.* **2**:281–300.

Almli, C. R., 1984, Early brain damage and time-course of behavioral dysfunction: Parallels with neural maturation, in: *Early Brain Damage:* Volume 2, *Neurobiology and Behavior* (S. Finger and C. R. Almli, eds.), Academic Press, New York, pp. 99–116.

Almli, C. R., 1985, Normal sequential behavioral and physiological changes throughout the developmental arc, in: *Neurological Rehabilitation* (D. Umphred, ed.), C. V. Mosby, St. Louis, pp. 41–71.

Almli, C. R., and Finger, S., 1984, *Early Brain Damage:* Volume 1, *Research Orientations and Clinical Observations,* Academic Press, New York.

Braun, J. J., 1978, Time and recovery from brain damage, in: *Recovery from Brain Damage: Research and Theory* (S. Finger, ed.), Plenum Press, New York, pp. 165–197.

Broca, P., 1865, Siege de la faculté de langage articulé dans l'hemisphere gauche du cerveau, *Bull. Soc. Anthropol.* **6**:377–393.

Cannon, W. B., and Rosenblueth, A., 1949, *The Supersensitivity of Denervated Structures,* Macmillan, New York.

Cotman, C. W., and Nadler, J. V., 1978, Reactive synaptogenesis in the hippocampus, in: *Neuronal Plasticity* (C. W. Cotman, ed.), Raven Press, New York, pp. 227–271.

Finger, S., 1978, Lesion momentum and behavior, in: *Recovery from Brain Damage: Research and Theory* (S. Finger, ed.), Plenum Press, New York, pp. 135–164.

Finger, S., and Almli, C. R., 1984, *Early Brain Damage:* Volume 2, *Neurobiology and Behavior,* Academic Press, New York.

Finger, S., and Almli, C. R., 1985, Brain damage and neuroplasticity: Mechanisms of recovery or development? *Brain Res. Rev.* **10**:177–186.

Finger, S., and Stein, D. G., 1982, *Brain Damage and Recovery: Research and Clinical Perspectives*, Academic Press, New York.

Gazzaniga, M. S., 1966, Visuomotor integration in split-brain monkeys with other cerebral lesions, *Exp. Neurol.* **16**:289–298.

Geschwind, N., 1974, Late changes in the nervous system: An overview, in: *Plasticity and Recovery of Function in the Central Nervous System* (D. G. Stein, J. J. Rosen, and N. Butters, eds.), Academic Press, New York, pp. 467–508.

Goldberger, M. E., 1972, Restitution of function in the CNS: The pathological grasp in *Macaca mulatta*, *Exp. Brain Res.* **15**:79–96.

Goldberger, M. E., 1974, Recovery of movement after CNS lesions in monkeys, in: *Plasticity and Recovery of Function in the Central Nervous System* (D. G. Stein, J. J. Rosen, and N. Butters, eds.), Academic Press, New York, pp. 265–337.

Goldstein, S. G., Kleinknecht, R. A., and Gallow, A. E., 1970, Neuropsychological changes associated with carotid endarterectomy, *Cortex* **6**:308–322.

Jennett, B., and Bond, M., 1975, Assessment of outcome after severe brain damage, *Lancet* **1**:480–484.

Johnson, D. A., and Almli, C. R., 1978, Age, brain damage, and performance, in: *Recovery from Brain Damage: Research and Theory* (S. Finger, ed.), Plenum Press, New York, pp. 115–134.

Land, P. W., and Lund, R. D., 1979, Development of the rat's uncrossed retinotectal pathway and its relation to plasticity studies, *Science* **205**:698–700.

Laurence, S., and Stein, D. G., 1978, Recovery after brain damage and the concept of localization of function, in: *Recovery from Brain Damage: Research and Theory* (S. Finger, ed.), Plenum Press, New York, pp. 369–407.

LeVere, T. E., 1980, Recovery of function after brain damage: A theory of the behavioral deficit, *Physiol. Psychol.* **8**:297–308.

Marshall, J. F., 1985, Neural plasticity and recovery of function after brain injury, *Int. Rev. Neurobiol.* **26**:201–247.

McMullen, N. T., and Almli, C. R., 1980, Serial lateral hypothalamic destruction with various interlesion intervals, *Exp. Neurol.* **67**:459–471.

Munk, H., 1877, Zur Physiologie der Grosshirnrinde, *Berl. Klin. Wochenschr.* **14**:505–506.

Schneider, G. E., and Jhavari, S. R., 1974, Neuroanatomical correlates of spared or altered function after brain lesions in the newborn hamster, in: *Plasticity and Recovery of Function in the Central Nervous System* (D. G. Stein, J. J. Rosen, and N. Butters, eds.), Academic Press, New York, pp. 65–110.

2

Neural System Imbalances and the Consequence of Large Brain Injuries

T. E. LeVERE

1. INTRODUCTION

It is reported that while addressing a group of graduating college students, Dr. Seuss related some advice from Uncle Terwillinger concerning the art of eating popovers. Dr. Seuss noted that it was Uncle Terwillinger's suggestion that "to eat these things, you must exercise great care. You may swallow down what is solid, . . . but you must spit out what is air!" Given the range of beliefs concerning the consequences of brain injury, it appears that students of recovery of function may also need to heed Uncle Terwillinger's sage counsel. For example, at one extreme there are those who believe that the organization of the nervous system is quite stable, so that the behavioral consequences of neural injury are determined by whether or not what is spared can function normally. At the other extreme there are those who believe that the nervous system is quite plastic, so that the behavioral consequences of neural injury depend on whether or not neural centers will reorganize and function vicariously. There should be little surprise that such extreme positions have set the boundaries of fertile soil in which considerable debate and controversy can germinate and be nurtured. And understandably so, since the consequences of brain injury will necessarily define the mechanisms of recovery of function as well as how we should best proceed to understand and ultimately facilitate these mechanisms.

Whereas the evidence that brain injury can, and will, induce neural reorganization and vicariation is vanishingly thin (Cotman and Nieto-Sampedro, 1982; LeVere, 1975), the evidence for neural plasticity and for neural stability is

T. E. LeVERE • Behavioral Neuropsychology Laboratory, Department of Psychology, North Carolina State University, Raleigh, North Carolina 27695-7801.

not. In fact, even though these latter positions appear to be mutually exclusive, convincing arguments for the veracity of each have been put forth, so that distinguishing what is solid and what is air is not a simple exercise. Yet, when reasonable researchers tender widely different answers to the same question, it may not be unreasonable to question the question. The question in this instance is typically phrased, "What is the consequence of brain injury?" The key words in this query are "What is the," and I suggest that it this singularity that is at the root of much of the controversy that entangles our understanding and generates needless debate. It may well be that I have missed some fundamental principle, but I believe that current evidence provides no single rule to guide us in understanding the effects of neural injury and what recovery may be expected. Rather, I believe that, at least for the present, we must hedge our answer to this seminal question with the qualifier that "it all depends." And I also believe that one of the principal factors that the consequences of brain injury depend on is the size of the lesion.

This is not to suggest a revival of K. S. Lashley's concept of mass action (Lashley, 1933). Quite the contrary. It is simply a suggestion that lesions of different sizes yield qualitatively different consequences in terms of both the behavioral dysfunctions that they produce and, necessarily, the mechanisms that may be involved in the reversal of these dysfunctions. To try to make this suggestion a palatable "solid," I first briefly review some of the data relative to subtotal lesions that spare some portion of the affected brain area. I then suggest some reasons why I believe these data are not related to the consequences of more complete injuries, which do not spare some portion of the affected brain area. Next, I discuss what may be a reasonable description of the behavioral dysfunctions produced by these more complete lesions and finally offer some data that are compatible with this position.

2. SUBTOTAL LESIONS

2.1. Normalization and Recovery of Function

When a brain lesion is subtotal and spares at least 10% or so of the affected brain area, the behavioral consequences of the injury appear to be a direct result of the lost tissue, and recovery of function may be accomplished by the nervous system's physiological response to the injury. Presently, there are two such physiological responses that are reasonable candidates for recovery of function in these circumstances, namely, compensatory neural sprouting to restore lost synapses and biochemical adaptations to modify the effectiveness of existing synapses.

As outlined by Finger and Stein (1982), the notion that neural sprouting may serve recovery of function is not new. For example, on the motor side,

Exner (1885) suggested that surviving nerves might grow and innervate those muscle fibers that had lost their efferent input. On the sensory side, early research by Weddle, Guttmann, and Gutman (1941) demonstrated that deafferented regions on the rabbit's skin could be innervated by sprouts from intact fibers and provide some recovery of cutaneous sensitivity. Later, the superior cervical ganglion was included in the list of nervous structures capable of remodeling their morphology in response to injuy (Murray and Thompson, 1957; Guth and Bernstein, 1961). However, although it became generally accepted that the peripheral and the autonomic nervous systems were capable of this sort of plasticity, it remained for Liu and Chambers (1958) to demonstrate that the central nervous system (CNS) was also so endowed. This CNS plasticity was later confirmed by Goodman and Horel (1966) in the visual system and by Raisman (1969) in the hippocampus of rat.

The other CNS response to injury that is a plausible candidate for recovery of function is the biochemical adaptation that can occur in synapses that survive the neural injury. These adaptations have sometimes been labeled denervation supersensitivity (Cannon and Rosenblueth, 1949; Trendelenburg, 1963) to describe the increased sensitivity of muscle and neural tissue when it is deprived of some of its afferent neural input. That this adaptation in synaptic efficiency can also mediate behavioral recovery has recently been suggested by the elegant investigations of Marshall and his group with respect to the reversal of somatosensory deficits produced by interrupting the mesostriatal dopaminergic projection (see Marshall, 1984, for review).

These data then demonstrate that the CNS is not without physiological resources to deal with the challenge of certain types of brain injury. Both morphological and neurochemical adaptations can occur, and either or both can potentially mediate behavioral recovery, although the evidence for behavioral recovery is less the evidence for the adaptations themselves (Finger and Stein, 1982; Harrell et al., 1983; McCough et al., 1958). Yet, notwithstanding the veracity of lesion-induced neural sprouting and synaptic adaptations, it must be kept in mind that there are several physiological and behavioral qualifications related to these phenomenon that limit their usefulness as a singular explanation of the consequences of brain injury and recovery of function.

2.2. Some Limitations

To begin, there are some specific "rules" or conditions that determine the occurrence of, and the progress of, compensatory collateral sprouting. These have been summarized by Cotman and Nieto-Sampedro (1982) as follows:

> The first rule is that the new synapses completely restore the synaptic input lost following partial denervation (Cotman et al., 1981). The second rule is that an afferent will reinnervate a denervated zone only if its terminal field overlaps that of the

damaged afferent. The third general rule is that reactive growth causes only a quantitative increase or rearrangement of previously existing connections. Qualitatively
new pathways are not created during lesion-induced synaptogenesis in the adult organism. . . . A fourth general rule may be formulated: when a neuron receives more
than one type of afferent, there is a definite hierarchy in the relative capacity of the
various afferents to grow in response to synapse loss. It seems that "like" afferents,
i.e., those from similar cell types, have growth preference (p. 375).

With regard to biochemical synaptic adaptations within existing synapses, there also appear to be some rules or conditions that must be met. First, there must be at least 5 to 10% sparing of the neural area affected by the lesion (Marshall, 1979; Kozlowski and Marshall, 1981). Second, the behaviors that are recovered are typically unlearned or simple behaviors such as locomotor activity or somatosensory reflexes. Third, and perhaps most importantly, the behaviors that recover in these circumstances recover spontaneously.

Unfortunately, brain injury is not always limited in scope. Nor does it always affect simple unlearned behaviors. Further, in some cases recovery may not occur without some sort of physiological or behavioral intervention (Bartus *et al.*, 1986; Davis and LeVere, 1982; Meyer *et al.*, 1963; Sprague, 1966). This is not meant to imply that lesion-induced neural repair and normalization are not responsible for behavioral recovery; under certain circumstances they very well may be. It is important, however, to remember that limited neural injury is but one type of neural injury and that the behavioral dysfunctions caused by these lesions are but one of many possible consequences of brain injury. There are other types of lesions, particularly those that are large and do not spare any significant portion of the affected brain area, and it is these lesions that we suggest are qualitatively different in terms of both their consequences and the reversal of these consequences.

3. COMPLETE LESIONS

3.1. Recovery without Normalization

Large brain injuries, of course, violate certain of the requirements necessary to realize behavioral recovery from the neural repair. Nonetheless, behavioral recovery can occur following complete lesions, and, more importantly, it appears to do so even though the lesion-induced neurophysiological pathologies remain unchanged. For example, Spear and his colleagues have demonstrated that cats can recover the ability to perform visual discriminations even after large lesions that completely destroy neocortical visual areas 17, 18, and 19 (Spear and Baumann, 1979; Baumann and Spear, 1977; Wood *et al.*, 1973). These same researchers also demonstrated that this recovery is dependent on another area of the cat's visual system, the lateral suprasylvian area (LS), since removal of LS subsequent to the recovery produces an irreversible visual discrimination deficit.

However, it is important to note that the recovery following the initial visual lesion occurs even though the aberrant electrophysiological activity that this lesion produces in LS remains unchanged. In other words, the behavioral recovery is not dependent on the normalization, repair if you will, of a neural area shown to be critical for the occurrence of the recovery.

Conceptually similar findings have been reported by Bartus *et al.* (1985) in an investigation of the effects of ibotenic acid lesions of the nucleus basalis magnocellularis (NBM) in rats. The NBM is a primary cholinergic projection to neocortex, and lesions to this area produce marked decreases in cortical choline acetyltransferase activity and high-affinity choline uptake. The behavioral consequence of the lesion is a severe disturbance of short-term memory as evidence by impaired radial-arm maze performance. Notwithstanding the severity of this behavioral dysfunction, the NBM-lesion animals are nonetheless able to gradually restore prelesion levels of performance if trained over a period of several weeks. However, similar to the data reported by Spear and his colleagues, this recovery occurs without any detectable evidence of morphological changes within NBM or neurochemical normalization within the cholinergic system. In fact, Bartus *et al.* were even unable to detect any change in the parallel cholinergic systems innervating the hippocampus or the olfactory bulbs. Thus, independent of the implications that cholinergic mechanisms may be important for certain aspects of memory (Murray and Fibiger, 1986), the normalization of this system does not appear responsible for reversing the short-term memory dysfunctions initially produced by the NBM lesion.

It appears from both of these investigations that the behavioral deficits following more complete lesions of some portion of a functional neural system are not a direct result of lost neural tissue and/or its functioning, since recovery is not dependent on the normalization of the affected brain areas. Again, it must be emphasized that this in no way denies that neurological normalization may mediate recovery of function in some circumstances. It is simply meant to suggest that larger, more complete brain injuries may have qualitatively different consequences, and if so, then the mechanisms of recovery also must be qualitatively different.

3.2. The Nature of the Behavioral Deficit

Yet, if the normalization of neural physiological mechanisms does not mediate recovery of function following more complete injuries, then what does? A good question, but one that first requires an answer to another question. This more fundamental question concerns the brain's response to large insults. That is to say, if the brain responds by attempting neurological repair given some sparing of the injured brain area, then what is the brain's response if there is no such sparing? Although the neurophysiological answer to this is still far from complete, it is possible to gain some appreciation of it from the behavioral consequences of such injuries. One particularly enlightening experiment was reported some time ago by Sprague (1966). It involved cats, large neocortical injuries, and the animal's ability to orient to visual stimuli.

It was Sprague's finding that unilateral lesions that involved virtually all neocortex associated with vision resulted in a severe and persistent contralateral hemianopia. Although this was an obvious result of neocortical injury of this magnitude, what was not so obvious was that the hemianopia could be reversed by a contralateral superior colliculus lesion. Sprague's interpretation of these data, later confirmed by the electrophysiological investigations of Goodale (1973), was that the initial unilateral neocortical injury caused an imbalance in the operation of the visual system by eliminating the normal corticofugal facilitation of the ipsilateral superior colliculus. Consequently, the superior colliculus ipsilateral to neocortical lesion could not produce effective visual orienting because of the normal inhibitory influence of the contralateral superior colliculus. Appropriate orienting in the hemianopic visual field, however, could be restored by correcting this imbalance through destruction of the contralateral superior colliculus or simply by removing its inhibitory influence by commissural section.

Although these data are obviously of great import to our understanding of the visual system, I believe that their ultimate contribution will be to our understanding of the brain as a whole and particularly the consequences of large brain injuries. And I believe so for two reasons. First, Sprague's data clearly indicate that a behavioral dysfunction following a large neocortical injury is not necessarily the result of the destruction of the neural mechanisms critically responsible for the behavior. Second, I believe that his interpretation provides an explanation of why there is a behavioral dysfunction even though the neural mechanisms critically responsible for the behavior are spared.

I suggest, in this regard, that a fundamental consequence of large brain injuries is a change in the normal balance among the individual functional neural systems that control behavior. For example, suppose that a large injury sparing little of some component of a functional neural system changes not only the balance within that system but also, and more importantly, the balance between that functional system and other functional systems. The most likely result of this imbalance would be that noninjured functional neural systems, or those sustaining less injury, would dominate behavioral control. This shift in what neural systems control behavior would, of course, give rise to behavioral deficits simply because any behavior primarily dependent on the injured functional system would not be expressed. But, and this is the critical aspect of this concept, the behavioral dysfunction would not necessarily be caused by a loss of the neural capacities required to accomplish the behavior. Rather, these neural capacities could well be spared and the behavior not expressed because the functional system with which they are associated is simply not utilized. This may seem something of a bizarre notion, since it suggests the possibility that a behavioral dysfunction may occur in the absence of any pathology of the neural mechanisms directly responsible for the behavior. Nonetheless, it is compatible with the data reported by Spear and by Bartus, and it is also able to accommodate a number of empirical findings that are otherwise difficult to understand.

3.3. Some Supporting Data

There are a number of studies that support the possibility that large brain injuries will induce compensatory shifts in which functional neural systems control behavior (see LeVere, 1980, for review). Many of these studies involve unilateral injuries and the observation that the affected behaviors will not be expressed unless the individual is forced to do so either by physical restraint (see Guth, 1974) or additional brain injury (Lashley, 1924, Sprague, 1966). More direct evidence, however, is obtained from experiments that involve training normal and brain-damaged animals with more than one type of discriminative cue.

For example, Gillespie and Cooper (1973) studied the effects of brain injury on transfer of training in a conditioned emotional response (CER) experiment. The CER paradigm typically involves assessing the affective value of a conditioned stimulus (CS) that has been paired with a noxious stimulus by determining whether the occurrence of the CS will suppress some ongoing behavior. In their experiment, Gillespie and Cooper initially paired the noxious stimulus with either a visual or an auditory CS in both normal and visual decorticate rats and then, after the CER was established, tested for transfer, respectively, to the auditory or visual CS. Neither of the two normal groups nor the visual decorticate group initially trained with the auditory CS demonstrated any transfer to the other stimulus. However, the visual decorticate rats initially trained with the visual CS showed perfect transfer to the auditory stimulus, which clearly demonstrates that the visual decorticate rat is more responsive to nonvisual information than is the normal rat.

Gillespie and Cooper next extended this finding in an experiment concerned with the phenomenon of blocking. The central question in this investigation was whether establishing the CER to a visual CS alone would block subsequently establishing the CER to an auditory stimulus that was simultaneously presented with the visual stimulus. Kamin (1967, 1969) had shown, of course, that normal rats would not associate the CER with the second superimposed stimulus, and Gillespie and Cooper replicated this result. Visual decorticate rats, on the other hand, did not show any such blocking and readily established the CER to the auditory stimulus as indicated by complete behavioral suppression when the auditory CS was subsequently tested alone. Conceptually similar results have more recently been reported by Nonneman and Warren (1977) with normal cats and cats deprived of their auditory neocortex. Our interpretation of all of these experiments is that they provide rather solid evidence that large brain injuries will induce shifts in what neural systems control behavior and that these shifts will favor those neural systems not affected by the injuries.

Another study that is compatible with this notion of lesion-induced compensatory shifts is one that was recently completed in our laboratory and demonstrates that large neocortical injuries can, under certain conditions, actually facilitate learning. The investigation began by training normal rats in a Yerkes Y maze with compound cue discriminations composed of visual cues and haptic cues (LeVere and LeVere, 1982). The normal animals learned the compound-cue problem after committing approximately eight errors, and subsequent transfer of training tests with only the visual component cues or only the haptic component cues indicated that each component cue contributed about equally to the original acquisition of the compound-cue discrimination. However, if the rats were first prepared with visual decortications, then the compound-cue discrimination was learned with but three errors, which is significantly better than the performance of the normal animals. Apparently, the large visual lesion facilitated the animal's ability to acquire the compound-cue discrimination. Yet, this is not quite correct, since the visual decorticate rats relied solely on the haptic component to solve the discrimination as demonstrated by the transfer test with just the visual component cues, where their performance fell to chance. This propensity of the visual decorticate rat to track the nonvisual aspects of the compound-cue discrimination was further demonstrated by the visual decorticate rat's performance when the nonvisual component cues were irrelevant. For example, if the haptic cues varied at random with the correct position that was indicated by the visual cues, then the visual decorticate rat committed nearly four times as many errors as the normal rat trained on the same discrimination. In short, a large neural injury that does not spare some portion of a component of a functional neural system will effectively relegate the entire system to second-class status with respect to behavioral control. And this is apparently so independent of whether or not what is spared of the injured system can ultimately sustain the behavior in question.

From the perspective of recovery of function, these investigations indicate that a behavioral deficit may be present even though the neural mechanisms necessary for the behaviors are spared. The practical problem then becomes one of how to express, or fully utilize, that which is spared. Compensatory strategies seem inappropriate and, in fact, could reasonably be considered the antithesis of realizing the full potential of the brain-injured individual. Perhaps it would be, at least from the animal research literature, more effective to "train the deficit" or at least to do so until it is clearly evident that the neural mechanisms that control the impaired behavioral capacity are lost.

3.4. The Chronic Consequences of Large Injuries

One might expect that once the effectiveness of spared neural capacities is reestablished, behavioral recovery following large lesions, like behavioral recovery following subtotal lesions, should be relatively complete and stable. Unfortunately, this is yet another difference between complete lesions and subtotal lesions. With complete lesions, there appears to be a chronic consequence of the injury that persists even after the individual has supposedly recovered.

For example, Lashley (1930), in an experiment designed to determine the differential brightness threshold in visual decorticate rats, noted that although the lesioned rats were able to discriminate brightnesses, they were nonetheless never able attain the performance level of normal rats. More recently, Cooper *et al.* (1967) and LeVere and Mills (1977) obtained similar results while investigating visual thresholds in visual decorticate rats. Moreover, this chronic visual dysfunction was present even though it was possible to show that the lesion did not change the rat's absolute or differential brightness threshold. Without exception, whether it was early in the postoperative training or late, whether the animals were trained to choose the brighter or dimmer discriminative cue, whether the question was the absolute brightness threshold or the differential brightness threshold, in no instance did the brain-injured animal's performance ever reach that of its normal counterpart. To be certain, there are visual behaviors that do not recover following complete lesions of visual neocortex, for example, the ability to discriminate oblique striped patterns (Lavond *et al.*, 1978). However, the concern here is with behaviors that are supposedly "recovered," and it is apparently the case that this recovery is not quite as complete as "postoperative training to criterion" would indicate or we might hope.

Notwithstanding the persistence of this chronic behavioral dysfunction, it is important to note that its severity may be modulated in ways that are consistent with our suggestion that large lesions induce compensatory shifts in the hierarchy of functional neural systems. The research that demonstrated this was recently carried out by Tom Chappell in our laboratory (Chappell and LeVere, 1988). Chappell's procedure utilized a shock-motivated Y maze in which he trained normal and visual decorticate rats 25 trials per day, 5 days per week, for a period of several months. Each animal was initially trained until its performance stabilized as defined by no more than a three-error difference in five consecutive 3-day running averages. After stabilization, each animal was trained for an additional 10 days on one of four experimental conditions. Following this, stabilized performance was reestablished, and then the animal was trained on a second experimental condition. This routine was continued until each animal had been trained on each of the four experimental conditions, with different animals experiencing the four conditions in different sequences to control for order effects.

The experimental conditions Chappell investigated were (1) quiet training, (2) distracting training, (3) massed training, and (4) spaced training. The quiet training condition involved eliminating the usual 10- to 20-db (A) masking noise in the chamber where the Y maze was located. The distracting training condition was created by replacing the masking noise with a tape recording of various laboratory, shop, and street noises as well as placing different olfactory cues under one arm of the maze. The massed training condition was accomplished by eliminating the usual 15-sec intertrial interval, whereas the spaced training condition involved lengthening the intertrial interval to 60 sec.

The results of this experiment are summarized in Fig. 1 with the performance of the normal rats in the top panel and that of the visual decorticate rats in the bottom panel. The bar at the far left is the pooled stabilization performance for each group, and the remaining bars give the average daily performance during each experimental condition. Two conclusions are immediately apparent from this figure. First, considering just the stabilization training, there is a clear and consistent impairment in the performance of the brain-injured rats. This is, of course, the chronic behavioral deficit noted by Lashley (1930), Cooper *et al.* (1967), and LeVere and Mills (1977) and shows that this deficit is not dependent on challenging the brain-injured rat with performances at or near its threshold capabilities. It also must be emphasized that Chappell observed no improvement in this deficit over the several months that the animals were trained. The second conclusion that is obvious from this figure is that the experimental conditions had absolutely no effect on the performance of the normal animals. However, for the visual decorticate rat, these manipulations had significant consequences in terms of both increasing and decreasing the severity of their chronic performance deficit.

I believe these data are compatible with the thesis that large lesions will induce compensatory shifts in the hierarchy of functional neural systems. It also

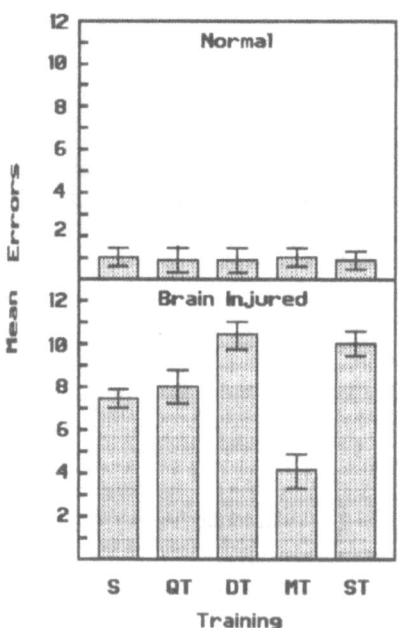

Figure 1. Extended training of normal rats (top panel) and visual decorticate rats (bottom panel) on a brightness discrimination task over a period of several months. The far left bar(s) is the pooled stabilization performance on the control training condition prior to each experimental training condition, and the bars to the right indicate the animal's performance during the four experimental training conditions as explained in the text. The vertical line associated with each bar is the 0.05 confidence interval. Abbreviations: S, stabilization performance during the control training condition; DT, distracting training condition; MT, massed training condition; QT, quiet training condition; ST, spaced training condition.

seems that the data extend our findings with the compound-cue discriminations by demonstrating that these shifts are not just associated with the acute consequences of large brain injuries, i.e., those that supposedly can be reversed. Rather, it appears that following large lesions there is a chronic tendency for the individual to attempt to compensate for the injury by preferentially utilizing noninjured functional neural systems. That is, if the "recovered" visual decorticate rat is given the opportunity to respond to nonvisual aspects of the training situation either by directly providing these sorts of cues (distracting training condition) or simply providing the time to do so (spaced training condition), then the rat will do so, and its performance will suffer. But if circumstances are arranged so that neither of these conditions is favored (massed training condition), then the lesioned rat's performance will significantly improve. I also suggest that the conditions that the rats found difficult are similar to the behavioral conditions found difficult in human cases of long-term brain injury (Lezak, 1983), particularly those cases of global amnesia described by Squire and Cohen (1984). In fact, the animal literature that has been discussed and the theoretical proposition that large lesions induce chronic compensatory shifts in the hierarchy of functional neural systems are quite compatible with much of the human clinical literature (see Miller, 1984).

4. CONCLUSIONS

The purpose of the present chapter is to suggest some parameters and present some evidence concerning one of the controversies within the general area of recovery of function. This controversy concerned "the" mechanism of recovery of function following brain injury, and it is the thesis of the present chapter that there should not be a controversy at all. Rather, I suggest that the dispute over whether recovery of function is a matter of collateral sprouting and synaptic adaptation or a matter of maximizing the functional significance of spared neural mechanisms is all somewhat silly. And it is so simply because the neurological and behavioral consequences of brain injury are not independent of the injury. By this I do not mean that visual, somatosensory, or frontal neocortex lesions do not have different behavioral effects. Obviously they have. That has proved the merit of the lesion method for investigating brain–behavior relationships. However, notwithstanding these obvious differences, I believe that there is a more fundamental consequence of different brain injuries that is independent of the particular brain area that is affected. This consequence is dependent on the size of the injury and, more importantly, on whether any significant portion of the affected brain area is spared.

If some significant portion of the affected brain area is spared, then demonstrated physiological responses to the injury may mediate behavioral recovery. These responses may be viewed as neural repair processes that serve to normalize the functional operation of the affected brain area. On the other hand, if there is no significant sparing of some component of a functional neural system, then the consequences of brain injury are qualitatively different. In this case, the behavioral dysfunction may not result from a loss of the neural capacity to perform the behavior. Rather, the behavioral dysfunction may arise from a compensatory shift in the neural systems that control behavior and from an underutilization of the total neural system affected by the lesion. Here, the neurobiology of neural repair is not an issue independent of the veracity of the phenomenon. Rather, the question is what controls the compensatory shift and how it may be reversed. At present, this question has few firm neurophysiological answers, but the behavioral parameters are beginning to be appreciated.

Thus, we do not at all dispute the counsel of Uncle Terwillinger. It is only that we believe that one would be well advised to consider first the size of a particular popover before swallowing anything!

REFERENCES

Bartus, R. T., Flicker, C., Dean, R. L., Pontecorvo, M., Figueredo, J. C., and Fisher, S. K., 1985, Selective memory loss following nucleus basalis lesions: Long term behavioral recovery despite persistent cholinergic deficiencies, *Pharmacol. Biochem. Behav.* **23**:125–135.

Bartus, R. T., Pontecorvo, M. J., Flicker, C., Dean, R. L., and Figueiredo, J. C., 1986, Behavioral recovery following bilateral lesions of the nucleus basalis does not occur spontaneously, *Pharmacol. Biochem. Behav.* **24**:1287–1292.

Baumann, T. P., and Spear, P. D., 1977, Role of lateral suprasylvian visual area in behavioral recovery from effects of visual cortex damage in cats, *Brain Res.* **138**:45–468.

Cannon, W. B., and Rosenblueth, A., 1949, *The Supersensitivity of Denervated Structures: A Law of Denervation*, Macmillan, New York.

Chappell, E. T., and LeVere, T. E., 1988, Recovery of function after brain damage: The chronic consequences of large neocortical injuries, *Behav. Neuroscience* (In press).

Cooper, R. M., Freeman, I., and Pinel, J. P. J., 1967, Absolute threshold of vision in the rat after removal of striate cortex, *J. Comp. Physiol. Psychol.* **64**:36–39.

Cotman, C. W., Nieto-Sampedro, M., and Harris, E. W., 1981, Synapse replacement in the nervous system of adult vertebrates, *Physiol. Rev.* **61**:684–784.

Cotman, C. W., and Nieto-Sampedro, M., 1982, Brain function, synapse renewal, and plasticity, *Annu. Rev. Psychol.* **33**:371–401.

Davis, N., and LeVere, T. E., 1982, Recovery of function after brain damage: The question of individual behaviors or functionality, *Exp. Neurol.* **75**:68–78.

Exner, S., 1885, Notiz zu der Frage von der Faservertheilung mehrerer Nerven in einem Muskel, *Pflugers Arch.* **36**:572–576.

Finger, S., and Stein, D. G., 1982, *Brain Damage and Recovery*, Academic Press, New York.

Gillespie, L. A., and Cooper, R. A., 1973, Visual cortical lesions in the rat and a conditioned emotional response, *J. Comp. Physiol. Psychol.* **83**:76–91.

Goodale, M. A., 1973, Cortico-tectal and intertectal modulation of visual responses in the rat's superior colliculus, *Exp. Neurol.* **17**:75–86.

Goodman, D. C., and Horel, J., 1966, Sprouting of optic tract projections in the brain stem of the rat, *J. Comp. Neurol.* **127**:71–88.

Guth, L., 1974, Axonal regeneration and functional plasticity in the central nervous system, *Exp. Neurol.* **45**:606–654.

Guth, L., and Bernstein, J. J., 1961, Selectivity in the re-establishment of synapses in the superior cervical sympathetic ganglion of the cat, *Exp. Neurol.* **4**:59–69.

Harrell, L. E., Barlow, T. S., and Davis, J. N., 1983, Sympathetic sprouting and recovery of spatial behavior, *Exp. Neurol.* **82**:379–390.

Kamin, L. J., 1967, Attention-like processes in classical conditioning, in: *Miami Symposium of the Prediction of Behavior: Adversive Stimulation* (M. R. Jones, ed.), University of Miami Press, Miami. pp. 9–31.

Kamin, L. J., 1969, Predictability, surprise, attention and conditioning, in: *Punishment and Aversive Behavior* (B. A. Campbell and R. M. Church, eds.), pp. 279–296, Appleton-Century-Crofts, New York.

Kozlowski, M. R., and Marshall, J. F., 1981, Plasticity of neostriatal metabolic activity and recovery from nigrostriatal injury, *Exp. Neurol.* **74**:318–323.

Lashley, K. S.. 1924, Studies of cerebral function in learning: V. The retention of motor habits after destruction of the so-called motor areas in primates, *Arch. Neurol. Psychiatry* **12**:249–276.

Lashley, K. S., 1930, The mechanism of vision: II. The influence of cerebral lesions upon the threshold of discrimination for brightness in the rat, *J. Gen. Psychol.* **37**:461–480.

Lashley, K. S., 1933, Integrative functions of the cerebral cortex, *Psychol. Rev.* **13**:1–42.

Lavond, D., Hata, M. G., Gray, T. S., Geckler, C. L., Meyer, P. M., and Meyer, D. R., 1978, Visual form perception is a function of the visual cortex, *Physiol. Psychol.* **6**:471–477.

LeVere, T. E., 1975, Neural stability, sparing, and behavioral recovery following brain damage, *Psychol. Rev.* **82**:344–358.

LeVere, T. E., 1980, Recovery of function after brain damage: A theory of the behavioral deficit, *Physiol. Psychol.* **8**:297–308.

Levere, N. D., and LeVere, T. E., 1982, Recovery of function after brain damage: Support for the compensation theory of the behavioral deficit, *Physiol. Psychol.* **10**:165–174.

LeVere, T. E., and Mills, J., 1977, Residual differential brightness thresholds following removal of visual neocortex in rats. *Physiol. Psychol.* **5**:490–496.

Lezak, M. D., 1983, *Neuropsychological Assessment*, Oxford University Press, New York.

Liu, C.-N., and Chambers, W. W., 1958, Intraspinal sprouting of dorsal root axons, *Arch. Neurol. Psychiatry* **79**:46–61.

Marshall, J. R., 1979, Somatosensory inattention after dopamine-depleting intracerebral 6-OHDA injection: Spontaneous recovery and pharmacological control. *Brain Res.* **177**:311–324.

McCough, G. P., Austin, G. M., Liu, C.-N., and Lui, C. Y., 1958, Sprouting as a cause of spasticity, *J. Neurophysiol.* **21**:205–216.

Meyer, P. M., Horel, J. A., and Meyer, D. R., 1963, Effects of *dl*-amphetamine upon placing responses in neodecorticate cats, *J. Comp. Physiol. Psychol.* **56**:402–404.

Marshall, J. R., 1984, Brain function: Neural adaptations and recovery from injury, *Annu. Rev. Psychol.* **35**:277–308.

Miller, E., 1984, *Recovery and Management of Neuropsychological Impairments*, John Wiley, New York.

Murray, C. L., and Fibiger, H. C., 1986, Pilocarpine and physostigmine attenuate spatial memory impairments produced by lesions of the nucleus basalis magnocellularis, *Behav. Neurosci.* **100**:23–32.

Murray, J. G., and Thompson, J. W., 1957, The occurrence and function of collateral sprouting in the sympathetic nervous system of the cat, *J. Physiol. (Lond.)* **135**:133–162.

Nonneman, A. J., and Warren, J. M., 1977, Two-cue learning by brain damaged cats, *Physiol. Psychol.* **5**:397–402.

Raisman, G., 1969, Neuronal plasticity in the septal nuclei of the adult brain, *Brain Res.* **14**:25–48.

Spear, P. D., and Baumann, T. P., 1979, Neurophysiological mechanisms of recovery from visual cortex damage in cats: Properties of lateral suprasylvian visual area neurons following behavioral recovery, *Exp. Brain Res.* **35**:177–192.

Sprague, J. M., 1966, Interaction of cortex and superior colliculus in mediation of visually guided behavior in the cat, *Science* **153**:1544–1547.

Squire, L. R., and Cohen, N. J., 1984, Human memory and amnesia, in: *Neurobiology of Learning and Memory* (G. Lynch, J. L. McGaugh, and N. M. Weinberger, eds.), pp. 3–64, Guilford Press, New York.

Trendelenburg, U., 1963, Supersensitivity and subsensitivity to sympathomimetic amines, *Pharmacol. Rev.* **15**:225–276.

Weddle, G., Guttmann, L., and Gutman, E., 1941, The local extension of nerve fibers into denervated areas of skin, *J. Neurol. Psychiatry* **4**:206–225.

Wood, C. C., Spear, P. D., and Braun, J. J., 1973, Direction-specific deficits in horizontal optokinetic nystagmus following removal of visual cortex in cat, *J. Comp. Physiol. Psychol.* **60**:231–237.

3

Bases of Inductions of Recoveries and Protections from Amnesias

DONALD R. MEYER

1. INTRODUCTION

In 1987, I celebrated my 40th anniversary as a student of inductions of recoveries of functions after injuries to the central nervous system. I think I know a fair amount about it. At least I have had a lot of practice with the subject and have only infrequently succumbed to temptations to give it up and study something else. I believe that I have come to some significant conclusions and propose in this discussion to say what they were and how they were finally arrived at. Then I contrast them with the views of other workers in our vineyard and do that by giving my reactions to statements that were made in a recent review of research on the neural mechanisms of memory.

I have had some disagreements with my colleagues. However, as best I can recall, we have had few public confrontations. Most of my arguments have been with study sections and with the editors of journals, whose judgments have only very rarely indicated that they understood what I was doing. But they were not completely to blame. Most of my work has been conducted in the eddies of the mainstream, and my reasons for doing it were often not completely clear to me. I am not ashamed to say that because, in my experience, a person who can give you all the reasons has either obtained them from somebody else or is planning a trivial pursuit.

My own style was formed, in large part at least, by the reaction of my teacher, Harry F. Harlow, to my plea for advice as to how I might select a topic for my Master's thesis. He replied that on the following Monday morning, I

DONALD R. MEYER • Department of Psychology, The Ohio State University, Columbus, Ohio 43210.   *Present address:* 476 Overbrook Drive, Columbus, Ohio 43214.

would have a group of monkeys to study, and that he expected me to keep him informed as to what kinds of things I had found out. He offered no suggestion as to what I should do, but it was clear that I should have to do something and that time was awasting. I cannot now remember what the study was about, but it taught me a most important lesson: if you watch animals, and watch them very closely, they will tell you what you ought to do next.

I was interested in studies of the brain. Such investigations were not Harlow's forte, although his laboratory at Wisconsin was among the leading institutions in that field. So in 1947 I went off to old Science Hall, took the elevator to the fourth floor where cadavers were kept in rooms whose walls were done in bright orange, and then climbed a narrow flight of stairs to an attic that housed an anatomic research laboratory being run by Paul H. Settlage. That day Paul and Henry Suckle, who was one of the city's neurosurgeons, were preparing a monkey with a huge unilateral cerebral lesion. The occipital and frontal lobes were amputated, and the rest of the dorsolateral cerebral cortex was topically destroyed except for the tissues on the banks of the fissures of Rolando.

Monkeys thus prepared were described by our group as semihemis. One of my first jobs in Harlow's shop was to study their behaviors and compare their performances with those of bilateral frontal subjects. Thereafter, over several years, the semihemis underwent surgeries that destroyed the contralateral frontal or temporal areas. I was thus introduced to what is now described as the method of serial ablations. The question being asked was whether such subjects, which underwent extensive and varied retraining between their first and second operations, would be able to perform tests that animals prepared with one-stage injuries would fail.

I learned several things from those investigations (Harlow *et al.*, 1952) that I think are still valid despite their having been conducted a long time ago. One is that impairments of delayed responding, the stablest of the deficits of monkeys that have suffered bilateral prefrontal injuries, are just as severe after serial ablations as they are after one-stage injuries. Another is that deficits of concept formation, which are frequently confused with impairments of discrimination learning (Meyer, 1971, 1972b), are just as severe after serial temporal injuries as they are after one-stage injuries. A few years ago, we reexamined the first of those conclusions and once again could find no reason to believe that impairments of the functions of the "uncommitted cortex" are subject to compensations if the injuries are inflicted in several operations (Meyer *et al.*, 1976).

However, while I still was a student at Wisconsin, I met and then worked with Harlow Ades. Ades was visitor in Harlow's laboratory, principally because the summers in Madison are cooler than they are in the deep South. He had a deep disdain for conventional opinions, which at that time included a presumption that the only kinds of neurobehavioral studies that are worth paying much attention to are those in which the injuries are bilaterally symmetrical and also are inflicted in one stage. For several years thereafter, he was virtually the only person, except of course for Harlow, who was carrying out controlled investigations of subjects with serial injuries to the brain.

Ades' work prompted my first investigations as a young professor at Ohio State. Ohio State's facilities were limited, and one worked with rats or not at all. Ades had observed that animals prepared with one-stage bilateral ablations of the auditory cortex will lose a conditioned avoidance response to a sound but will not if the injuries are serially inflicted and several days elapse between the surgeries. The subjects were not given formal interoperative practice, which suggested that the compensations were spontaneous in the sense that they occurred as a function of time, and time alone (Stewart and Ades, 1951). I was curious as to whether that was so, but because our laboratory was a very noisy place, I was forced to approach the question through a study of performances of visual conditioned avoidance responses (CARs).

2. TRAINING EFFECTS

Essentially, we trained rats to jump across a barrier whenever a light came on (Meyer *et al.*, 1958). Our surgical target was the visual neocortex, which we also term the posterior cortex. As Ades had noted with auditory injuries, we found that bilateral posterior injuries suppress retention of a CAR unless the injuries are serially inflicted and several days elapse between the surgeries. However, we also found that the effect depends on whether the animals are kept in the light between successive operations, and our question then became one of why that should be the circumstance. But I failed to come up with an interesting approach to the question, and instead of pursuing the matter any further, I went back to Madison on leave to study the functions of the temporal lobes of monkeys.

A few years later, several articles appeared in which rats with serial posterior ablations were trained in a shock-avoidance version of the classical two-choice discrimination learning apparatus (Thompson, 1960; Petrinovitch and Bliss, 1966; Petrinovitch and Carew, 1969). One of the questions addressed in those studies was whether a serial effect could be observed for performance on a "visual brightness" task if the subjects were not given training on the task between the first and second operations. The findings were mixed, and when our CAR results were included on the grounds that the two tasks were measures of the same thing, two investigations had said that it could and the other two had said that it could not.

The difference between the two sets of outcomes was, of course, attributed to variations in the scopes of the injuries to the brain. However, I knew that such an explanation could not account for my results, and so I built a version of the "Thompson box" (Thompson and Bryant, 1955) to see for myself what the facts were. I found that for injuries of the scopes that I had previously inflicted, a

serial effect is not observable for habits of responding to a white and not to a
black door unless the animals are given interoperative practice (Kircher *et al.*,
1970). The observation, when taken by itself, was not particularly impressive,
but it prompted inquiries that kept me very busy for a period of nearly 30 years.

I think that the reasons for my enamorment with the black–white discrimination problem can best be perceived if we ponder the question as to just what
would be the minimal ingredients for systematic studies of the bases of serial
effects. There are five if we decide to explore the behaviors of subjects with two-
stage ablations. We can either elect to train or not to train the subjects before
their first surgeries; we can also elect to train or not to train the subjects between
their surgeries; and then we must retrain, or reretrain, the subjects following their
second operations. The other two procedures are the surgeries, and of course all
five must be conducted at times that have previously been rationalized with data.

That basic paradigm cannot be simplified if we wish to be able to distinguish the causes of impairments of performance. At least that is so if we prefer
facts to guesses, which are easier to come by than facts. Also, if we try to employ
it with a task that a subject will rapidly forget, we are sunk at the outset because
by the time the various maneuvers are completed, the ultimate test of retention of
the habit may take place many weeks following the animal's initial exposure to
the task. Therefore, the fact that a normal animal will relearn the black–white
discrimination problem in two or three trials after layoffs of more than a month
(Glendenning, 1972) is one of the features that commends the use of the method
in studies of recoveries.

We also discovered that performance of the problem by operated subjects is
governed by some very simple rules. To illustrate, except for one result that has
not been replicated (Gray and Meyer, 1981), it doesn't matter how much training
on the task a subject receives prior to surgery. Nor, within broad limits, does it
matter a great deal how much interoperative training a subject with two successive injuries receives before its second injury is inflicted (Meyer and Meyer,
1977). Also, we have found that serial effects are easy to obtain with the
procedures but that such effects can only be observed if the animals are trained
both before and after their first and second operations (Bodart *et al.*, 1980).

The black–white problem is a visual task, and hence we have had to pay
attention to the question of whether impairments of remembering the problem are
caused by the perceptual effects of injuries to the cortex. At the time we began
our inquiries, it was generally believed that mammals prepared with posterior
injuries can still detect differences in flux but are permanently form-blind. We
examined those conclusions for ourselves and found them to be acceptable
(Horel *et al.*, 1966), but we also observed that posterior preparations can learn to
discriminate between visual patterns if the patterns are different with respect to
amounts of visual contour. Thus, such subjects can readily discriminate between
the two sides of visual cliffs (P. M. Meyer, 1963); between visual alleys that are
either deep or shallow (Braun *et al.*, 1970); between perspective drawings of

deep and shallow alleys (Wetzel, 1969); and between arrays of small and large visual figures that present the same amounts of visual flux (Dalby *et al.*, 1970). Hence, we concluded that posterior subjects still have a form of spatial vision but that they are permanently unable to discern the spatial orientations of contours (Ritchie *et al.*, 1976; Lavond *et al.*, 1978; Lavond and Dewberry, 1980).

For a time, we believed that it was possible to think that animals with posterior injuries fail to remember the black–white problem because, prior to surgery, they had learned the problem by responding to visual-detail cues. We were not alone in that opinion (e.g., Cooper *et al.*, 1972). We thought it was sound when we compared the rates of learning of the problem by normal and posteriorly decorticated subjects, which Lashley (1935) had found to be exactly the same in his pioneering studies of the question. We confirmed his results, but we also observed that his finding had been artifactual in that normal rats are inferior to or equal to operated subjects as a function of increasingly more stringent criteria of performance (Horel *et al.*, 1966). Hence, insofar as we could see at that point, postoperative recovery of performance of the task is a substitutive process in which the animals must form new habits to replace the more efficient habits they had learned prior to surgery.

3. DRUG EFFECTS

However, at about the same time, my wife was engaged in a study that dealt with the behaviors of cats with enormous cerebral cortical ablations (P. M. Meyer, 1963). As had been expected from the work of Bard and Brooks (1934), the animals exhibited persistent impairments of placing to visual surfaces. For a very long time, I had kept in mind a finding that treatments with amphetamine will reinstate the righting responses of pontine preparations (Maling and Acheson, 1946), and when she had completed her investigation, I suggested to Pat that she had nothing to lose if we tested her subjects' responses to treatments with the drug.

We learned a lot from that investigation (P. M. Meyer *et al.*, 1963). Our main observation was that the treatments served to reinstate visual placing even though the animals had shown no signs of recovery of placing for many months following their surgeries. But we also observed that the treatment's effects had nothing to do with the drugs' activating properties. Thus, if you attempt to test a treated cat a few minutes after the injections, all that you will get are growls, spits, hisses, and scratches. However, when the animal has calmed down, it will place as nicely as a normal cat, presumably by using the kinds of visual cues that our other experiments have shown that such subjects still detect.

Those results for cats prompted us to study the effects of treatments with amphetamine on recovery of performance of the black–white discrimination problem (Braun *et al.*, 1966; Jonason *et al.*, 1970). The subjects, rats, were

bilateral posterior preparations that either had learned the problem prior to operation or had not learned it prior to operation. The treatments had no effect on the
latter group but facilitated relearning of the task by subjects with prior experiences. Hence, it was apparent that preoperatively established memories for the
black–white problem are neither erased by posterior ablations nor silenced because the subjects have perceptual impairments.

We next asked if treatments with amphetamine will yield a serial effect if
the drug is given interoperatively to rats prepared with two-stage posterior ablations (Kircher *et al.*, 1970). The treatments had no effect whatever. Hence, it
appeared that the action of amphetamine, whatever its molecular basis may be,
facilitates recoveries if and only if the treatments are given in conjunction with
practice on a task. I know of no exception to the rule, which in our investigations
was also shown to hold for reinstatements of placing by decorticated rats (Braun,
1966). Moreover, a few years later we observed that if the combination is
employed for treating cats with chronic impairments of placing, the animals will
then continue to place for many days following withdrawal of the drug if given
frequent tests for the reactions (Ritchie *et al.*, 1976).

Until very recently, the findings just described were looked on as bother-
some because they contradicted the venerable theory that placing is a "cortical
reflex" (Bard and Brooks, 1934). For about 10 years, we were the only workers
who were actively studying the effects. But then they were confirmed by an
elegant study in which the drug was used to induce reversals of impairments of
somesthetic placing (Amassian *et al.*, 1972). We have recently been heartened
by the fact that Feeney and his colleagues have found them interesting and have
also observed that the effects are contingent on a combination of the treatment
and experience with a task (Feeney *et al.*, 1985). At present, the group is
exploring the use of the method in treatment of human hemiplegic patients
(Fenney and Sutton, 1986). As a British sergeant-major would put it, their work
is a jolly good show.

4. MEMORY AND REMEMBERING

By the early 1970s, we had come to regard impairments of placing as being
forms of retrograde amnesias and not, in the main at least, as caused by perceptual impairments (cf. Meyer, 1972a). Hence, we looked around for yet another
task that cannot be performed by decorticated subjects unless special methods are
employed. We had recently observed that rats prepared with one-stage ablations
of the entire neocortex are incapable of learning a shuttle-box avoidance response
(P. M. Meyer *et al.*, 1968). However, the nature of the deficit was not at once
apparent. The classical notion that the engrams, or traces, of memories are stored
as modifications of the cortex suggested that the rats had failed to learn the
problem because their memory banks had been destroyed. However, we ob-

served that if the cortical injuries were combined with ablations of the septum, the animals not only learned the task very promptly but at a somewhat faster rate than subjects whose brains were intact (P. M. Meyer *et al.*, 1970).

The latter observation showed that the subjects still had a place to put their memories. Hence, we were prompted to ask if the failures of animals with injuries to the cortex by itself were impairments of remembering memories. We began by preparing two groups of bilaterally decorticated rats (Beattie *et al.*, 1978). One group was given saline, and the other was given low doses of amphetamine. Both groups failed to learn the problem within 10 days of training. However, we then reversed the treatments and observed that the rats that were first given saline performed the task at higher levels than naive normal animals when treated with amphetamine. From that it seemed plain that they had learned the task at a time when they could not perform the task and that their memories were completely latent before they were treated with the drug. That result reminded us of what we had observed with respect to the black–white problem, for once again it seemed that at the doses we were using, the drug facilitated remembering of memories but not the acquisition of memories.

Significantly, we had looked very carefully at three impairments that were widely believed to be perceptual impairments and/or memorial impairments. We had shown that at least the greater part of those impairments were reversible with methods that could not have been effective unless those components were impairments of remembering or retrieval. Hence, we were prompted to suggest to clinicians that the retrograde amnesias of victims of strokes were not very likely to represent destruction of their memories and hence should be treatable with more efficient methods than the ones presently in use (Meyer and Meyer, 1977). Thereafter, we reviewed the general problem of whether the engrams of any kinds of memories whatever are stored as modifications of the cortex. We concluded that a century's worth of studies of the question had failed to provide us with as much as one argument in favor of that hoary proposition (P. M. Meyer and Meyer, 1982).

I next began to spend a fair proportion of my time in examinations of a curious set of findings with respect to performance of the black–white problem. In my well-spent youth I had been a statistical guru but had grown very tired of the classical presumption that a properly designed experiment must have its own experimental and control groups. Hence, I decided to standardize ·our black–white procedures and to include in each successive study a replicative group so that we could be assured of their comparabilities (Meyer and Meyer, 1977). When the program was concluded in 1985, the strategy had given us a cross-validated bank of information for approximately 2000 subjects, of which the great majority had been subjected to more than one injury to the brain. The data in the bank now enables us to say, within a trial or two and at the most within about 10% what the mean performances of many kinds of subjects with two-stage injuries to the cerebral cortex will be.

The curious set of findings to which I have referred was first described by me to my graduate students as the "pseudomathematics of the cortex." In the early days of our program, we had focused our attention on the consequences of injuries to the posterior cortex. However, we had also studied many rats with injuries to the extravisual cortex, which by our definition consists of the tissues that lie anterior to the bregma. We found that, in general, ablations that destroy a quadrant of the cortex have the same effects on the retention of the problem regardless of whether the injuries are anterior or posterior. Moreover, we found that ablations that destroy two of the quadrants of the cortex have twice the effects, as measured by mean trials, as injuries to one of the quadrants. Thus, if the methods permitted a one-quadrant subject to relearn the problem in 8.5 trials, a similarly treated two-quadrant subject would relearn it in 17 trials (Meyer and Meyer, 1977).

There was one exception to the rule. We observed that animals prepared with one-stage bilateral posterior ablations required about half again as many trials to relearn the problem after surgery. To illustrate, although bilateral anterior rats, or rats with an injury to one posterior quadrant and the ipsilateral or contralateral anterior quadrant, required 17 trials, animals whose injuries bilaterally destroyed the posterior cortex required about 25 trials. For a number of years, I was unable to see that the eight-trial difference was a measure of the cost of an injury that destroys form perception. However, I hope to be forgiven on the grounds that such a hypothesis implied that the regionally nonspecific cost of an injury to the "visual" neocortex was about twice the cost of the animals' perceptual handicaps.

Now we can say, and with considerable assurance, that impairments of remembering, as contrasted with impairments of perceiving, are the principal deficits of animals prepared with injuries to the posterior cortex. Thus, through the use of serial procedures, we have shown that the specific eight-trial impairment can be dissociated from a larger deficit that happens to be equal to deficits of animals with anterior injuries of the same scopes (Cloud *et al.*, 1982; Meyer *et al.*, 1985). That is, insofar as postoperative performance of the black–white problem is concerned, the locus of the injury does not matter in the slightest provided that the injury has neither destroyed nor completed the bilateral destruction of the visual neocortex. In the latter circumstance, an animal's impairment will be equal to the cost of the rest of its injuries plus eight trials regardless of the kinds of injuries it sustains in addition to one-stage or two-stage posterior injuries.

The impairments proved to be dissociable because retraining on the problem after first-stage injuries conveys a protection against further deficits except for the eight-trial deficit. To illustrate, a hemidecorticated subject will relearn the problem in approximately 17 trials. If it then is prepared with a second-stage ablation of the contralateral anterior cortex, it will rerelearn the task in about three trials, as quickly as a subject that was trained on the task and was subsequently tested for forgetting of the problem without having undergone any kind of cerebral operation (Glendenning, 1972). In contrast, if a hemidecorticated subject is prepared with a second-stage ablation of the posterior cortex, it will take eight trials to relearn the problem because the second injury completes the bilateral destruction of the visual neocortex (Cloud et al., 1982). That is just one example of our findings with respect to subjects with varying kinds of first-stage and second-stage injuries to the cortex, and the rule just described has been amply confirmed for all of the conditions we have studied (Meyer et al., 1985).

Notably, if subjects with anterior injuries are given retraining on the problem, they exhibit protection of retention if prepared with second-stage posterior injuries. Hence, the compensation does not depend on a preexisting injury to the visual mechanisms of the cortex. Nor is it a function of the scope of the first-stage ablations, as is nicely illustrated by the scores of animals with two-stage injuries whose final operations leave them with only one posterior quadrant of the cerebral cortex. Such a subject will relearn the problem as quickly as a normal animal and do so regardless of whether its first-stage injury was larger or smaller than its second (Cloud et al., 1982).

The protection afforded by the compensation is not observed unless the subject has a memory to remember and is then reminded of the memory through retraining following an injury to its brain (Horel et al., 1966; Howarth et al., 1979; Meyer and Meyer, 1977; Hata et al., 1980). We have unpublished data that suggest that a very small injury will permit initiation of the process irrespective of whether the injury is to the frontal or the visual neocortex. We have argued that the finding has a bearing on the fact, as was first pointed out by Hughlings Jackson (1873), that injuries to the cortex that develop very slowly can become very large before a person thus afflicted is compelled to seek advice from a physician. That is, when the lesion is still very small, it will not have a marked effect on remembering of memories, and memories that are used while they are still usable will not be forgotten as the lesion continues to expand.

In summary, then, the principal fruits of our studies of inductions of recoveries of functions include a denial that the cerebral cortex serves as memory storage bank, a proof that the cortex functions as a whole in retrieval of memories from storage, and, perhaps above all, a mass of evidence that demonstrates that engrams, or traces of memories, are almost incredibly resilient. The secondary fruits, which were not discussed in this review, have included explorations of recoveries from early injuries, a scientific rationale for the use of electroconvulsive therapies, and studies that suggest that motives are determinants of how a memory will be stored.

5. CONTROVERSIAL ISSUES

Thus far, none of our principal conclusions has been challenged in a public forum. Hence, by definition, they are not controversial, for it takes two to tango and at least two to have a controversy. But now I shall present an invitation to the dance by giving my assessments of statements that appear in a very recent article by Squire (1986). I found Squire's paper to be remarkable because there was virtually nothing in it with which I could comfortably agree.

In his opening paragraph, Squire asserts that

> One powerful strategy for understanding memory has been to study the molecular and cellular biology of plasticity in individual neurons and synapses, where the changes that represent stored memory must ultimately be recorded (p. 1612).

The statement confuses promise with fulfillment. It is not plain to me that the molecular approach has taught us very much in the 30 years or so that it has now been actively pursued. At least, I am not aware of solid proofs of the existence of "memory molecules," nor of proofs that changes in the brain that are induced by experiential factors are the engrams or substrates of memories. Why, for example, should we jump to a conclusion that changes in the structure of the cerebral cortex are putative engrams if the evidence at hand suggests that the cortex is simply not a memory storage bank?

Next, Squire defines the questions with which his review is explicitly concerned: "Where is memory stored? Is there one kind of memory or many? What brain processes or systems are involved in memory, and what jobs do they do?" (p. 1612). He asserts that

> In recent years, studies of complex vertebrate nervous systems, including studies in humans and other primates, have begun to answer those questions (p. 1612).

The first statements are difficult to fault because the questions are of central importance in neuropsychological science. But the process of studying them has been underway for a long time and can only be said to have "recently begun" if we think of 1880 as "recent."

A little later on, Squire asserts that

> Memory is stored as changes in the same neural systems that ordinarily participate in perception, analysis, and processing of information to be learned. For example, in the visual system, the inferotemporal cortex (area TE) is the last in a sequence of visual pattern-analyzing mechanisms that begins in the striate cortex. Cortical TE has been proposed to be not only a higher-order visual processing mechanism but also a repository of the visual memories that result from this processing (p. 1612).

There are several things wrong with the statement. First, it is not at all apparent to me that the striate cortex is the *sine qua non* of the brain's visual form perceptual system. The evidence at hand suggests, instead, that the principal honors should be given to the bands that surround it (e.g., Hughes, 1977; Lavond and Dewberry, 1980). Moreover, the evidence at hand suggests that the so-called memorial impairments of TE subjects are disorders of attending and that if these are corrected the animals will learn a visual discrimination problem just as quickly as subjects whose brains are intact (Meyer, 1972b; P. M. Meyer *et al.*, 1986).

Thereafter, Squire says that

> The idea that information storage is localized in specific areas of the cortex differs
> from the well-known conclusion of Lashley's classic work that memory is widely and
> equivalently distributed throughout large brain regions (p. 1612).

He fails to mention that Lashley believed that memories for the black–white discrimination problem are stored within the visual cortex if the cortex is intact. However, that belief no longer matters, for as we and others have shown very clearly, the impairment is a failure of remembering of a memory that is stored somewhere else (Braun *et al.*, 1966; Jonason *et al.*, 1970; LeVere and Morlock, 1973).

Squire then says,

> In his most famous study, Lashley showed that when rats relearned a maze problem
> after a cortical lesion, the number of trials required for relearning was proportional to
> the extent of the lesions and was unrelated to its location. Yet Lashley's results are
> consistent with the modern view if one supposes that the maze habit depends upon
> many kinds of information (for example, visual, spatial, and olfactory) and that each
> kind of information is separately processed and localized (p. 1612).

That view is hardly modern; it amounts to a restatement of Hunter's (1930) classical critique of Lashley's theory. In the studies that led us to propose that the cortex is equipotentially involved in the process of remembering, we found that Hunter's theory was unable to account for the fact that different kinds of first-stage injuries will permit the induction of protection of remembering after various kinds of second-stage injuries. Thus, in the rat, the anterior neocortex is principally composed of "somatosensorimotor areas" (Woolsey *et al.*, 1950), and Hunter could have argued that animals prepared with first-stage ablations of anterior or posterior quadrants exhibit impairments that are different even though the animals relearn the problem at the same rates. But after either injury, retraining on the problem induces a protection of performance of the problem following a second injury to the converse quadrant. That finding shows that a common mechanism was affected, even though it also is unquestionably true that localized ablations of the cerebral cortex have regionally specific consequences.

A little further on, Squire asserts that

> Animal models for human amnesias have been recently developed for the monkey and
> rat. Animal models make it possible to identify the specific neural structures that when
> damaged produce the syndrome, and they set the stage for more detailed biological
> studies (p. 1613).

The statements suggest a new departure. That is simply not so, for I have now been engaged in the development of animal models for amnesias for approximately a quarter of a century. The statements also overlook the fact that models can be useful for assessments of approaches to treatments for diseases. Although we have emphasized that feature of our work (e.g., Meyer and Meyer, 1977; P. M. Meyer and Meyer, 1982; Meyer, 1984), it is not plain to me that the "recent models'' have cast much light on the question of how a clinician might try to help his patients.

Next, Squire asserts that

> It has been known for nearly 100 years that memory is impaired by bilateral damage to either of two brain regions, the medial aspect of the temporal lobes and the midline of the diencephalon (p. 1613).

If that statement is so, then I have been asleep for twice as long as Rip van Winkle. Furthermore, the statement does not address the question of why a patient might be forgetful. Failures of storage can occur for a variety of reasons, and failures of retrieval can be so severe as to lead one to think that the patient has memorial losses. Eventually those problems will be sorted out, but so far as I can see, it will be a long time before they are.

Next, Squire asserts that

> The capacity for long-term memory requires the integrity of the medial temporal and diencephalic regions, which must operate in conjunction with the assemblies of neurons that represent stored information (p. 1613).

From that I presume that he regards the former structures as routes to and from memory storage and, from his previous statements, that the cortex would be a good place to look for the assemblies. But as we have shown, the latter idea has no empirical support, and it seems to me at least that there is not much support for the former. We have known for many years that bitemporal monkeys, that is, with resections of both lobes, can learn discriminations between visual objects as rapidly as naive normal animals (Riopelle *et al.*, 1953). Of course, it could be argued that the latter operation, if combined with a medial diencephalic injury, would block both storage and retrieval, but I know of no experimental evidence for such a proposition.

Squire then observes that

> Amnestic patients demonstrate intact learning and retention of certain motor, perceptual, and cognitive skills and exhibit intact priming effects: that is, their performance, like that of normal subjects, can be influenced by recent exposure to stimulus materials (p. 1614).

I am thoroughly familiar with priming effects in animals with retrograde amnesias. To us, they are the outcomes of tests we must conduct to reinstate placing by decorticated cats that are treated with amphetamine and of the interoperative training that our studies have shown will induce protection from retrograde amnesias after second-stage cerebral ablations. In the latter instance, and importantly I think, retraining is most effective if given to subjects that have been treated with a drug that prevents initial learning of the task (LeVere and Fontaine, 1978; Davis and LeVere, 1979). The fact that priming works for both animal and human amnestics is consistent with our theory, but it certainly is not with theories that propose that amnesias are memorial impairments.

A little further on, Squire asserts that

> Memory is not fixed at the moment of learning but continues to stabilize (or consolidate) with the passage of time (p. 1615).

That is a venerable idea. However, as we pointed out a dozen years ago, there was then no support for such a concept, and so far as I can see, none has been developed since that time (Meyer and Beattie, 1977). Elsewhere, we have argued that Squire's own data for ECT patients (his Fig. 4B) do not support his theory because it would predict a difference at 3–4 years (Meyer, 1984). In our own investigations, we have shown that remembering of old and new memories is equally affected by ECS treatments provided that the subjects, at the time that they are treated, have the same motives as they had when they were trained on particular discrimination problems (Robbins and Meyer, 1970; Howard and Meyer, 1971; Howard et al., 1974). That observation is no longer an isolated finding. Thus, investigations with the black–white reversal paradigm have shown very clearly that preoperative memories will not interfere with postoperative learning of the black–white discrimination problem unless the same motives are employed (LeVere and Davis, 1977; see also LeVere et al., 1984a,b). Hence, we have interpreted the findings of Squire as supportive of our theory that electroshock treatments are effective because they selectively affect retrieval of the memories that were learned by the patients after they were already ill. In passing, I note that the theory implies that the treatments should never be given to a patient at a time when he is feeling just fine, for under such conditions they are certain to affect retrieval of happy memories.

In his next section, Squire asserts that

> Monkeys with bilateral lesions of the amygdala and hippocampal formation, which included perirhinal cortex and parahippocampal gyrus, exhibited severe memory impairment (p. 1617).

The statement implies an uncritical acceptance of the notion that a casual inspection of a task is all that is required for an assessment of the function it measures. Although I find it interesting that medial–temporal monkeys are grossly impaired in their performance of delayed nonmatching to sample, a person who believes that the deficits are memory impairments must be prepared to speak to the question of why a bitemporal monkey is the equal of a naive normal monkey in learning discrimination problems (Riopelle et al., 1953).

I must stress that my differences with Squire, and with many of the people whom he cites, are not monkish quarrels but, instead, are concerned with fundamental questions in the science of cerebral organization. For example, although I share the beliefs of my colleagues that molecular approaches to the problem of the nature of memorial traces will bear fruit, the time frame is going to be a long one unless the people who practice those particular arts begin to look for engrams where engrams are likely to be. I think I have shown that the cerebral cortex is not a storage site for memories, and I know of no convincing evidence that memories are stored within the cerebral hemispheres. Hence, by exclusion, it seems to me at least that the traces are formed within the brainstem and that structural changes in cerebral organs that are brought about by training are reflections of their roles in retrieval of memories from storage.

I close this discussion by asserting, once again, that the substrates of memories are tough. There is no proof whatever that memories decay, nor is there any proof that any cerebral injury, however inflicted, will destroy the trace of any form of memory. In contrast, remembering is fragile, but, happily for persons with memorial disorders, the impairments are often treatable. My wife Pat and I regard that as being our main scientific legacy, and we both look forward to the day when its potentials will be fully understood by clinicians.

REFERENCES

Amassian, V. E., Ross, R., Wertenbaker, C., and Weiner, H., 1972, Cerebello-thalamocortical interrelations in contact placing and other movements in rats, in: *Cortico-Thalamalic Projections and Sensorimotor Activities*, (T. L. Frigyesi, E. Rinvik, and M. D. Yahr, eds.), Raven Press, New York, pp. 395–444.

Bard, P., and Brooks, C. M., 1934, Localized cortical control of some postural reactions in the cat and rat together with evidence that small cortical remnants may function normally, *Res. Pub. Soc. Nerv. Ment. Dis.* 13:107–157.

Beattie, M. S., Gray, T. S., Rosenfield, J. A., Meyer, P. M., and Meyer, D. R., 1978, Residual capacity for avoidance learning in decorticate rats: Enhancement of performance and demonstration of latent learning with *d*-amphetamine treatments, *Physiol. Psychol.* 6:279–287.

Bodart, D. J., Hata, M. G., Meyer, D. R., and Meyer, P. M., 1980, The Thompson effect is a function of the presence or absence of preoperative memories, *Physiol. Psychol.* 8:15–19.

Braun, J. J., 1966, The neocortex and visual placing in rats, *Brain Res.* 1:381–394.

Braun, J. J., Meyer, P. M., and Meyer, D. R., 1966, Sparing of a brightness habit in rats following visual decortication, *J. Comp. Physiol. Psychol.* 61:79–82.

Braun, J. J., Lundy, E. G., and McCarthy, F. V., 1970, Depth discrimination in rats following removal of visual neocortex, *Brain Res.* 20:283–291.

Cloud, M. D., Meyer, D. R., and Meyer, P. M., 1982, Inductions of recoveries from injuries to the cortex: Dissociation of equipotential and regionally specific mechanisms, *Physiol. Psychol.* 10:66–73.

Cooper, R. M., Blochert, K. P., Gillespie, L. A., and Miller, L. J., 1972, Translucent occluders and lesions of posterior neocortex in the rat, *Physiol. Behav.* 8:693–697.

Dalby, D. A., Meyer, D. R., and Meyer, P. M. M., 1970, Effects of occipital neocortical lesions upon visual discriminations in the cat, *Physiol. Behav.* 5:727–734.

Davis, N., and LeVere, T. E., 1979, Recovery of function after brain damage: Different processes and facilitation of one, *Physiol. Psychol.* 7:233–240.

Feeney, D. M., and Sutton, R. L., 1986, Catecholamines and recovery of function after brain damage, in: *Pharmacological Approaches to the Treatment of Brain and Spinal Cord Injury* (B. Sabel and D. G. Stein, eds.), Plenum Press, New York, pp. 121–139.

Feeney, D. M., Sutton, R. L., Boyeson, M. G., Hovda, D. A., and Dail, W. G., 1985, The locus coeruleus and cerebral metabolism: Recovery of function after cortical injury, *Physiol. Psychol.* 13:197–203.

Glendenning, R. L., 1972, Effects of training between two unilateral lesions of visual cortex upon ultimate retention of black–white habits by rats, *J. Comp. Physiol. Psychol.* 80:216–229.

Gray, T. S., and Meyer, D. R., 1981, Effects of mixed training and overtraining on recoveries from amnesias in rats with visual cortical ablations, *Physiol. Psychol.* 9:54–62.

Harlow, H. F., Davis, R. T., Settlage, P. H., and Meyer, D. R., 1952, Analysis of frontal and posterior association syndromes in brain-damaged monkeys, *J. Comp. Physiol. Psychol.* **45**:419–429.

Hata, M. G., Diaz, C. L., Gibson, C. H., Jacobs, C. E., Meyer, P. M., and Meyer, D. R., 1980, Perinatal injuries to extravisual cortex enhance the significance of visual cortex for performance of a visual habit, *Physiol. Psychol.* **8**:9–14.

Horel, J. A., Bettinger, L. A., Royce, G. J., and Meyer, D. R., 1966, Role of neo-cortex in the learning and relearning of two visual habits by the rat, *J. Comp. Physiol. Psychol.* **61**:66–78.

Howard, R. L., and Meyer, D. R., 1974, Motivational control of retrograde amnesia in rats: A replication and extension, *J. Comp. Physiol. Psychol.* **74**:37–40.

Howard, R. L., Glendenning, R. L., and Meyer, D. R., 1974, Motivational control of retrograde amnesia: Further explorations and effects, *J. Comp. Physiol. Psychol.* **86**:187–192.

Howarth, H., Meyer, D. R., and Meyer, P. M., 1979, Perinatal injuries to the visual cortex enhance the significance of extravisual cortex for performance of a visual habit, *Physiol. Psychol.* **7**:163–166.

Hughes, H. C., 1977, Anatomical and neurobehavioral investigations concerning the thalamocortical organization of the rat's visual system, *J. Comp. Neurol.* **175**:311–336.

Hunter, W. S., 1930, A consideration of Lashley's theory of the equipotentiality of cerebral action, *J. Gen. Psychol.* **3**:455–468.

Jackson, J. H., 1873, Lectures on the diagnosis of tumors of the brain, *Med. Times Gaz.* **2**:139.

Jonason, K. R., Lauber, S., Robbins, M. J., Meyer, P. M., and Meyer, D. R., 1970, The effects of d-amphetamine upon discrimination behaviors in rats with cortical lesions, *J. Comp. Physiol. Psychol.* **73**:47–55.

Kircher, K. A., Braun, J. J., Meyer, D. R., and Meyer, P. M., 1970, Equivalence of simultaneous and successive neocortical ablations in production of impairments of retention of black–white habits in rats, *J. Comp. Physiol. Psychol.* **71**:420–425.

Lashley, K. S., 1935, The mechanisms of vision: XII. Nervous structures concerned in the acquisition and retention of habits based on reactions to light, *Comp. Psychol. Monographs* **11**:43–79.

Lavond, D., and Dewberry, R. G., 1980, Visual form perception is a function of the visual cortex. I. The rotated horizontal–vertical and oblique-stripes pattern problems, *Physiol. Psychol* **8**:1–8.

Lavond, D., Hata, M. G., Gray, T. S., Geckler, C. L., Meyer, P. M., and Meyer, D. R., 1978, Visual form perception is a function of the visual cortex, *Physiol. Psychol.* **6**:471–474.

LeVere, T. E., and Davis, N., 1977, Recovery of function after brain damage: The motivational specificity of spared neural traces, *Exp. Neurol.* **57**:883–899.

LeVere, T. E., and Fontaine, C. W., 1978, A demonstration of the importance of RNA metabolism for the acquisition but not performance of learned behaviors, *Exp. Neurol.* **59**:444–449.

LeVere, T. E., and Morlock, G. W., 1973, The nature of visual recovery following posterior decortication in the hooded rat, *J. Comp. Physiol. Psychol.* **83**:62–67.

LeVere, T. E., LeVere, N. D., Chappell, E. T., and Hankey, P., 1984a, Recovery of function after brain damage: On withdrawals from the memory bank, *Physiol. Psychol.* **12**:275–279.

LeVere, T. E., Chappell, E. T., and LeVere, N. D., 1984b, Recovery of function after brain damage: On deposits to the memory bank, *Physiol. Psychol.* **12**:209–212.

Maling, H. M., and Acheson, G. R., 1946, Righting and other postural activity in low decerebrate cats after d-amphetamine, *J. Neurophysiol.* **9**:379–386.

Meyer, D. R., 1971, The habits and concepts of monkeys, in: *Cognitive Processes of Non-Human Primates* (L. E. Jarrard, ed.), Academic Press, New York, pp. 83–102.

Meyer, D. R., 1972a, Access to engrams, *Am. Psychol.* **27**:124–133.

Meyer, D. R., 1972b, Some features of the dorsolateral frontal and inferotemporal syndromes in monkeys, *Acta Neurobiol. Exp.* **32**:235–260.

Meyer, D. R., 1984, The cerebral cortex: Its roles in memory storage and remembering, *Physiol. Psychol.* **12**:81–88.

Meyer, D. R., and Beattie, M. S., 1977, Some properties of substrates of memory, in: *Neuropeptide Influences on Brain and Behavior* (L. Miller, C. Sandman, and A. Kastin, eds.), Raven Press, New York.

Meyer, D. R., and Meyer, P. M., 1977, Dynamics and bases of recoveries of functions after injuries to the cerebral cortex, *Physiol. Psychol.* **5**:133–165.

Meyer, D. R., Isaac, W., and Maher, B., 1958, The role of stimulation in spontaneous reorganization of visual habits, *J. Comp. Physiol. Psychol.* **51**:546–548.

Meyer, D. R., Hughes, H. C., Buchholz, D. J., Dalhouse, A. D., Enloe, L. J., and Meyer, P. M., 1976, Effects of successive unilateral ablations of principalis cortex upon performances of delayed alternation and delayed response by monkeys, *Brain Res.* **108**:397–412.

Meyer, D. R., Gurklis, J. A., and Cloud, M. D., 1985, An equipotential function of the cerebral cortex, *Physiol. Psychol.* **13**:48–50.

Meyer, P. M., 1963, Analysis of visual behavior in cats with extensive neocortical ablations, *J. Comp. Physiol. Psychol.* **56**:397–401.

Meyer, P. M., and Meyer, D. R., 1982, Memory, remembering, and amnesia, in: *Expression of Knowledge* (R. L. Isaacson and N. E. Spear, eds.), pp. 179–212, Plenum Press, New York.

Meyer, P. M., Horel, J. A., and Meyer, D. R., 1963, Effects of dl-amphetamine upon placing responses in neodecorticate cats, *J. Comp. Physiol. Psychol.* **56**:402–414.

Meyer, P. M., Yutzey, D. A., Dalby, D. A., and Meyer, D. R., 1968, Effects of simultaneous septal–visual, septal–anterior, and anterior–posterior lesions upon relearning a black–white discrimination, *Brain Res.* **8**:281–290.

Meyer, P. M., Johnson, D., and Vaughn, D., 1970, The consequences of septal and neocortical ablations upon learning a two-way avoidance response, *Brain Res.* **22**:113–120.

Meyer, P. M., Meyer, D. R., and Cloud, M. D., 1986, Temporal neocortical injuries in rats impair attending, but not complex visual processing, *Behav. Neurosci.* **100**:845–851.

Petrinovich, L., and Bliss, D., 1966, Retention of a learned brightness discrimination following ablations of the occipital cortex in the rat, *J. Comp. Physiol. Psychol.* **61**:136–138.

Petrinovich, L., and Carew, T. J., 1969, Interaction of neocortical lesion size and interoperative experience in retention of a learned brightness discrimination, *J. Comp. Physiol. Psychol.* **68**:451–454.

Riopelle, A. J., Alper, R. G., Strong, P. N., and Ades, H. W., 1953, Multiple discrimination and patterned string performance of normal and temporal-lobectomized monkeys, *J. Comp. Physiol. Psychol.* **46**:145–149.

Ritchie, G. D., Meyer, P. M., and Meyer, D. R., 1976, Residual spatial vision of cats with lesions of the visual cortex, *Exp. Neurol.* **53**:227–253.

Robbins, M. J., and Meyer, D. R., 1970, Motivational control of retrograde amnesia, *J. Exp. Psychol.* **84**:220–225.

Squire, L. R., 1986, Mechanisms of memory, *Science* **232**:1612–1619.

Stewart, J. W., and Ades, H. W., 1951, The time factor in reintegration of a learned habit lost after temporal lobe lesions in the monkey (*Macaca mulatta*), *J. Comp. Physiol. Psychol.* **44**:479–486.

Thompson, R., 1960, Retention of a brightness discrimination following neo-cortical damage in the rat, *J. Comp. Physiol. Psychol.* **53**:212–215.

Thompson, R., and Bryant, J. H., 1955, Memory as affected by activity of the relevant receptor, *Psychol. Rep.* **1**:393–400.

Wetzel, A. B., 1969, Visual cortial lesions in the cat: A study of depth and pattern discrimination, *J. Comp. Physiol. Psychol.* **68**:580–588.

Woolsey, C. N., Settlage, P. H., Meyer, D. R., Sencer, W., Pinto-Hamuy, T., and Travis, A. M., 1950, Patterns of localization in precentral and "supplementary" motor areas and their relation to the concept of a premotor area, in: *Patterns of Organization in the Central Nervous System*, Proceedings of the Association for Research in Nervous and Mental Disease, William & Wilkins, Baltimore, pp. 238–264.

4

Neural Spare Capacity and the Concept of Diaschisis
Functional and Evolutionary Models

ROBERT B. GLASSMAN and AARON SMITH

1. INTRODUCTION

1.1. Intimations of Spare Capacity

Instances of extensive recovery from brain damage suggest that the brain has spare capacity. In a follow-up of 50 infantile hemiplegics who sustained surgical removal of all neocortex of one hemisphere for intractable seizures or other injuries, Wilson (1970) reported that all but one developed normal speech or recovered it completely irrespective of which hemisphere had been removed (Wilson, 1970, p. 166). Smith and Sugar (1975) carried out a comprehensive neuropsychological follow-up on a patient at ages 21 and 26 who had had left hemispherectomy for seizures as a 5½-year-old boy. He demonstrated superior language and intellect, including WAIS verbal IQ of 126 and performance IQ of 102, had graduated from a university, and was working as a traffic controller. Normal psychological function also was observed in 279 cases of hydrocephalus with onset before the end of the first year of life (Berker *et al.*, 1983). Most remarkable is one young man in whom a CAT scan shows ventricular dilatation occupying over 95% of the intracranial space. When tested on the Michigan Neuropsychological Battery at age 25, he had graduated from Sheffield University with honors in mathematics, had a verbal IQ of 140 and performance IQ of 130, and had been successfully employed for several years.

ROBERT B. GLASSMAN • Department of Psychology, Lake Forest College, Lake Forest, Illinois 60045.   AARON SMITH • University of Michigan NHR Project, Ann Arbor, Michigan 48103.

1.2. Do Large Ablations More Readily Reveal Spare Capacity?

In a provocative but largely unrecognized paper, B. Campbell (1960) argued that there is a "factor of safety in the nervous system," as in other organs. Some of Campbell's comments suggested greater recovery from hemispherectomy in cats than was observed by Glassman and Glassman (1977) following lesser sensorimotor cortical ablations. Summarizing the results from 12 human adults hemispherectomized for tumors, Burklund (1969) commented that "The quality and quantity of movement appear to be much greater following removal of an entire cerebral hemisphere, including the basal ganglia, than in excisions involving only a portion of the cerebral hemisphere, especially lesions of the sensorimotor strip" (p. 25). Consistent with this conclusion are observations of kittens by Villablanca and Olmstead and associates (e.g., Villablanca et al., 1984). Because earlier results with lesions restricted to frontal cortex or caudate nuclei revealed long-term sensorimotor deficits similar to those that followed lesions made in the adult, they were surprised to observe special sparing of sensorimotor behaviors when they removed an entire hemisphere (neocortex plus neostriatum and hippocampus).

Reminiscent of Burklund's comments about a possible role of the red nucleus in motor recovery, Villablanca et al. reported anomalous but well-organized projections to the red nucleus from the remaining hemisphere. Lesion-induced growths may willy-nilly contribute to behavioral recovery, be irrelevant, or cause harm because they obey "narrow-minded" embryological programs. For example, this generalization applies to the principle that axons tend to conserve their total arborization (Devor and Schneider, 1975). Because of the narrow focus of compensatory processes there may sometimes be a better outcome when there is a larger ablation; this may preclude local imbalances while engaging global developmental processes. Markowitsch (1985) lists other observations of more extensive ablations causing lesser deficits.

1.3. Evolutionary and Functional Puzzle of Spare Capacity

Why should begrudging natural selection bestow features beyond those barely adequate for survival and reproduction? We propose three hypotheses. (1) The brain breaks information down into simple qualities represented in "diffuse domains"; each such domain has no inner differentiation beyond a one-dimensional range of values. This hypothesis is intimately related to the notion of diaschisis. (2) Provisions for subsystem reliability are related to system complexity, with recoverability from brain damage as a byproduct. (3) Evolutionary paths to large brains may involve preadapting factors having nothing to do with information processing. We explain how these hypothetical principles work individually and together.

2. THE ELEMENTS OF BRAIN INFORMATION PROCESSING ARE DIFFUSE DOMAINS

2.1. Unrealistic Aspects of Machine Metaphors and the Bugaboo Mosaic

Analytically it is necessary to see any organized whole in terms of components, but components of what sort? It is naive to reify psychological functions as discrete machinelike parts of the brain, each bounded by a closed surface. Although evolution entails a degree of modular packaging of functions (Glassman and Wimsatt, 1984), natural selection also seizes expedients, and this suggests that the physical components of the brain must have mixed functions.

The idea that the brain comprises a mosaic is parsimonious but simplistic; yet for as long as this conception has been set aside on empirical or logical grounds, it has continued to resurrect itself (e.g., Sherrington, 1941, p. 228). For example, studies of thousands of aphasics have failed to confirm the existence of discrete language syndromes as claimed by Geschwind (1970, p. 941). Smith (1978) presented a historical review of the "mosaic issue" with respect to language functions, and recently Markowitsch (1985) has reviewed the literature on mnemonic information processing in attempting to define a sensible position between localizationist and nonlocalizationist views of memory.

2.2. Diffuse Domains Are Adequate for Maintaining Distinctions: A Metaphor of the Brain as an Immense Set of Counters

We propose an extension of Campbell's (1960) hypothesis that diffuse domains of neural excitation and inhibition help explain how a factor of safety might work in the brain. Beginning with the conservative premise that the "bottom line" of brain functions is the mediation of appropriate distinct outputs for distinct inputs, we suggest that a system of diffuse domains can hold a great deal of such discriminative information. Without wishing to demean the remarkable capabilities of the brain, we first offer a brief sampling of behavioral evidence of parsimony in discriminative processes:

1. Images. Pictures match what we would see monocularly only when viewed from a particular angle and distance, yet they look realistic from many angles and distances (Hochberg, 1972).
2. Memory. Modest capacity of everyday memory is evident in the inability of most people to reconstruct the correct configuration of the Lincoln-head penny (Nickerson and Adams, 1979/1982). Also, eyewitness identification is surprisingly poor and is coupled with lack of awareness of fallibility (Loftus and Loftus, 1980).
3. Probability judgments. Despite long experience, gamblers sometimes

mistake independent events for connected ones in assuming that a
"lucky break" is overdue. Also, the stock market is famous as an arena
of misperceived patterns; for example, data about Dow–Jones Industrial
Average fluctuations often taken to support the notion of a summer
rally—part of Wall Street lore—equally well suggest a summer crash in
each of the same years (Andrew, 1984).

4. Motivational confusions. Nisbett and Wilson (1977) review many studies in which subjects misattributed the sources of their own motives;
these social psychology findings are reminiscent of the literature on
salesmanship. For example, in the "committing question" technique,
the salesman gradually builds friendly small talk into an affirmative
mood fog that suffuses perception of his product (Linden, 1979; Frank
and Lapp, 1959).

With this sampling of humbling examples in the background, let us consider how
the brain might manage its input–output relationships.

2.2.1. What Is a Diffuse Domain?

The information capacity of an associative system can be characterized in
terms of simple component modules; each merely summates all the inputs that
converge on it. Such a module might be a single neuron or part of a dendritic
field, a neuronal pool, the regional concentration of a single neurotransmitter, a
large region whose electrotonic potential varies in a unitary way, or some highly
distributed, irregularly shaped subsystem that combines these things and acts as a
unit. Thus, a diffuse domain may occupy a tiny region, a broad mass of the
brain, or anything in between. In any case, the subset of brain substance functioning as a module has no internal differentiation as far as that function is
concerned. A single modular function whose anatomic substrate encompasses a
substantial volume of the brain merely summates excitatory and inhibitory influences over that volume without otherwise using its internal spatial or chemical
differentiation. At the same time that one aspect of a large region's activity
comprises a diffuse, unitary modular function, other aspects may be composed of
more differentiated functions of smaller modules.

In one version of neural network theory, Wood (1982) proposed input lines
broadly converging to and diverging from a set of associative modules; each
module summated its weighted inputs. Computer-simulated ablations of subsets
of these inputs sometimes left the system's ability to discriminate nearly intact.
Even in brain systems such as those involved in fairly precise sensorimotor
functions, point-to-point topographic mappings can do no better a discriminative
job than diffuse overlapping area-to-area axonal connections, with the map regis-
tering local average excitation levels rather than punctiform excitations (Churchland, 1986). Overlapping fiber distributions may have the additional advantage

of allowing for minor adjustments in the mapping function between two layers of brain tissue.

Brain organization can be considered a result of the way the variously weighted outputs of such modules act in combination on other modules. The behavior of one module is a one-dimensional variable. To make this issue easier to think about, each module in our metaphor has a range of discrete integer values. This is biologically unrealistic, but the logic is basically the same as for subsystems that vary continuously. The informational capacity of a set of several modules is simply the number of possible combinations of activation levels of the set (Glassman, 1985). This is easiest to think about when the ranges of all modules are the same. For convenience, let us suppose each module has a range of ten values.

This hypothetical system then becomes analogous to a counter, and each module of our metaphorical system can be thought of as a single digit that counts from 0 to 9. A neural system of one-dimensional modules thus "keeps its books" merely by reserving a numerical label for each distinct input–output relationship. The business of the whole nervous system, in any one instant of time, is to maintain throughout its substance a large array of these sets of digits.

As noted above, one type of diffuse domain might be a portion of a dendritic field. Indeed, the dendritic arborization of a neuron may easily be thought of as analogous to a five-digit counter because of the phenomenon of input segregation. Grinnell (1977) lists several regions in the brain in which there are on the order of five different categories of inputs contacting different parts of the dendritic tree; the functional significance is that inputs nearer the soma have stronger excitatory or inhibitory influences. Here, we hypothesize additionally that the electrotonic influences of segregated input categories do not blend continuously with one another but form "quantum levels" that act analogously to digits.

2.2.2. Shared Subsystems

There are two senses in which modules may be shared among different brain functions: (1) functions may exclusively reserve portions of the range of a module (or set of modules functioning as a larger-range unit), or (2) more than one function may use the same portions of range, with the distinction between this and other functions set by an additional module. For an example of the latter, suppose that function F and function G both use the joint range of modules J, K, and L from values 000 to 398 (i.e., $J = 0$ to 3, $K = 0$ to 9, and $L = 0$ to 8). But a fourth module, I, distinguishes between functions F and G by taking on two alternative values, say $I = 5$ for function F and $I = 6$ for function G. We define such additional modules as "higher" or more significant digit places. In general, the simplest brain functions use only a segment of the range of a single digit-place module, whereas the greater differentiation in complex functions requires the joint range of several adjacent digit places.

Modules containing higher-place or more significant digits represent more stable conditions of the brain. We can think of any number of levels of more significant digits as setting contexts for all the lower-place digits. (Whereas we ordinarily write lower places to the right on a horizontal line, the brain must have a variety of other conventions for setting hierarchies of this sort.) In ordinary counting, a small change in a higher place drastically changes the meaning of numbers shown in lower places.

2.2.3. Large Information Capacity of the Abstract Model

The digit-sets model suggests that the brain has an immense informational capacity even where there is poor spatial differentiation. This implication emerges from the basic property of combinations of multivalued elements, that there is an exponential increase in values with the addition of new elements. To continue the illustration using modules that are like digit places having a range of ten possible values, each time another module is added to a set, the base range of 10 is raised to one higher power. Thus, five digit places contain 100,000 possible values (0 to 99,999), and six places have one million possible values, and so on.

2.2.4. Information Capacity of the Anatomicochemical Brain

Some possible memory substrates are as follows: regional expansions of dendritic fields, expansion or multiplication of postsynaptic receptor sites (Lynch and Baudry, 1984) or presynaptic autoreceptor sites, proliferation of glia, enlarged somas supporting persistent neuronal firing, and local changes in amounts of neurotransmitter-regulating enzymes. In all such cases there should be corresponding changes in protein concentrations (see Davis and Squire, 1984).

In one view, the brain's great information capacity requires highly differentiated spatial patterns of dendrites and axon terminals. Researchers who seek memory correlates in tissue homogenates have not been able to get this thorn out of their side. Could there be sufficient room in a brain if each item of memory information had to be embodied in a diffuse change of a sort that would be evident in a homogenate? Where could organized complexity possibly reside other than in fine spatial structure?

Perhaps memory is embodied in changes in the distribution of a large complement of brain proteins and peptides. Conceivably, what is important is some regional average concentration of these substances in each "diffuse domain." Thus, the significance of neurons' shapes may be in protein-typical aspects of spatial conformations that are influenced by learning variables. The regional type of conformation, rather than a computer-chip-like precision in the smallest spatial details of unique individual meetings of axons and dendrites, may be what is functionally significant. The greater the number of chemical "dimensions" of the brain, the greater is the possibility that a combinatorial effect of these dimensions can code a large number of distinct memories. Our digit-places model makes this point clear. In other words, it is possible that mass action and equipotentiality are properties of all regions of brain tissue for some functions (see Frommer, 1978; Meyer, 1984; Meyer *et al.*, 1985; Watanabe *et al.*, 1984 for discussions of mass action).

2.3. Von Monakow's Concept of Diaschisis

Our diffuse-domains hypothesis can be linked to a number of historical sources, including Gowers' comment that a single functional center "may consist of elements that are anatomically distant" (Gowers, 1885, pp. 4–5). However, it is von Monakow's concept of diaschisis that is most intimately related to the hypothesis that the brain maintains information in diffuse domains. We review the concept of diaschisis (see also the chapter in Finger and Stein, 1982) and show its importance for our model.

Like Hughlings Jackson, von Monakow (1914) recognized the need to distinguish between deficits caused by some active process, e.g., irritation, and deficits of a more passive sort. Von Monakow emphatically defined diaschisis as a passive phenomenon, contrasting this form of shock with the hypothetical forms emphasized by other workers. Yet in his definition of diaschisis, von Monakow allowed for imbalances, some of which involve release phenomena as in spinal shock (see Sherrington, 1947, for corroborating testimony) and some cases of aphasia.

According to von Monakow, diaschisis represents a shock of distinct neural connections and a "ceasing of operations," . . . "limited in time," where "the initial symptoms blend fluently with the residual ones." "The temporary local symptoms can be separated into legitimate and illegitimate ones," where the latter "are connected with pathological concomitant symptoms (hemorrhaging, clogged vessel, . . . , edema, . . . , etc.)," and the former involve a "suspension of excitability (cessation of function)" whose ramifications follow the normal neural connections as far as they extend "as a result of disruption of the brain substance. . . ." (all quotes from von Monakow, 1914).

Von Monakow's massive tome includes a conception of what we think of today as feature analysis (". . . nerve cells react . . . in . . . organized groups only to stimuli of a special quality . . ."), and perhaps it is this conception that allowed him to say that diaschisis involves a loss of excitability in some regions to a "physiologically well-defined parameter of stimulation" from the damaged region, rather than a complete loss of function. At the same time, "this parameter does not coincide with the usual physiological paths of innervation which extend from the center." We assume he meant by this that when a given region of the brain excites the other regions with which it connects, it typically does so

as a function of excitability characteristics that neural network theorists now call
"weights." In diaschisis, these weights are generally reduced, but the different
connections do not have their excitability diminished by a single common factor.

Von Monakow's concept of diaschisis as a passive phenomenon was motivated by his perception of the immensity of the problem of neural reorganization
demanded by the notions of vicariation and compensation. He rejected the possibility of "excess capacity" when conceived as subsystems, ordinarily held in
abeyance or used for other functions, which must make new, intelligently organized compensations following brain damage.

> . . . When one invests the central nervous system with the capability . . . to act
> functionally in "several fashions," not much room is left for "overcapacity" in the
> sense of compensation while the brain parts in question [are] under pathological
> conditions. . . . And as far as "vicariating hypertrophy" is concerned, with the
> increase of [a previously] secondary capacity, [one would expect these] residual
> symptoms: upset of chronological sequence, spurious merging of successive acts in
> simultaneous ones. . . . Even that hypothesis much loved by clinicians of the vicariat-
> ing onset of other parts of the brain for the damaged one is lastly traced back to the
> work of Munk. . . .

Munk might have realized more clearly that in recovery from spinal shock
following cord transections, the " 'new constitution' (in the severed section of
the spinal cord) . . . is more damaging than useful to the original purpose."
Moreover, in considering dogs' partial recovery of locomotion following removal of the cerebellum, Munk should have given more explanatory weight than
he did to diaschisis, because a global loss of locomotion follows cerebellectomy
even though "after this trauma the extracerebellatory locomotion centers remain
whole anatomically and are not inhibited by the neighborhood of the wound."

> It seems arbitrary to me when Munk regards only the spinal motoric centers as
> accessible to "shock effects" . . . it seems unreasonable to expect that it should
> [thus] limit itself [and bypass] those segments of the oblongata, of the midbrain, etc.
> which are connected to the cerebellum fiber masses. . . .

He asks how can "those apparatuses, which allow the animal without a cerebellum to locomote [respond to] demands placed on them in considerably 'increased' fashion" when the anatomic loss must have reduced the ability of these
subsystems to communicate with other regions.

Although the following two quotes, in their emphasis on competition and
compensation, might be thought to contain hints of hypothesized intelligence or
social system metaphor at the level of neural subsystems, the passages are better
interpreted in the more passive terms of von Monakow's idea that when neural
circuits suffer a loss of nerve cells, there may be elevated "internal resistance"
of an entire region.

> Since in every nervous capacity . . . uncounted stimuli of periodic and locally differ-
> ent form compete with each other, the end result of such a match should correspond to
> the respective power relationships of the innervation components in question. The

more elements within a combined function are placed in diaschisis condition . . . the more this could lead to progressive diaschisis . . . and eventually arrest the work along the entire line. . . .

In a similar vein is his analysis of a speech deficit:

If . . . the stimulus circle X is lost temporarily to [a subsystem] usually briskly exchanging stimuli with stimulus circle Y, indeed it can transmit its inactivity even to stimulus areas usually active in association and so paralyze the entire area of function of certain physiological valuation in its activity. This entire occurrence frequently has the character of an acute, shocklike interference with equilibrium (diaschisis), which in a healthy brain is usually balanced in a short time by corrective interference of other . . . functioning components of the cortex or of deeper parts of the brain as well.

All these quotations point to the idea that except for a simple excitability adjustment, each region of a recovered brain, after the dissipation of diaschisis and other reversible pathophysiological influences, comes back to doing qualitatively exactly what it always did.

Language function has been a favorite topic of superficial reasoning about extreme localization of psychological function; for example, the dichotomania in popularizations of the split-brain literature has reflected back into the scientific literature. It has therefore been widely assumed that recovery implies vicariation. A few additional remarks about the logic of studying language representation will be useful at this point. The literature contains numerous cases of recovery from aphasia following left hemisphere lesions in adults, and Bastian, Gowers, Goldstein, Russell, and others attributed the recovery of language functions wholly or partly to the capacities of the intact right hemisphere (Burklund and Smith, 1977).

However, the restoration of language attributed to the compensatory capacities of the right hemisphere might be consistent with any of three possible interpretations: (1) the right hemisphere structures had not previously been involved in the language functions that were lost; (2) the right hemisphere had participated to a limited degree, and following damage to the left hemisphere structures, the role of the right hemisphere structures was augmented; or (3) the lost language functions had been initially acquired by the structures in both hemispheres, but with cerebral maturation the role of the right hemisphere was inhibited; as a result of damage to the left hemisphere, this inhibition of the right was reduced or eliminated (A. L. Campbell et al., 1981, p. 506).

The notion that language is "taken over" by an intact area that had not previously participated, however, would require that the new structures "learn" language anew, including grammar, syntax, etc. The rate of recovery of language in some adult aphasics, however, is far too rapid. Comprehension of speech and reading and, to a lesser degree, even writing is demonstrated to varying degrees not long after the entire left "dominant" hemisphere has been removed (left hemispherectomy for tumor), in far less time than it would take for an adult to learn a new language.

2.4. *Experimental Studies of Diaschisis*

There have been few direct tests of the diaschisis idea. Some empirical research has demonstrated shocklike effects, manifested as reduced sensory evoked potentials and slowed EEG, in the vicinity of a brain wound (Glassman and Malamut, 1976), but electrophysiological evidence of diaschisis was not observed extending along functional pathways in the hippocampal system (West *et al.*, 1976; Markowitsch and Pritzel, 1978). Kempinsky (1958) inferred that diaschisis was responsible for the depressed EEG and evoked potentials in the intact visual cortex following a lesion in the other hemisphere, because this effect was prevented by sectioning of the corpus callosum several weeks in advance. However, other studies of the visual system have not revealed diaschisis (reviewed by Marshall, 1984).

Feeney *et al.* (1985) present evidence that catecholaminergic hypoactivity underlies at least some forms of diaschisis or "remote functional depression." These authors describe a postlesion window of opportunity for enhancing recovery by boosting catecholaminergic function. Because catecholaminergic influences are widespread, it is arguable whether the interesting results of Feeney *et al.* represent the thesis that diaschisis extends along specific functional pathways.

Isaacson (1975), taking a very skeptical view of vicariation, lists 16 simpler alternative sources of change in vegetative effects following brain injury. He cites numerous instances of failure to find enhanced recovery from early brain damage and instances of overstatement of positive results reported in the literature. Although there is no question that organisms do show remarkable behavioral plasticities, it is possible that authors have indeed usually exaggerated the degree of brain plasticity for many of the functions studied. Perhaps the marvel of the brain is somehow to amplify at the behavioral level distributions of subtle parametric changes at the neuronal level. In this regard, our model preserves von Monakow's essential point that after recovery all parts of the brain are qualitatively doing exactly what they did before, but in some parts to a greater degree.

2.5. *Diaschisis in the Model*

More severe losses of function after brain insult—or diaschisis—will occur insofar as the damaged functional systems serve as more significant digit places. In some cases, even slight damage and only the most minute loss of information will cause a broad spectrum of functional loss because of the change in the setting of a subsystem having the property we describe here as a higher digit place. Regarding memory functions, lost function without much loss of information has been spoken of as a deficit in "retrieval." Our model provides a new way of considering this idea.

A loss or alteration of a given size in a subsystem serving as a higher digit place causes a greater functional disturbance for two reasons. First, higher places are components of a greater number of digit sets. In anatomic terms, the neural embodiment of a higher-place digit covers a large expanse of the brain, or, alternatively, many lower-place subsystems by direct or indirect connections refer to the same small region setting a higher-place value.

Second, it is an interesting property of digit sets that shrinkage by a given amount of range in any one digit place causes an equal overall loss of values in the whole set as does the same shrinkage in any other digit place; however, lower-place losses cause a scattered pattern of gaps in the digit-set values, whereas higher-place losses are associated with more global gaps. For example, a loss of the numerals 8 and 9 either in the "units" digit place of a five-digit counter or in the "thousands" digit place means that one-fifth of the 100,000 possible values from 00,000 to 99,999 can no longer be registered. But the losses caused by the "deficit" in the thousands place are in large chunks of 2000 consecutive values (e.g., 88,000 through 89,999 and 98,000 through 99,999), as opposed to groups of two consecutive values with the deficit in the units digit (e.g., 86278–86279 and 86288–86289). Although residual functions might somehow interpolate over scattered "units-type" neural system losses, interpolation cannot occur with global losses of value settings.

2.6. Implications of the Model for Understanding Early Brain Damage

This model can be fitted to cases of early brain damage in which there is unusually great sparing and to the finding that older patients show more evidence of diaschisis than do younger ones (von Monakow and Mourgue, 1928). For example, in one sample of patients with verified lesions restricted to one hemisphere (Berker and Smith, 1981), all six of the oldest (57–65 years old) showed bilateral sensory and/or motor deficits, whereas only two of the six youngest (25–30 years old) did so.

Ontogeny of brain functions may involve a hierarchical sequence of establishing parameter settings (Glassman and Wimsatt, 1984). In terms of the present model, higher-place digits are generally the first to be set into correspondence with behaviors as the brain matures. The first settings later become most refractory to change, because developmental organizing processes have passed them by. Normally, this is fortunate, because subsequent settings of lower-place digits retain their correct code values only in the context of stable and accurate higher-place digits. Were it not for sequential developmental ossification of most-significant parameters, dramatic recovery from brain damage might occur without the need to relearn radically or reconstruct lost information. Further theoretical and empirical research may suggest ways in which simple therapeutic interventions could cause the latent information to blossom forth.

3. ERROR AND RELIABILITY WHEN LARGE NUMBERS OF SUBSYSTEMS INTERACT

3.1. Introduction to Neuroeconomics: Costs and Benefits in the Natural Selection of Spare Neural Capacity

Our model suggests a ready evolutionary pathway to larger brains. Whereas reorganizational change is always a difficult problem, diffuse domains taken one at a time are inherently simple; therefore, evolutionary random variation might easily increase their size or number. But will such additional loads be passed by natural selection? Are such loads useful?

With every human birth a large head meets the limited size of the pelvic passageway. The birth canal size itself is constrained evolutionarily by the skeletal requirements of upright-postured locomotion. Studies of human neonates (reviewed by Berker and Smith, 1983) have revealed blood in the CSF, indicating intracerebral hemorrhage, possibly in as many as 10 or 15% of unselected consecutive births (including both full-term and premature). Moreover, almost all neonates show some evidence of asphyxia. Notwithstanding these injuries, only 2–3% show neurological signs (Ford, 1960). Thus, in these data there is both evidence of evolutionary pressure against large head size and evidence of spare capacity.

In this section we propose that the evolutionary foundation of safety margins is in the probabilistic nature of physical processes and the consequent vulnerability of complex systems. Our argument is not that safety margins evolved to protect against brain trauma but that they underlie reliability of normal function. More specifically, spare capacity should be present in proportion to the number of different subsystems that must be coordinated, to the duration of time over which they must act, and to the importance for survival and reproduction of the behaviors in which the subsystems participate. In the case of highly social species such as human beings, the likelihood that natural selection sometimes focuses on units larger than the individual (e.g., Glassman et al., 1986) suggests an especially high brain safety factor.

A close look at some aspects of behavior indicates semiautonomous subsystems having built-in backups. As a prototype for regulatory and motivational functions, Lorenz (1969, 1981) pointed to the coupled pacemaker subsystems of the heart. The atrioventricular node's natural rhythm is slightly longer than the pace of the sinoatrial node, which usually does the driving; if the SA node fails, the AV may substitute as a secondary pacemaker (Schmidt and Thews, 1983, p. 360). Under the heading of "tool activities," ethologists have documented the partial autonomy and backing-up of animals' motivational subsystems (Eibl-Eibesfeldt, 1975). Just as there is a lag before a physiological or behavioral backup engages, so seems to be the case with developmental timing. For example, accelerated catching up has been observed in the development of brain-damaged children (Smith, 1984).

3.2. Two Types of Safety Factor: Reiteration (Redundancy) and Aiming High

Backing up subsystems is one of two ways in which a safety factor logically may be attained. Alternatively, evolution may "aim high" in the "specifications" for a subsystem. The random variation component of evolution suggests that both of these routes have been exploited as opportunities have fortuitously arisen. Evolutionary "aiming high" has been studied in nonneural biological systems. For example, bones are subject to both accidental and fatigue fractures; they generally have a safety factor between two and five (Alexander, 1984). Using limpet shells as a model system, Lowell (1985) has argued that the safety factor difference between actual mean strength and the mean maximum load likely to be encountered is directly related to the variability of strength in the population. Considering the great variability of psychological traits in a human population, Lowell's finding suggests that we may have a very large brain safety factor.

There is ample nonneural evidence that evolution has sometimes taken the first safety factor route of backing up, or reiterating, subsystems; the ability to survive normally with only a single kidney is one such example (Campbell, 1960). It might be argued that bilateral duplication of kidneys, liver mass, and lungs is a mere epiphenomenon of the basic embryology of bilaterally quasisymmetrical organisms, but because all parts of the body imply an additional burden of maintenance and additional possibilities of injury or disease, we might expect natural selection to keep the size of organs, bilateral or otherwise, to a minimum. That this has not occurred is suggestive of natural selection for safety factor. The mass of kidneys, liver, or lungs arises largely from reiterations of the same basic elements, and a small region of any of these organs performs its role in essentially the same way as almost any other region. This may be analogous to some instances of mass action in parts of the brain.

We do not know the degree to which reiterations of behaviorally significant brain subsystems are concentrated in local, specialized cohesive masses or interleaved through large extents of the brain's mass. Marshall (1984) gives some examples of the former in sensory and motor systems and in the nigrostriatal path; the papers about mass action, cited earlier, are also relevant. Studies of initial and later effects of hemispherectomy for tumor in adults and for lateralized epileptogenic lesions in children indicate that early in life, the two cerebral hemispheres provide a safety factor akin to the pairing of other human organs. However, with cerebral maturation and increasing specialization of hemispheric functions, the safety factor on this global level is diminished considerably (Smith, 1972, 1981).

With reiterated subsystems there is a basic problem of coordination. Each reiterated subsystem type must participate in the functions of the larger systems of which it is a part only insofar as it is needed. There must not be too many cooks. We can think of five ways in which such coordination may be accomplished. (1) Certain privileged iterations may always have priority. (2) A function may cycle among a group of iterated subsystems, which share the load over time. (3) Large numbers of subsystems may work in a summative fashion, swamping small malfunctions in a few of the connections. (4) The system might always be active in excess of what is needed, with an occlusionlike process (Ochs, 1965) rendering harmless any outputs beyond the adequate signals. (5) There may be some sort of regulatory monitoring function that involves feedback control and takes as its signal some concomitant of normal functioning of one of the iterated subsystems. However, for some brain functions, perhaps the mode of regulating subsystem participation is analogous to the ways in which simpler organs such as kidneys, liver, and lungs manage to do their massive jobs without individual monitoring of the level of participation of each nephron, liver cell, or alveolus.

Some evidence indicates that the brain has more afferent than efferent redundancy. Jackson (Taylor, 1958) called attention to the differential sensitivities of afferent and efferent functions to brain insults. He believed that sensory or afferent functions were more widely represented or rerepresented than motor functions. Today we know that there are many more sensory neurons than motor neurons, about 2,000,000 versus 350,000 in humans (Sinsheimer, 1971). Sensory cortex greatly expands on afferents arriving through the bottleneck of the sensory nerves; for example, in the human there are approximately 538 million visual cortical cells, and 100 million auditory cortical cells elaborate information from only 30,000 auditory nerve fibers (Worden, 1971). In arthropods there are a few dozen motor axons for each segmental ganglion but several thousand sensory axons (Bullock *et al.*, 1977, p. 411). Smith (1981, 1984) has reviewed neuropsychological evidence of the greater vulnerability of motor function to relatively small brain insults.

3.3. Numerical Demonstration of the Importance of the Reiterative Safety Factor

Consider a hypothetical, mathematically convenient creature ("HYMACC") having the following characteristics: (1) it has to live 10 years if it is to reproduce and raise its young, (2) it has ten behaviors in its repertoire, (3) each behavior must successfully occur ten times a year, and (4) each behavior involves coordinated activity of ten different kinds of neural subsystems. Each such subsystem must perform once in order for the behavior to occur. Inevitably, such a subsystem will occasionally malfunction; let us say that there is a one in 1000 chance of failure for each of the subsystems, each time it is called into play. Suppose that whenever there is such a subsystem failure, there is a one in ten chance that the HYMACC will die.

Multiplying these figures, in the lifetime of the HYMACC there must be 10^4 = 10,000 successful subsystem operations; there is a $1/1000 \times 1/10 = 0.0001$ chance of death, and therefore a comfortable-in-the-short-run 0.9999 chance of living, associated with each subsystem operation. However, when the figure 0.9999 is raised to the 10,000th power to get the likelihood of success over a lifetime, the result is only 0.37. Thus, almost two-thirds of the HYMACCs will leave no progeny.

Evolution might exploit such a situation in natural selection of a tendency for larger numbers of offspring. Insects reproduce prolifically because externally imposed mortal accidents are likely, but the solution to the problem of accidental failure, whether of external or internal origin, is not carried to insectlike extremes with birds and mammals.

If natural selection happened to reiterate each of the HYMACC's subsystems with a single duplicate, then the chances of failure of both members of that subsystem pair in any one operation would be greatly reduced, from $1/1000$ to $(1/1000)^2 = 1/1,000,000$. Recalculating as above, this improves the lifetime reliability from 0.37 to 0.999. Pursuing a few more numerical examples, suppose the chances of subsystem failure were not $1/1000$ but $1/100$. Then (recalling the $1/10$ chance of death in each failure) there would be a $1/1000$ chance of death with each subsystem operation. This leads to a 0.999 chance of living through each subsystem operation and only a 0.000045 lifetime prospect. Fewer than one in 10,000 of this strain of HYMACCs will leave progeny. In this case, duplicating each subsystem results in a fair 0.90 lifetime prospect, and triplicating each subsystem would raise this to a safe 0.999. If each subsystem had a poor $1/10$ failure rate, to reach the same good lifetime prospect of 999 chances in 1000 of survival and reproduction, there would have to be six iterations of each subsystem. These simple examples illustrate the importance of neural safety factor when many subsystems interact. (See Glassman, 1987, for further development of these mathematical arguments. Soviet research on reliability theory in application to biological systems is summarized in Koltover, 1983.) Reliability arguments have also been used in attempted explanation of redundancy in the genome (Bernstein *et al.*, 1984).

3.4. Implications of the Numerical Demonstration for Ablation Research

Natural selection acts with respect to the entire reproductive lifetime of organisms and thus on a scale far larger than the usual laboratory measures. Proponents of strict localizationist thinking, among them physiologists whose fascination is with regional microscopic properties of the brain, argue that sufficiently refined laboratory measures should always reliably reveal a clear behavioral deficit following brain damage; all we must ostensibly do is to discover the appropriate test. However, the foregoing argument clearly shows that no laboratory test within reason is sensitive enough to reveal some evolutionarily significant unreliabilities, even where half the brain is missing. Tiny, occasional behavioral malfunctions do not satisfy a laboratory observer but may have significant survival significance across a population.

As an example of a behavior for which this issue has arisen, there is apparently complete recovery from tactile discrimination and orientation deficits in cats having experimental damage to the entire surface of the anterior ectosylvian gyrus containing the SII representation of the body so long as the subjacent orbital–anterior sylvian gyrus is undamaged (Glassman and Glassman, 1977). Perhaps some more sensitive test would reveal a reliable small, residual threshold elevation, but perhaps not. Consider the life-and-death situations that a domestic cat faces daily if it has to live in an alley, fighting, preying on small animals without being bitten, and evading dogs, juvenile delinquents, and disgruntled bird-lovers. An occasional small error in tactile discrimination or orientation could significantly reduce the chances of survival. Perhaps SII in cats and other mammals evolved under these pressures and analogous ones. Although the degree of anatomic and physiological organization of SII might suggest it was sculpted by a persistent evolutionary process for its own special function, it is possible that SII largely comprises reiterations of functional elements present in other areas.

Let us carry these arguments further. Considering the variability of populations and recalling Lowell's (1985) safety factor argument, as cited above, there may actually be cases of complete recovery with no residual deficits at all. Like many other characteristics, the extents of reiterative and aiming-high safety factors in various systems are probably normally distributed across different individuals. Thus, an individual who recovers from damage may still happen to be in better shape than most of the rest of the population if he had an unusually high safety margin in the brain systems that were damaged.

However, the majority of people who are brain damaged in youth or adulthood do show deficits, at least on some testing occasions (Smith, 1984). Stuss et al. (1985) observed what they interpreted as subtle deficits in dividing attention in the majority—but not all—of a sample of closed-head injury patients who tested normally on most of the usual mental measures. This finding fits our reliability argument if we additionally hypothesize that different task components sometimes select brain subsystems from a common pool. A reduced number of reiterated subsystems would mean fewer are available for reliable simultaneous operation; so a deficit would be seen when task components cannot be queued.

3.5. Reiterations Are Unlikely to Comprise Large, Complex Units

Taken together, the foregoing arguments suggest a reason for a safety factor at the level of the most elementary or diffuse-domain functional units of the brain. Reiterative safety factor is unlikely to have evolved with regard to more complex functional systems in a manner that would support an analogy with a "spare tire." It seems unlikely that safety factor evolved in response to a selection pressure attending instances of gross accidental brain trauma to whole functional systems because the circumstances of virtually all brain traumas in nature must be deadly (also see Finger and Almli, 1985; Glassman, 1974; Glassman and Malamut, 1977). Therefore, natural selection of a facility for recovery from brain trauma never has a chance to get off the ground. [However, the ultrasocial human species (D. T. Campbell, 1983) may be an exception to this rule. Mutual caring and nursing to recovery from illnesses seems an essential part of the human adaptation; this social behavior pattern must go well back into prehistory.]

Spare capacity might protect against too-early debilitation from gradual, anatomically distributed losses of brain tissue occurring with aging. Any natural selection relevant to lengthening the life span should thus increase safety factor. In contrast to the present argument, Finger and Almli (1986) have mobilized a considerable review of findings implying that gradual loss of brain tissue in aging or disease cannot be an important evolutionary factor underlying recoverability. These authors give figures suggesting that the vast majority of wild animals and—until recent history—human beings have died long before a natural loss of neurons could be relevant to natural selection. However, as aging and probably postreproductive hominids ourselves, we believe Finger and Almli underestimated the important contributions of our age group to advanced biosocial adaptations (e.g., deWaal, 1982; Glassman, 1984; Glassman et al., 1986).

3.6. Relevance of Research on Brain Size for the Safety Factor Hypothesis

In mammals, brain size is positively correlated with longevity and with body size; a high neonatal index of encephalization is correlated with small litter size and long gestation (Hofman, 1983a,b; also see Calder, 1983). These facts may be interpreted in terms of the evolutionary strategy just described. That is, members of prolifically reproducing, small-brained species (insects are the most extreme examples) have the strategy of casting their genes on the world in small, short-lived packages. Although each package is relatively unreliable, the aggregate has a good chance of including a member who happens to survive and reproduce prolifically. The deaths of some gene-package bearers have no effect on the others. Thus, the species as a whole persists reliably, but no individual has this quality.

More reliability than this must be built into each relatively expensive large individual from small-litter species if such a species is to survive. Instead of reiterated, uncoupled, independently peripatetic packages of genes, there is a massive reiteration of functions within each individual.

As proposed earlier, the existence of a reiterative safety factor implies a need for a special coordinative function to avoid the too-many-cooks problem. The more reiteration within and among subsystem types, the greater is this coordination problem. Depending on other assumptions, this consideration may imply either a positive or negative within-species correlation between brain size and intelligence. In a brief but broad-ranging review, Van Valen (1974) concluded that for humans there is a positive coefficient of correlation, which may be as high as 0.3.

3.7. Implications when There Is Additional Loss of Tissue

Occasionally, in some individuals, a misfortune of ontogeny could combine with a fortune of trauma to the brain in such a way that the individual may be better off after the loss of tissue has either reduced the brain's problem of coordinating too many reiterated subsystems or fortuitously adjusted a "higher digit place" parameter that was poorly set during early development.

However, since the average brain-damaged individual has less tissue than the average brain-intact individual, the former should generally show greater vulnerability to subsequent loss of brain cells. In a number of Kennard's famous monkeys that had shown special sparing after damage to motor cortex in infancy, there was an unusual onset of spasticity with aging. Other examples of apparent deterioration were observed in late neuropsychological follow-up measurements of schizophrenics (Smith, 1959) who had been topectomized many years earlier, and for whom no increment in cognitive impairment had then been reported. Similarly, late deterioration of cognitive function has been reported for boxers, who presumably have diffuse brain damage (Smith, 1984). A "postpolio syndrome," comprising recurrence of motor weaknesses 30–40 years after the disease, has been observed in a number of clinics (Spencer, 1985).

4. FIVE POSSIBLE NONNEURAL PREADAPTATIONS FOR SAFETY FACTOR

Any of the following five hypothetical processes might work in conjunction with those discussed above. An important implication of such an interaction is noted at the end.

4.1. Developmental Heterochrony

A cephalocaudal developmental order is apparently dictated embryologically, but other evolutionary factors may exaggerate this tendency. One mode of evolutionary change comprises developmental acceleration or deceleration. The limiting case of deceleration is fixation of an infantile characteristic in the adult; this is called "neoteny" or "pedomorphic descendence" (DeBeer, 1958; Gould, 1977). For example, kangaroo rats show greater encephalization than others of their lineage. Concomitantly with the enlarged brain, these animals also possess the other features of the pedomorphic syndrome, including enlarged eyes, poorly ossified skeleton, weakly fused cranial sutures, and retarded development of motor roots (Hafner and Hafner, 1984).

Such evolutionary changes might be a response to any of a variety of factors, with only one component of the change being the adaptive focus of selection. The others are carried as costs. But any of these components of excess baggage then remain available for possible later exploitation. Thus, large brain size may have originated as a secondary factor, only later turning out to be useful for systems reliability or intelligence.

4.2. The Head as a Releaser of Imprinting at Birth

Although mortality during childbirth selects against large head size, another selection factor, related to imprinting, may favor an effortful birth scenario. Textbooks usually emphasize affectional imprinting in infancy; however, also important is the later formation of discriminative affectional attachments on the parent's part. In sheep, the stimulation of the birth process sets a critical period for learning in the mother; the lamb with which she spends the next 30–120 min is the one she adopts as her own, and others are rejected (Keverne et al., 1983).

Considering the extreme sociality of the human adaptation, it is conceivable that there has been natural selection for a particularly difficult birth process in humans via the simple evolutionary expedient of large head size. The childbirth event is an important human ritual in which family and community implicitly communicate values, roles, and an intention to build the new individual into the social system. The mother, pressed into a highly expressive role, is the best source of a timing signal for these social interactions. As ethologists have pointed out, communication signals work best when they are vigorous, conspicuous, and prolonged (Eibl-Eibesfeldt, 1975).

4.3. The Visual Proportions of Infants as Affectional Releasers

The appearance of children, including relatively large head, disproportionately large eyes, small nose, etc., is so attractive that humans generalize their nurturing emotions to the young of other species and even to toy dolls or cartoon characters having supernormally exaggerated proportions. In the evolutionary process of ritualization, a morphological or behavioral feature acquires signaling characteristics. Once a feature becomes significant as a communication, subsequent evolution typically leads it to become more exaggerated, conspicuous, and distinct (Eibl-Eibesfeldt, 1975).

4.4. A Large Head on a Large Body Is Fearsome rather than Cute

A visual size advantage in dominance contests is accomplished in various ways in different species, e.g., broadside display or erected fur, fins, or feathers (Lorenz, 1966). The antlers of deer, lion's mane, and orangutan's fleshy facial prominences have analogous social effects. In the human species, deference has been elicited by the headgear or hats worn by soldiers, kings, Indian chiefs, and captains of industry; perhaps these customs all play on a biologically evolved emotional responsiveness.

As a further speculation, perhaps such learned simulations of large heads at the dawn of humankind relieved some of the natural selection pressure for a large skull. In turn, this would have relaxed the selection pressure for a large pelvis, allowing further evolution of upright posture. So long as there remained a particular proportion of pelvis to natal head size, the consideration in Section 4.2 would still obtain.

4.5. Surface/Volume Ratio in Thermoregulation

Endothermic temperature regulation is challenging in colder climates. A three-dimensional object of constant shape decreases in the ratio of its surface to volume as its size grows larger. This decrease implies an energy cost savings because heat is lost at the skin surface. The brain demands a disproportionate amount of the body's metabolism (20%), presumably because of the critical homeostatic requirement of this most highly organized of organs (Hofman, 1983b). The ratio of brain weight to body weight in birds and mammals deviates from linearity, with the brain comprising a higher proportion of the body in smaller species, but this deviation disappears when body weights are adjusted for rates of oxygen consumption (Armstrong and Bergeron, 1985). This might be because in smaller animals more brain mass is actively involved in energy processing, because large size provides thermal "inertia," protecting against sudden perturbations, and/or because of smaller surface/volume ratio.

5. AN IMPORTANT IMPLICATION OF NONNEURAL NATURAL SELECTION FACTORS FOR NEURAL INFORMATION PROCESSING, DIASCHISIS, AND RECOVERY

Nonneural natural selection pressures cannot have been the only causes of brain enlargement, because the simplest and most probable evolutionary route to filling a large skull would merely be to expand the volume of the fluid-filled ventricles. However, if any of the above arguments about nonneural selection factors are valid, then the large human skull may have provided something of a free evolutionary playground for the brain, a preadaptation or concurrent adaptation helping to pave the way for the development of a reliable information-processing system.

There is an additional possibility. As Jackson recognized, the brain evolves higher functions in part by inhibiting more primitive functions (Taylor, 1958). Whatever the possible functional advantages or disadvantages of this course, it is an easier evolutionary route because relatively little change is involved simply in diffusely inhibiting a system—in contrast to what would have to be involved in evolutionary extirpations and reconnections to achieve the most efficient reorganization. If natural selection is permissive or even encouraging of large head size for nonneural reasons, this implies that some of the neural stuff enclosed in our large skulls is evolutionary detritus. Some of the living brain mass may comprise fossilized accretions of bygone eras; these accretions engage in little or no differentiated organizing activity, but they sit there always inhibited. Alternatively, small systems that once were differentiated may have merged and slowed down in their electrochemical dynamics to form large masses that sit dully in dynamic electrotonic or chemical balance with each other as well as with parts of the brain critically necessary for information processing. Though serving little use, these hypothetical balanced masses would make for vulnerability to diaschisislike effects. This speculation suggests a new hope for remediation of brain damage if the hypothesized systems can be clearly identified.

6. SUMMARY

Cases of extensive recovery from brain damage imply that brains have spare capacity, but we doubt that brain subsystems have the sort of self-contained, local intelligence needed for vicariation. Spare capacity as we envision it requires a simpler form of organizing available resources. We hypothesize that the elementary components of neural information processing each have one dimension of dynamic range. We call these elements "diffuse domains," a term that encompasses single neurons, groups of neurons acting as unitary excitable masses for some function, and other possibilities. Using the metaphor of a digital counter, we show how a set of such subsystems might hold a great deal of discriminative information and how it might participate in more than one function. Such subsystems are subject to the sort of noninsightful, continuous adaptive modification that must be the currency of evolution, ontogeny, and recovery, via adjustments in degree of excitability or number. This hypothesis unifies the concept of diaschisis with the notion of spare capacity and implies that profound deficits sometimes occur with only a slight loss of relevant information.
Spare capacity or safety factor probably evolved for reliability in complex, long-lived organisms with undamaged brains. Logically, reliability may be based on backup subsystems and/or on high "specifications" of a given subsystem. Evolution has probably led to both of these modes and to readjustments taking place at low levels of system organization. Possible evolutionary routes to large brains having spare capacity involve preadapting factors not relevant to neural information processing. This implies that the large human brain may include obsolescent masses of tissue that make it exceptionally vulnerable to diaschisis.
Together, these hypotheses suggest a hope of therapies capable of restoring functions to some damaged brains by means of relatively minor interventions to restore appropriate settings and balances.

REFERENCES

Alexander, R. McN., 1984, Optimum strengths for bones liable to fatigue and accidental fracture, *J. Theor. Biol.* **109**:621–636.

Andrew, J., 1984, Legendary summer rally of stocks is sighted but it could be a mirage, *Wall Street Journal* **64**(177):25,45 (June 25).

Armstrong, E., and Bergeron, R., 1985, Relative brain size and metabolism in birds, *Brain Behav. Evol.* **26**:141–153.

Berker, E., and Smith, A., 1981, Specific site and diaschisis in Raven Coloured Matrices and other performances in 41 adults with acute focal lesions, *Int. J. Neurosci.* **22**:225–226.

Berker, E., and Smith, A., 1983, "Cryptogenic" epilepsy, reading, learning and related disorders of childhood in light of intracerebral hemorrhage in neonates, *INS Bull.* Oct:56–57.

Berker, E., Lorber, J., and Smith, A., 1983, Influences of extra-cerebral factors on cerebral development of 289 patients with varying degrees of hydrocephalus, in: *11th Annual Meeting, International Neuropsychological Society*, Mexico City.

Bernstein, H., Byerly, H. C., Hopf, F. A., and Michod, R. E., 1984, Origin of sex, *J. Theor. Biol.* **110**:323–351.

Bullock, T. H., Orkand, R., and Grinnell, A., 1977, *Introduction to Nervous Systems*, W. H. Freeman, San Francisco.

Burklund, C. W., 1969, Cerebral hemisphere function in the human: Fact versus tradition, in: *Drugs, Development, and Cerebral Function* (W. L. Smith, ed.), Charles C. Thomas, Springfield, IL, pp. 8–36.

Burklund, C. W., and Smith, A., 1977, Language and the cerebral hemispheres: Observations of verbal and non-verbal responses during eighteen months following left ("dominant") hemispherectomy, *Neurology (Minneap.)* **27**:627–633.

Calder, W. A., 1983, Body size, mortality, and longevity, *J. Theor. Biol.* **102**:135–144.

Campbell, A. L., Bogen, J. E., and Smith, A., 1981, Disorganization and reorganization of cognitive and sensorimotor functions in cerebral commissurotomy, *Brain* **104**:493–511.

Campbell, B., 1960, The factor of safety in the nervous system, *Bull. Los Angeles Neurol. Soc.* **25**:109–117.

Campbell, D. T., 1983, The two distinct routes beyond kin selection to ultrasociality: Implications for the humanities and social sciences, in: *The Nature of Prosocial Development: Theories and Strategies* (D. Bridgeman, ed.), Academic Press, New York, pp. 11–41.

Churchland, P. M., 1986, Cognitive neurobiology: A computational hypothesis for laminar cortex, *Biol. Philos.* **1**:25–51.

Davis, H. P., and Squire, L. R., 1984, Protein synthesis and memory: A review, *Psychol. Bull.* **96**:518–559.

deBeer, G., 1958, *Embryos and Ancestors*, 3rd ed., Oxford University Press, New York.

Devor, M., and Schneider, G. E., 1975, Neuroanatomical plasticity: The principle of conservation of total axonal arborization, in: *Aspects of Neural Plasticity* (F. Vital-Durand and M. Jeannerod, eds.), INSERM, Paris, pp. 191–200.

deWaal, F., 1982, *Chimpanzee Politics: Power and Sex among Apes*, Harper & Row, New York.

Eibl-Eibesfeldt, I., 1975, *Ethology: The Biology of Behavior*, Holt, Rinehart, and Winston, New York.

Feeney, D. M., Sutton, R. L., Boyeson, M. G., Hovda, D. A., and Dail, W. G., 1985, The locus coeruleus and cerebral metabolism: Recovery of function after cortical injury, *Physiol. Psychol.* **13**:197–203.

Feldman, J. A., and Ballard, D. H., 1982, Connectionist models and their properties, *Cog. Sci.* **6**:205–254.

Finger, S., and Almli, C. R., 1986, Brain damage and neuroplasticity: Mechanisms of recovery or development? *Brain Res. Rev.* **10**:177–186.

Finger, S., and Stein, D. G., 1982, *Brain Damage and Recovery: Research and Clinical Perspectives*, Academic Press, New York.

Ford, F. R., 1960, *Diseases of the Nervous System in Infancy, Childhood, and Adolescence*, Charles C. Thomas, Springfield, IL.

Frank, W. W., and Lapp, C. L., 1959, *How to Outsell the Born Salesman*, Collier Books, New York.

Frommer, G. P., 1978, Subtotal lesions: Implications for coding and recovery of function, in: *Recovery from Brain Damage: Research and Theory* (S. Finger, ed.), Plenum Press, New York, pp. 217–280.

Geschwind, N., 1970, The organization of language and the brain, *Science* 170:940–944.

Glassman, R. B., 1974, Equipotentiality and sensorimotor function in cats, *Neurosci. Res. Prog. Bull.* 12:246–249.

Glassman, R. B., 1984, A sociobiological examination of management theory Z, *Hum. Relat.* 37:367–392.

Glassman, R. B., 1985, Parsimony in neural representations: Generalization of a model of spatial orientation ability, *Physiol. Psychol.* 13:43–47.

Glassman, R. B., 1987, An hypothesis about redundancy and reliability in the brains of higher species: Analogies with genes, internal organs, and engineering systems, *Neurosci. Biobehav. Rev.* 11:275–285.

Glassman, R. B., and Glassman, H. N., 1977, Distribution of somatosensory and motor behavioral function in cat's frontal cortex, *Physiol. Behav.* 18:1127–1152.

Glassman, R. B., and Malamut, B. L., 1976, Recovery from electroencephalographic slowing and reduced evoked potentials after somatosensory cortical damage in cats, *Behav. Biol.* 17:333–354.

Glassman, R. B., and Malamut, B. L., 1977, Does the brain actively maintain itself? *Biosystems* 9:257–268.

Glassman, R. B., and Wimsatt, W. W., 1984, Evolutionary advantages and limitations of early plasticity, in: *Early Brain Damage, Volume 1: Research Orientations and Clinical Observations* (C. R. Almli and S. Finger, eds.), Plenum Press, New York, pp. 35–58.

Glassman, R. B., Packel, E. W., and Brown, D. L., 1986, Green beards and kindred spirits: A preliminary mathematical model of altruism toward nonkin who bear similarities to the giver, *Ethol. Sociobiol.* 7:107–115.

Gould, S. J., 1977, *Ontogeny and Phylogeny*, Harvard University Press, Cambridge.

Gowers, W. R., 1885, *Diagnosis of Disease of the Brain and of the Spinal Cord*, William Wood, New York.

Grinnell, A., 1977, Structural basis of connectivity, in: *Introduction to Nervous Systems* (T. H. Bullock, R. Orkand, and A. Grinnell, eds.), W. H. Freeman, San Francisco, pp. 97–127.

Hafner, M. S., and Hafner, J. C., 1984, Brain size, adaptation and heterochrony in geomyoid rodents, *Evolution* 38:1088–1098.

Hochberg, J., 1972, The representation of things and people, in: *Art, Perception, and Reality* (E. H. Gombrich, J. Hochberg, and M. Black, eds.), Johns Hopkins University Press, Baltimore, pp. 47–94.

Hofman, M. A., 1983a, Evolution of brain size in neonatal and adult placental mammals: A theoretical approach, *J. Theor. Biol.* 105:317–332.

Hofman, M. A., 1983b, Energy metabolism, brain size and longevity in mammals, *Q. Rev. Biol.* 58:495–512.

Isaacson, R. L., 1975, The myth of recovery from early brain damage, in: *Aberrant Development in Infancy* (N. G. Ellis, ed.), John Wiley & Sons, New York, pp. 1–26.

Kempinsky, W. H., 1958, Experimental study of distant effects of acute focal brain injury: Study of diaschisis, *Arch. Neurol. Psychiatry* 79:376–389.

Keverne, E. B., Levy, F., Poindron, P., and Lindsay, D. R., 1983, Vaginal stimulation: An important determinant of maternal bonding in sheep, *Science* 219:81–83.

Koltover, V. K., 1983, Theory of reliability, superoxide radicals and aging, *Uspekhi Sovremennoi Biologii* (Progress in Contemporary Biology) 96(4):85–100.

Lewin, R., 1986, Punctuated equilibrium is now old hat, *Science* 231:672–673.

Linden, E., 1979, *Affluence and Discontent: The Anatomy of Consumer Societies,* Viking, New York.

Loftus, E. F., and Loftus, G. R., 1980, On the permanence of stored information in the human brain, *Am. Psychol.* **35**:409–420.

Lorenz, K. Z., 1966, *On Aggression,* Harcourt, Brace and World, New York.

Lorenz, K. Z., 1969, Innate bases of learning, in: *On the Biology of Learning* (K. H. Pribram, ed.), Harcourt, Brace and World, New York, pp. 13–93.

Lorenz, K. Z., 1981, *The Foundations of Ethology,* Springer-Verlag, New York.

Lowell, R. B., 1985, Selection for increased safety factors of biological structures as environmental unpredictability increases, *Science* **228**:1009–1011.

Lynch, G., and Baudry, M., 1984, The biochemistry of memory: A new and specific hypothesis, *Science* **224**:1057–1063.

Markowitsch, H. J., 1985, Hypotheses on mnemonic information processing by the brain, *Int. J. Neurosci.* **27**:191–227.

Markowitsch, H. J., and Pritzel, M., 1978, Von Monakow's diaschisis concept: Comments on West *et al., Behav. Biol.* **22**:411–412.

Marshall, J. F., 1984, Brain function: Neural adaptations and recovery from injury, *Annu. Rev. Psychol.* **35**:277–308.

Meyer, D. R., 1984, The cerebral cortex: Its roles in memory storage and remembering, *Physiol. Psychol.* **12**:81–88.

Meyer, D. R., Gurklis, J. A., and Cloud, M. D., 1985, An equipotential function of the cerebral cortex, *Physiol. Psychol.* **13**:48–50.

Monakow, C. von, 1914, *Localization in the Cerebrum and the Degeneration of Functions through Cortical Sources,* J. F. Bergmann, Wiesbaden.

Monakow, C. von, and Mourgue, R., 1928, *Introduction Biologique a l'Etude de la Neurologie et de la Psychophysiologie,* Libraire Felix Alcan, Paris.

Nickerson, R. S., and Adams, M. J., 1979/1982, Long-term memory for a common object, in: *Memory Observed* (U. Neisser, ed.), W. H. Freeman, San Francisco, pp. 163–175.

Nisbett, R. E., and Wilson, T. D., 1977, Telling more than we can know: Verbal reports on mental processes, *Psychol. Rev.* **84**:231–259.

Ochs, S., 1965, *Elements of Neurophysiology,* John Wiley & Sons, New York.

Schmidt, R. F., and Thews, G., 1983, *Human Physiology,* Springer-Verlag, New York.

Sherrington, C. S., 1941, *Man on his Nature,* Macmillan, New York.

Sherrington, C. S., 1947, *The Integrative Action of the Nervous System,* 2nd ed., Yale University Press, New Haven.

Sinsheimer, R. L., 1971, The brain of Pooh: An essay on the limits of mind, *Am. Sci.* **59**:20–28.

Smith, A., 1959, Changes in psychological test performances of brain-operated schizophrenics after 8 years, *Science* **129**:149–150.

Smith, A., 1968, Ambiguities in concepts and studies of "brain damage" and "organicity," *J. Nerv. Ment. Dis.* **135**:311–326.

Smith, A., 1972, Dominant and nondominant hemispherectomy, in: *Drugs, Development and Cerebral Function* (W. L. Smith, ed.), Charles C. Thomas, Springfield, IL, pp. 37–68.

Smith, A., 1978, Lenneberg, Locke, Zangwill, and the neuropsychology of language and language disorders, in: *Psychology and Biology of Language and Thought* (G. Miller and E. Lenneberg, eds.), Academic Press, New York, pp. 133–149.

Smith, A., 1981, Principles underlying human brain functions in neuropsychological sequelae of different neuropathological processes, in: *Handbook of Clinical Neuropsychology* (S. Filskov and T. Boll, eds.), John Wiley & Sons, New York, pp. 175–226.

Smith, A., 1984, Early and long-term recovery from brain damage in children and adults: Evolution of concepts of localization, plasticity and recovery, in: *Early Brain Damage,* Volume 1: *Re-*

search Orientations and Clinical Observations (C. R. Almli and S. Finger, eds.), Academic Press, New York, pp. 299–324.

Smith, A., and Sugar, O., 1975, Development of above normal language and intelligence 21 years after left hemispherectomy, Neurology (Minneap.) 25:813–818.

Spencer, J., 1985, Postpolio: Decades after the "cure" victims are suffering again, Chicago Tribune, Section 5, December 18, pp. 1, 3.

Stuss, D. T., Ely, P., Hugenholtz, H., Richard, M. T., LaRochelle, S., Poirier, C. A., and Bell, I., 1985, Subtle neuropsychological deficits in patients with good recovery after closed head injury, Neurosurgery 17:41–47.

Taylor, J., ed., 1958, Selected Writings of John Hughlings Jackson, Staples Press, London.

Van Valen, L., 1974, Brain size and intelligence in man, Am. J. Phys. Anthropol. 40:417–424.

Villablanca, J. R., Burgess, J. W., and Sonnier, B. J., 1984, Neonatal cerebral hemispherectomy: A model for postlesion reorganization of the brain, in: Early Brain Damage, Volume 2: Neurobiology and Behavior (S. Finger and C. R. Almli, eds.), Academic Press, New York, pp. 179–210.

Watanabe, S., Hodos, W., and Besette, B. B., 1984, Two eyes are better than one: Superior binocular discrimination learning in pigeons, Physiol. Behav. 32:847–850.

West, J. R., Deadwyler, S. A., Cotman, C. W., and Lynch, G. S., 1976, An experimental test of diaschisis, Behav. Biol. 18:419–425.

Wilson, P. J. E., 1970, Cerebral hemispherectomy for infantile hemiplegia, Brain 93:147–180.

Wood, C. C., 1982, Implications of simulated lesion experiments for the interpretation of lesions in real nervous systems, in: Neural Models of Language Processes (M. A. Arbib, D. Caplan, and J. C. Marshall, eds.), Academic Press, New York, pp. 485–509.

Worden, F., 1971, Hearing and the neural detection of acoustic patterns, Behav. Sci. 16:20–30.

5

Kurt Goldstein and Recovery of Function

GABRIEL P. FROMMER and AARON SMITH

1. INTRODUCTION

In the first half of this century, Kurt Goldstein developed a holistic or organismic model of cerebral functioning that has important implications for conceptions of recovery. The term "holistic" is frequently used rather casually, so it needs some clarification. It refers to the fundamental assumption that it is impossible to consider a biological phenomenon in isolation; it must be considered in the context of the biological system, the environment, and the history of the whole organism. For Goldstein, holism was a specific response to the epistemological problem of what one can usefully discover about biology through the process of "taking apart" the organism to generate "a multitude of isolated facts" (Goldstein, 1939, p. 7). Goldstein acknowledged that analysis through isolation is the necessary first step in the scientific study of living organisms, but dissatisfaction with the usefulness of the results of this procedure for medical practice led him to his holistic approach (Goldstein, 1939, p. 8; 1942, p. 11). The dangers of drawing unwarranted conclusions based on reductionistic approaches, analysis of phenomena in isolation, or "decomposition" were also emphasized by other neurologists of the time, notably Henry Head (1926).

Goldstein drew primarily on two sources in developing his model. He repeatedly acknowledged his heavy debt to the writings of "antilocalizational" neurologists, especially Constanin von Monakow in Switzerland and John Hughlings Jackson in England. More importantly, Goldstein obtained a rich body of empirical data from his own work with brain-damaged patients, especially brain-injured German veterans of World War I in a hospital established

GABRIEL P. FROMMER • Department of Psychology, Indiana University, Bloomington, Indiana 47405.   AARON SMITH • University of Michigan NHR Project, Ann Arbor, Michigan 48105 specifically for their rehabilitation. Two kinds of clinical observations were particularly important: detailed qualitative investigations of the patients' psychological deficits, which Goldstein found to be common to diverse symptoms of cerebral injury, and especially continuing follow-up studies with careful attention to the processes underlying the patients' recovery.

. A central feature of Goldstein's holistic model was its conception of localization of function in the cerebral hemispheres. Like Hughlings Jackson before him, Goldstein distinguished between localization of defect or symptom, which can be established statistically by clinicopathological correlations, and localization of psychological performance or function, which can be identified from detailed qualitative analysis of individual patients' performances. On the basis of such analyses, Goldstein concluded that "localization of performance no longer means to us an excitation in a certain place, but a dynamic process which occurs in the entire nervous system . . . and which has a definite configuration for each performance" (Goldstein, 1939, p. 260).

Goldstein's main empirical work was published in German, but the major findings were summarized in English in *Aftereffects of Brain Injuries in War* (Goldstein, 1942) and in *Language and Language Disturbances* (1948). His major theoretical work, *Der Aufbau des Organismus* (Goldstein, 1934), written in a "time of enforced leisure" in the Netherlands during his flight from Nazi Germany, was published with only minor revisions in English translation as *The Organism* (Goldstein, 1939). Other important new writings appeared in English following his arrival in the United States in 1935 (Meiers, 1968).

It is difficult for most neuroscientists and psychologists, especially those trained in the North American scientific tradition during the second half of the 20th century, to understand and assimilate Goldstein's writings and concepts. Goldstein, whose first published work appeared in 1903, came from the Central European tradition of scholarship, heavily steeped in German philosophy and literature. His posthumous autobiographical paper (Goldstein, 1967) provides an interesting and valuable intellectual testament. His theoretical work, *The Organism* (1934/1939), will intrigue the modern reader. Goldstein drew on remarkably diversive and unusual sources to present a general theory of the organism, its biological nature, and its capacities as variously altered in its relationship to its immediate environment and individual circumstances. *The Organism* may be unique among major scientific works of the 20th century in its extensive citation of Goethe, whom Goldstein deeply admired, as a scientific writer. The book also presents problems to the modern reader. Goldstein attempted to present a general theory of organismic function; hence, much of the book presents closely reasoned arguments on diverse theoretical matters. The modern neuroscientist or psychologist, who is frequently ignorant of the origins, history, and logic of his/her discipline, may find much that appears only remotely relevant to the problem of brain organization and behavior. Furthermore, the holistic position Goldstein developed seems at first to be incompatible with today's dominant modular, connectionistic conception of cerebral organization. Finally, Goldstein's prose could have benefited greatly from skilled editing.

In his vigorous exposition of his conceptions of neural functioning, Goldstein made a number of strong claims. Although many have withstood the test of time, others have been shown to be incorrect, or at least incomplete, by a half century of additional research. This would not have surprised Goldstein. When one of us (A.S.), Goldstein's student in the 1950s, showed Goldstein results that failed to confirm his view that brain-damaged patients would be unable to shift criteria on the Goldstein–Weigl Sorting Test, Goldstein sighed and said, "Ja . . . Aaron, we was wrong. We must rewrite the book." On another occasion Goldstein said, "Aaron, if the patient does not agree with the book, throw the book away. The patient can never be wrong." Compare this with the remark, "The animal is always right," attributed to another proponent of detailed investigation of single subjects, B. F. Skinner.

Goldstein is considered one of the major originators of the neuropsychological tradition of applying psychological concepts and methods to the study of neurological disorders (Luria, 1966). He was one of the first to use the term "neuropsychology" (Goldstein, 1939, p. 365) and is credited with introducing formal psychological testing to the study of functional deficits following brain damage (Teuber, 1966), albeit in a form that is quite different from neuropsychological evaluation as it is often practiced today, at least in the United States. The journal *Neuropsychologica* published a memorial issue on his death in 1965, in which Teuber (1966), among others, wrote a warm appreciation of Goldstein and his writings and acknowledged the great influence Goldstein had on his (Teuber's) own development as scientist and scholar.

Despite the warm tributes Goldstein received from leaders of the neuropsychological establishment, direct evidence of his influence on theoretical and methodological questions in modern neuropsychology often is not immediately obvious. His name appears frequently in the indices of recent works on neuropsychology, but usually for his writings on specific neuropsychological phenomena. His methodological approach and his theoretical conception of cerebral function usually receive little or no elaboration. Yet these represent Goldstein's most important contributions, and they are fundamental to an understanding to his conception of recovery of function.

2. METHODOLOGICAL ASSUMPTIONS AND EMPIRICAL ORIGINS

Goldstein (1939) presented three methodological postulates. First, the investigator must consider, at least initially, all phenomena the organism (in the present context, the brain-damaged patient) presents, "giving no preference . . . to any special one" (p. 21). This postulate is intended to warn the investigator against the common practice of focusing attention on a prominent symptom, which may or may not not be of primary significance, to the exclusion of less striking ones, which may turn out to be crucial to the understanding of the functions underlying the observed behavioral phenomena. Second, the investigator must do more than simply describe behavioral phenomena; he/she must try to discover the functional causes of success or failure underlying various performances by use of appropriate qualitative tests. For Goldstein the scores a patient earned on tests were much less informative than were investigations undertaken to determine the specific nature of the the correct or incorrect performance, or how those scores were earned. Third, all behavioral phenomena must be considered "with . . . reference to the organism concerned and to the situation in which it appears" (p. 25). This postulate reflects Goldstein's affinity to Gestalt theory and his firm insistence on the artificial and misleading character of observations made on isolated (experimentally or conceptually) systems.

Goldstein repeatedly returned to the problem of isolation for its methodological and also its theoretical implications. Although he acknowledged that the process of analysis is a necessary first step in the scientific approach to biology, he subjected models of living systems based on such analyses, such as the reflexology current in his time, to intense criticism. He believed that the process of isolation introduces serious biases at the levels of both empirical observation and theoretical interpretation. Goldstein (1939, pp. 84*ff.*) argued that reflexes are abstractions created by the process of isolation and that the response elicited in reflexes depends on the condition of the system as well as the eliciting stimulus. Sherrington, as Goldstein noted, would have no difficulty with these points. However, Goldstein further argued that various attempts to interpret complex behavior in terms of integration of reflex activities are inadequate because they require that the direction of behavior come from without. Sherrington's summative conception of the integrative action of the nervous system accounts for the order seen in behavior through the total stimulation from the environment (p. 89). But Goldstein argues that order in behavior in fact comes from within the organism. He points out that an organism's environment is defined by the order inherent in the biological nature of the organism itself (see also Smith, 1986).

Surprisingly, an explicit statement of his alternate approach is difficult to find beyond the three methodological postulates described above, perhaps because it is difficult to express concisely. Goldstein's approach reflects characteristic features that are important but may be overlooked:

1. It emphasizes investigation of specific neural and behavioral phenomena in relation to the overall neural and behavioral activity of the organism.
2. It stresses detailed qualitative observation over extended periods using diverse methods to discover the specific nature of the underlying psychological processes on which normal and disturbed performance are based.

3. It distinguishes between the quantitative knowledge of natural science and the qualitative knowledge of biology.

The phenomena of recovery of function following brain damage were especially important in leading Goldstein to his holistic conception of cerebral organization because recovery is incompatible with a strict localization of psychological function. For example, patients with unilateral calcarine lesions show a reorganized visual field, which Goldstein believed did not map simply on the organization of the retinal projection on the remaining visual cortex. Patients with severe motor speech disturbance relearned the multiplication tables by substituting visual imagery for the motor speech series on which this performance had originally been based.* Variability of performance, which is quite common in patients with brain damage, has similar implications. For example, severely aphasic patients can in certain situations "utter" appropriate phrases that they cannot "say" or repeat on request. Reise (1960) has argued that because such complex behavioral processes can reappear under appropriate circumstances in brain-damaged patients, they cannot be related exclusively to the specific damaged cerebral areas.

Modern research on recovery of function has greatly extended the body of data against which models of cerebral organization can be evaluated (Finger and Stein, 1982). Smith (1979, 1981a,b, 1983, 1984, 1986) has reviewed literature particularly relevant to a holistic conception of cerebral organization. If one carefully considers such factors as the "momentum" of the lesion (the rate at which a lesion develops; Finger, 1978), the difference between "resolving" and "evolving" lesions, the time since damage (see also Braun, 1978), and functional condition of the remaining cerebral tissue, this body of data forms a coherent pattern that fails to support strict localization of function as presented in many textbooks. A dramatic example of this failure is the normal development or recovery of functions conventionally attributed to a cerebral hemisphere following its surgical removal for infantile hemiplegia or for cerebral tumors in adults. The recovery in adults can be greater than that following more circumscribed lesions. Evidence also indicates that recovery following cerebral damage depends on the condition of the remaining cerebral tissue in the same as well as the opposite hemisphere and on the integrity of the cerebral commissures between

*Contemporary reports have described similar phenomena. Dimond et al. (1976) have recently reported a striking shift in subjective central vision in normal subjects viewing test material through slits in opaque contact lenses and spectacles that prevented viewing with central vision. In this viewing condition, subjects had a perimetrically defined visual field restricted to a small oval area offset about 20° from the foveal representation, which was itself completely obscured. After removing these lenses, the subjects were surprised to discover "that they had not been stimulated at the centre of their vision" (p. 691). Kashiwaga et al. (1987) found that aphasic patients relearned the multiplication tables much more effectively using the visual modality than the verbal rhyming procedure by which Japanese children normally learn them.

them (Campbell *et al.*, 1981; Smith, 1981a,b, 1983, 1984). [Goldstein (1948, pp. 51–53) had anticipated this finding many years earlier.] Furthermore, massive destruction of both cerebral hemispheres in certain forms of hydrocephalus can have surprisingly meager psychological and behavioral consequences. The development of not only normal but even superior adult intelligence has been documented in patients with extreme congenital hydrocephalus exhibiting ventricular expansion occupying over 90% of the intracranial cavity (Berker, 1985).

3. THEORETICAL APPROACH

The holistic conception of neural organization implies that an intervention or modification in any one part of the brain has widespread influences on other parts of the brain; hence, the functional consequences that follow reflect the adaptation of the whole brain, indeed of the whole organism, to the demands of the environment. Every reaction represents a "figure" that acts against the "ground" of the activity of the rest of the nervous system and body. For example, execution of a specific movement involves not only the muscles of the particular limb directly involved in that movement but an adjustment of the whole body. The movement is a "figure" on the "ground" of the postural adaptations in the rest of the body. Following amputation of a limb, both vertebrate and invertebrate species immediately adjust to the loss by a compensated gait, which is very effective for locomotion. To achieve this compensation, the organism must adjust the whole body (the "ground") as well as the gait itself (the "figure").

However, areas of the brain are not equally involved in all behavioral and psychological processes. Although the whole brain is influenced by a given stimulus, the effects usually appear primarily in one more or less extended area (Goldstein, 1939) because of the filtering effect of the neural network interpolated between sensory input and motor output. The area of heightened activity can be considered as a figure imposed against the ground of activity in the rest of the nervous system. This idea and many other statements in his writings show that Goldstein did not dispute the obvious differentiation of the brain and its functional significance as seen from anatomy and symptomatology following brain lesions. What he did dispute was the interpretation that this differentiation reflected specificity of psychological function. Failure to make this distinction has led to a common misinterpretation of Goldstein's conception of cerebral function.

4. LOCALIZATION

One feature of Goldstein's work presents a particular problem for the modern reader. Goldstein went about his task very differently from most modern neuropsychologists. Description of the location of the lesion was often an after-thought, if it appeared at all, when he described the disturbance of psychological function following brain damage. Whereas Teuber and others (e.g., Teuber, 1959; Grafman *et al.*, 1986) capitalized on the availability of detailed operative notes to localize the cerebral lesions in studies of brain-injured veterans, Goldstein in his rehabilitation work largely ignored clinicopathological correlations and localizing signs, which seem so central to modern investigations of the psychological deficits following brain damage. Of course, Goldstein did not have the modern armamentarium of technology for identifying the location of lesions, but the real difference from current practice is in the interest, or lack of it, that Goldstein showed in the problem.

As an experienced and skilled clinical neurologist, Goldstein fully appreci-ated that lesions in different parts of the cerebral hemispheres resulted in differ-ent patterns of symptoms. He wrote a 240-page chapter on localization in the 25 volume *Handbuch* edited by Bethe in the late 1920s (Goldstein, 1927). In this chapter he systematically reviewed the localizing significance of symptoms after lesions and of responses to stimulation in each cerebral area. He summarized this material in *The Organism* (Goldstein, 1934, 1939) and clarified his position on localization in a paper specifically devoted to the subject (Goldstein, 1946). In his volume on aphasia, Goldstein (1948) discussed extensively the anatomic correlates of different general classes of aphasia. Geschwind (1964) was sur-prised that a presumed opponent of localization like Goldstein should present such detailed descriptions of specific symptomatology following brain damage in different locations. What Geschwind apparently did not recognize was that in Goldstein's mind this symptomatology did not reflect in a simple way the cere-bral organization underlying deficits in psychological performance that followed brain injury. As Luria (1966) has pointed out,

> Goldstein has made a heroic attempt to overcome the conflict of strict localization and the mentalistic approach of the noetic school . . . which tried to prove that an analytic approach to the brain functions is impossible. . . . [A]lthough he highly sympathized with the latter, [he] remained to the end of his life a brilliant representative of classical neurology with its analytic approach. The conflict of the two traditions was the basic content of his life; the attempt to construct a new neurology which had to include the truth of both was his endeavour (p. 311).

Goldstein followed Hughlings Jackson in differentiating between localiza-tion of defect or symptom and localization of function (Finger and Stein, 1982). The former represents clinical observation; the latter is a theoretical interpreta-tion and does not flow directly from the former. Melzack and Wall (1981) make an analogous distinction in the study of pain between specialization and specifici-ty. The former represents the physiological fact that most neurons studied in the somatosensory system are highly selective in their sensitivity to somatic stimuli; the latter is a theoretical explanation of the relationship between this physiologi-cal fact and the psychological phenomena of pain. By analogy, localization of function is a theory about the relationship between cerebral anatomy and physiology on the one hand and psychological function on the other. Localization of defect represents empirical correlations between physiological and anatomic disorganization and defects in behavioral performance.

Establishing localization of defect presented major technical difficulties (by no means completely overcome with modern techniques) as well as conceptual difficulties even in simpler cases. The theoretical concept of localization of function presents an additional set of problems (Goldstein, 1946). There are negative cases (lesions without the expected symptoms and symptoms without the expected lesions). There are differences related to the momentum and dynamics of the lesion (Finger, 1978). There are difficulties in assessing accurately the location of lesions even in histologically prepared autopsy material. There are remote effects of damage in the undamaged nervous system, a concept he adopted from von Monakow (Finger and Stein, 1982; Glassman and Smith, Chapter 4, this volume). There is the uncertainty of the relationship between histological findings and functional integrity. There are differences in premorbid personality. There are differences in individual patients' functional adaptation to the defect produced by the lesion.* For all of these reasons, Goldstein concluded that it is difficult to pick out from the multitude of symptoms that a brain lesion can produce those that depend specifically on the location of the lesion. Although enormous technical and empirical advances have been made since Goldstein's time, his criticisms remain valid.

Despite the difficulties in establishing correlations between location of cerebral damage and psychological function, Goldstein recognized that the great heterogeneity of the cortex implied complex functional organization. Different areas do not contribute equally or in the same way. Goldstein differentiated between "the periphery of the cortex," equivalent to primary sensory and motor areas, and "the central sector, which comprises the parietal, the Insula Reili, and particularly the frontal lobe" (Goldstein, 1939, p. 251). The latter is relatively loosely connected to the peripheral inputs and outputs and differs from the former in microscopic structure as well. These differences are reflected in psychological functioning.

Goldstein summarized his conception of localization thus:

> A particular locality in the brain matter is characterized by the influence which the structure of this locality exercises on the total process, by the contribution of the excitation of this process—as effect of its particular structure. Thus, for instance, the area striata contributes to this process something which is necessary for the experience of vision, the frontal lobes something which is the presupposition for the mental

---

*The following illustrate the last two points. Grafman *et al.* (1986) found that premorbid IQ was a major predictor of generalized cognitive functioning following head injury in their sample of Viet Nam veterans. Goldstein (1948, p. 327) found that patients may show one of two reactions to an irreversible defect. They may yield to its constraining effects on performance, showing simpler, more restricted, but more stable performance, or they may try to compensate for the loss. In the latter case, the performance is more variable and more likely to lead to catastrophic reactions.

phenomenon we call abstract attitude, etc., but, to reiterate, we may not localize corresponding functions in these parts of the brain. . . . Such a rejection [of localization of function] is not in contradiction to an assumption that to each performance corresponds an excitation of definite structure . . . widespread over the whole cortex differently in each performance (Goldstein, 1948, p. 50).

Goldstein was primarily interested in the underlying psychological deficit that resulted from cerebral damage to help design therapeutic interventions. Knowing the location of the lesion was secondary to understanding the psychological deficit it produced, because his qualitative tests indicated that the fundamental psychological consequences of the brain damage were often similar despite variation in the form of their expression as symptoms. In contrast, Teuber's early work (e.g., Teuber, 1959) on World War II and Korean War veterans emphasized identifying the functional differences between different areas of the cerebral hemispheres by statistical analysis of scores on carefully controlled tests from patients grouped on the basis of location of lesion. Grafman *et al.* (1986) adopted a similar approach to study Viet Nam veterans. This approach reflects the dominant theme in much of American neuropsychology: to correlate location of lesions with the resulting pattern of deficits on tests. To this end Teuber applied the double-dissociation method (Milner and Teuber, 1967), which has become something of a standard for assessing brain function despite limitations of its rigid application (Glassman, 1978, pp. 16–18) and its underlying assumptions (Smith, 1983, 1984).

5. PSYCHOLOGICAL DEFICITS FOLLOWING BRAIN DAMAGE

According to Goldstein, cerebral lesions disturb performance in a pattern of predictable ways. Performance deficits are not restricted to a single "performance field,"* the changes in different performance fields show the same underlying psychological disturbance, and a single performance field does not drop out completely. Performance in affected fields undergoes a process Goldstein called "dedifferentiation." Dedifferentiation appears as an impairment in the ability to execute separate actions and to execute responses discretely and selectively, to respond selectively to single features, to segregate figure from ground, etc. It also appears as elevated threshold and reduced acuity, defective localization of stimuli, decreased ability to differentiate between sensory qualities, etc. Following lesions of the central portions of the cerebral cortex, dedifferentiation is often the consequence of isolation of different cerebral areas from each other. Modern research provides a number of examples of what appear to be dedifferentiation following cerebral lesions (Kinsbourne, 1981).

---

*Recent claims of category-specific deficits may reflect incomplete analysis and interpretation of the observations (Humphry and Riddoch, 1987).

Another fundamental consequence of cerebral lesions is the "catastrophic reaction," which occurs when any organism, intact as well as brain damaged, is confronted with environmental demands that are beyond that organism's capacities. "Catastrophic reactions . . . are not only inadequate but also disordered, inconstant, inconsistent, and embedded in physical and mental shock" (Goldstein, 1939, p. 37), and their disturbing aftereffects can affect performance on tasks that the patient solves readily under other circumstances. Brain-damaged patients make a persistent effort to adapt to their disturbed and limited capacities by returning to an ordered condition. This is achieved by restructuring the situation in which the brain-damaged individual must operate and by learning round-about procedures to accomplish the adequate performance to compensate for the loss of normal, direct procedures. The motivating process behind these adaptations is self-actualization, a term later popularized by Maslow (1954). It is the general drive that originates from "that state of tension . . . which enables and impels the organism to actualize itself in further activities according to its nature" (Goldstein, 1939, p. 197). This "tension" results from the organism's effort to maintain its inherent ordered state of organization. Goldstein contrasted self-actualization, which does not discharge the tension from which it originates, to traditional concepts of separate drives, which are aimed at discharging special tensions resulting from abnormal isolation of parts of the organism.

Goldstein believed that the basic psychological disturbance common to performance deficits of diverse specific forms is a decrease in the patient's ability to use the "abstract attitude." More recent investigations indicate that this is not a necessary consequence of cerebral damage, even in the frontal lobes, damage to which results most frequently in the loss of the abstract attitude (Smith, 1962, 1975). The abstract attitude permits the individual to deal with stimuli as representing one or more classes or categories. Features of this attitude include, among others, the ability to shift reflectively from one aspect of a situation to another, to plan ahead ideationally, to consider what might be possible, and to think or act symbolically. The loss of the abstract attitude requires him/her to depend more and more on the "concrete attitude." It forces the individual to deal with stimuli only as specific events or in terms of specific actions or responses one makes to them. Goldstein and Sheerer (1941, p. 3ff.) presented a detailed summary of this distinction.

Some specific examples will clarify these concepts. In his studies of "amnesic aphasias," Goldstein (1939, 1948) concluded that the deficit in these patients is not in the ability to associate a word with its referent object. Rather, it represents the patient's inability to use a word to represent an abstract idea or class of object. The circumlocutions that such patients employ reflect the use of the concrete attitude to guide their behavior. By using a concrete example or application of the stimulus object or a concrete association with it, the patient is able to make an appropriate response that does not depend on the abstract concept of class membership. Thus one patient responded to an umbrella with a phrase like "that is a thing for rain" instead of saying "umbrella" directly. Her subsequent spontaneous statement "I have three umbrellas at home" demonstrated that the patient knew the word "umbrella" and had access to it under appropriate circumstances. Similarly, in performing a sorting test patients often put together colored skeins that are almost identical in all aspects rather than accept moderate variation in hue or saturation within a color class. When such patients were asked to name a color put before them, they often were unable to answer. If they did, their responses often included a concrete referent having that color, e.g., strawberry red or sky blue. These performances reflected responding to some specific constant feature of the stimulus rather than a more abstract classification in which variations of irrelevant attributes are ignored. Goldstein and Sheerer (1941) provided many more examples. Studies of thousands of patients by one of us (A.S.) since studying with Goldstein have indicated that this specific defect described as characteristic of patients with brain damage is not as common as Goldstein had suggested and could occur following lesions in any part of the brain, not just the frontal lobes. Unfortunately, most of the tests and techniques designed for qualitative studies of patients with brain damage have not been administered to normal control subjects. Thus, the extent to which patients lose abstract attitude or depend on "concrete approaches" as a function of age, education, or other individual factors independent of the brain insults has never been determined (Smith, 1983, 1984).

6. PSYCHOLOGICAL TESTING OF BRAIN-DAMAGED PATIENTS

For Goldstein, the underlying psychological deficit following brain damage can be identified only by careful analysis of the symptoms by systematically varying psychological tests to identify how the patient solves or fails to solve them. The correctness or incorrectness of performance on a test is much less informative than is the way the solution (or failure) was achieved. In this, Goldstein's position resembles Luria's and differs markedly from the quantitative psychometric tradition, which usually emphasizes quantitative comparison of a patient's performance to established norms and is the basis of much of modern Western neuropsychological testing. Kohn (1986) recently summarized Goldstein's position on qualitative versus quantitative approaches to assessment of behavioral deficits following brain damage and its significance to modern practice of neuropsychological assessment.

Goldstein and Sheerer (1941) and Goldstein (1948) described several qualitative sorting tests and other methods for investigating the psychological disturbances that result from brain damage. The protocols provide detailed instructions for administering the tests, with branching instructions based on the subject's performance. The examiner must record in detail what the patient does and says, especially in response to the examiner's requests for the subject to explain his performance. The examiner is encouraged to follow up the standard protocol with variations intended to explore the basis of the patient's performance.

Goldstein argued that conventional psychometric methodology is not applicable to testing brain-damaged patients because they approach the test problems differently than do normal individuals. Thus, the large variations in test performance often observed in brain-damaged patients from day to day or under slightly varying testing conditions do not represent errors of measurement. Rather, they can represent important clues to the underlying deficit. According to Goldstein, brain-damaged patients are strongly influenced by extraneous circumstances because they are more sensitive to fatigue, because their catastrophic reactions to previous failures of performance affect performance on subsequent testing, and because slight changes in the testing task can prevent (or permit) solution of the test by the concrete attitude (Goldstein, 1942).

7. RECOVERY AND REHABILITATION

There are a number of different models of recovery of function (Finger and Stein, 1982), some with a long history, others based on evidence obtained with modern techniques. Goldstein's conception of recovery through retraining fits best what Finger and Stein called behavioral substitution and response and cue substitution theories. Finger and Stein carefully considered the conceptual problems with this model, especially in differentiating it from a vicariation model in which undamaged parts of the nervous system in some fashion take over the specific functions of the damaged areas. The central question is how well we can understand the "function" of a localized area of the cerebral cortex.

Goldstein (1942) identified three processes that contribute to "recovery" of performance following brain damage: "restitution of the damaged substratum" resulting from healing processes acting on the damaged brain area itself; simplifying the environment to avoid encountering situations with which the brain-injured individual is unable to cope; and relearning a disturbed performance by using other systems that remain functional following brain damage. The first process is the basis of the spontaneous recovery that patients show over the first weeks or months following brain injury. Such recovery is assumed to reflect the progressive clearing of pathological processes resulting from the damage. The diminution or disappearance of certain *"symptomes temporaires"* is related to von Monakow's concept of diaschisis (Finger and Stein, 1982; Glassman and Smith, Chapter 4, this volume). The second process, creating a simplified situation, permits the patient's spared concrete attitude to be more effective in meeting the challenges of everyday functioning following impairment of the abstract attitude. It also prevents the occurrence of catastrophic reaction, which occurs when the patient is faced with a challenge he/she cannot meet. Such catastrophic reactions severely disturb other performances that the patient ordinarily can do successfully.

The third process, recovery through retraining, can take two forms: relearning a performance capacity in the same way that it was executed before injury (restitution) and relearning by compensation with other performances (substitution) (Goldstein, 1942, 1948). If the brain substratum for a performance is damaged such that its organization as a whole is preserved, then retraining may be possible "in about the same way as it (the performance) was trained in childhood" (Goldstein, 1942, p. 148). Such a replication of ontogeny in recovery following brain damage has been demonstrated experimentally following lateral hypothalamic lesions in animals (Teitelbaum *et al.*, 1969). If the brain damage results in irrevocable loss of a performance, then whatever recovery that can occur must be achieved "by building up other, compensatory performances . . ." (p. 148). Such compensatory recovery depends on the interaction of functionally intact cerebral tissue in the same as well as the opposite hemisphere and on the commissures between them (Campbell *et al.*, 1981; Smith, 1981a,b, 1983, 1984). Although training of this sort can produce performance that resembles normal performance, the performance relearned by substituting intact processes for the ones normally used must be deficient in some aspect. The distinction between restitution and substitution is of basic importance in the process of recovery and its facilitation by the therapist. Different underlying defects require different approaches to the process of treatment. Unfortunately for the purposes of practical treatment and theoretical understanding, it is often very difficult to determine which process takes place (Goldstein, 1948, pp. 51*ff.*).

Goldstein's (1942, pp. 149*ff.*) treatment of a case of very severe visual agnosia illustrates the process of recovery through retraining. Following injury "at the back of the head" from "mine splinters," this patient was unable to recognize objects visually even though his acuity and fields were essentially normal. However, the patient noticed that even though he was unable to recognize letters themselves, he could recognize their meaning. Careful observation of the patient revealed that he achieved this ability by tracing the letters with eye movements, although he was quite unaware of his technique. The patient eventually learned to "read" (though the level of reading was not specified) following treatment based on this substitute performance. This case also illustrates the importance Goldstein placed on using clues the patient gives to devise successful methods of treatment. It also emphasizes the importance of differentiating between effects of damage to specific modalities (input and/or output) and higher "cognitive" functions involved in tests used in studies of patients with brain damage. This distinction, which Goldstein frequently emphasized, led one of us (A.S.) to develop a neuropsychological test battery based on such differentiations (Smith, 1975, 1983, 1984).

Goldstein's (1942, pp. 152*ff.*) approach to the treatment of "amnesic aphasia" is instructive in its demonstration of the relationship between the treatment
strategy and the identification of the underlying deficit. Patients with this disorder are disturbed in finding words, especially names of concrete objects, even
though they readily recognize the correct word when it is given to them. Extensive training by associating words with pictures of their referents produces very
little gain, because the patients' underlying deficit is an inability to use words as
representative of a class of objects. This deficit is, according to Goldstein, the
result of an inability to adopt the abstract attitude, that is, to use words to
represent classes. These patients are often able to find a word if it is part of a
rhyme, a series, or some other imposed structure. Such performance depends on
the concrete attitude, which is less likely to be affected by localized brain
damage than is the abstract attitude. Improvement in word finding can be
achieved by building on these remaining skills.

The usefulness of Goldstein's emphasis on careful evaluation of the psychological deficit underlying disturbed performance is illustrated in its capacity to
differentiate among patients with superficially similar symptoms. Goldstein
(1942, p. 159*ff.*) described a patient whose difficulty finding words resembled
"amnesic aphasia." His underlying deficit turned out, on careful testing, to be
quite different from the deficit in the patients described above. For him, the
ability to represent the idea of a word as a set of sound patterns was disturbed.
Unlike the patients described above, he easily learned to associate spoken words
with their referent objects, but he was quite unable to use the words in commu-
nicative speech. Thus, the form of treatment was necessarily quite different,
concentrating on the consolidation of the representation of ideas as words. This
was done by intensive practice using specific words in their various forms and
transformations to relearn that they were all derived from the same root and
reflected the same idea.

8. SIGNIFICANCE FOR NEUROPSYCHOLOGY

Goldstein's conception of deficit and recovery provides a theoretical basis
for common practice among many workers in cognitive rehabilitation. Therapists
emphasize finding surviving concrete cognitive processes, external cues, specific
events, etc. on which to build the therapeutic interventions. They try to identify
the specific nature of the psychological defect underlying the performance deficit
to find the most efficient intervention, and they recognize that improvements in
performance usually are achieved by processes that differ from those used before
brain damage. They create simplified environments in which the patients are able
to operate. All of these are directly parallel to the rehabilitation program that
Goldstein established in the 1920s.

There are, however, important differences between Goldstein's approach and that of modern workers. These differences appear especially in methods of assessment and interpretation of symptoms and deficits. A careful consideration of Goldstein's approach may provide useful insights both for the practice of cognitive rehabilitation and for the theoretical understanding of the psychological consequences of brain damage and the processes of recovery.

Goldstein's qualitative assessment procedures were intended to identify the underlying psychological defect resulting from cerebral damage. For the purposes of cognitive rehabilitation, location of damage, *per se*, is irrelevant, except that it may help predict on an actuarial basis the nature of the defect and the probable course of recovery. An increased emphasis on trying to understand the psychological deficit with neuropsychological batteries is a very useful goal to establish. It should improve the therapist's ability to design remediation programs more accurately aimed at the underlying defect. The emphasis on the careful analysis of the underlying psychological disturbance caused by brain damage is perhaps the most important practical contribution that a reassessment of Goldstein's work can provide.

The theoretical difference between Goldstein and modern students of recovery is more difficult to spell out clearly. It appears to originate in asking a different question and/or adopting a different level of analysis. Goldstein attempted to understand the psychological functions mediated by the cerebral cortex. He believed that there are common psychological deficits underlying diverse symptoms that follow cerebral damage. In contrast, modern neurologists and neuropsychologists tend to approach the consequences of brain damage the same way workers in experimental animal neuropsychology do. These workers ordinarily design studies that isolate the brain and behavioral systems under consideration. This powerful simplifying technique lends itself to the demonstration of specificity, but this specificity may mislead by appearing to localize a particular behavioral function (see Finger and Stein, 1982) and by distorting the phenomenon under investigation (Goldstein, 1948). Valenstein (1973, 1975) has described this problem in the application of findings from animal experiments to clinical interventions in human patients.

Modern workers also tend to adopt a connectionistic, modular model of psychological organization to parallel the connectionistic, modular features of cortical anatomy and physiology. This is not a necessary relationship. The facts of precise anatomic connections and physiological specialization in the cerebral cortex do not require a simple psychological parallel. We have presented above one example of this distinction between anatomic and physiological specialization and psychological specificity in the writings of Melzack and Wall (1981) on pain. Other phenomena in sensory systems suggest a similar dissociation. Dykes and Metherate (Chapter 15) have reviewed the literature on variation in the cortical somatotopic representation induced by altering the afferent input or internal state of the subject. These phenomena appear contrary to strict localization. Nevertheless, most neurons retain their functional attributes over extended periods, as localization requires. Dykes and Metherate proposed a model to account for these apparently contradictory properties.

More generally, Erickson (1984) has shown that his distributive account of representation of sensory information can be applied to the representation of behavioral processes in general. He argued that sensory information and behavioral responses are represented by specific complex patterns of activation of populations of these highly specialized neurons. Recent investigations of motor function have demonstrated a distributed representation of the coding of specific movements. Georgopoulos *et al.* (1986) have found that a specific movement of a monkey's arm is best represented by a weighted average of the activity of a population of units in the precentral arm representation. Other recent work has suggested that the cerebral cortex is divided into highly specialized units, the operations of which may have an underlying uniformity. Thus, individual cognitive processes may be represented by individual complex patterns of activation in populations of these highly specialized, but uniformly operating, "elements." Such "elements" have been proposed based largely on anatomic and physiological observations (e.g., Edelman and Mountcastle, 1978; Phillips *et al.*, 1984). Elsewhere in this volume, Glassman and Smith (Chapter 4) develop a simple numerical model exhibiting such properties.

Finally, a recent paper (Eaton and DiDomenico, 1987) has analyzed the relationship between the physiological and behavioral functions of command neurons in simple nervous systems. Activation of these neurons is held to be necessary and sufficient for specific behavioral reactions. For example, the Mauthner cell in the goldfish brainstem is usually held specifically to mediate the brisk escape reaction characteristic of the species. This definition of the command neuron represents an extreme case of specific correspondence between anatomic and physiological specialization and a specific behavioral function. Eaton and DiDomenico provide empirical and theoretical evidence that command neurons do not and cannot operate as defined in any but the simplest circuit.

REFERENCES

Berker, E. A., 1985, Principles of brain function in neuropsychological development of hydrocephalics (Doctoral dissertation, University of Michigan, 1985), *Dis. Abstr. Int.* **46**:1373b.

Braun, J. J., 1978, Time and recovery from brain damage, in: *Recovery from Brain Damage* (S. Finger, ed.), Plenum Press, New York, pp. 165–197.

Campbell, A. L., Bogen, J. E., and Smith, A., 1981, Disorganization and reorganization of cognitive functions in cerebral commissurotomy: Compensatory roles of the forebrain commissures and cerebral hemispheres in man, *Brain* **104**:493–511.

Dimond, S. J., Farrington, L., and Johnson, P., 1976, Differing emotional reactions from right and left hemispheres, *Nature* **261**:690–692.

Eaton, R. C., and DiDomenico, R., 1985, Command and the neural causation of behavior: A theoretical analysis of the necessity and sufficiency paradigm, *Brain Behav. Evol.* **27**:132–164.

Edelman, G. M., and Mountcastle, V. B., 1978, *The Mindful Brain*, MIT Press, Cambridge, MA.

Erickson, R. P., 1984, On the neural basis of behavior, *Am. Sci.* **72**:233–242.

Finger, S., 1978, Lesion momentum and behavior, in: *Recovery from Brain Damage* (S. Finger, ed.), Plenum Press, New York, pp. 135–164.

Finger, S., and Stein, D. G., 1982, *Brain Damage and Recovery*, Academic Press, New York.

Georgopoulos, A. P., Schwartz, A. B., and Kettner, R. E., 1986, Neuronal populations coding of movement direction, *Science* **233**:1416–1419.

Geschwind,.N., 1964, The paradoxical position of Kurt Goldstein in the history of aphasia, *Cortex* **1**:214–224.

Glassman, R. B., 1978, The logic of the lesion experiment and its role in the neural sciences, in: *Recovery from Brain Damage* (S. Finger, ed.), Plenum Press, New York, pp. 4–31.

Goldstein, K., 1927, Die Lokalisation in der Grosshirnrinde, in: *Handbuch der Normalen und Pathologischen Physiologie*, Volume 10 (A. Bethe, G. v. Bergmann, G. Embden, and A. Ellinger, eds.), Springer, Berlin, pp. 600–842.

Goldstein, K., 1934, *Der Aufbau des Organismus*, Martinus Nijhoff, The Hague.

Goldstein, K., 1939, *The Organism*, American Book Co., New York.

Goldstein, K., 1942, *Aftereffects of Brain Injuries in War*, Grune & Stratton, New York.

Goldstein, K., 1946, Remarks on localization, *Confin. Neurol.* **7**:25–34.

Goldstein, K., 1948, *Language and Language Disturbances*, Grune & Stratton, New York.

Goldstein, K., 1967, Kurt Goldstein, in: *A History of Psychology in Autobiography*, Volume 5 (E. G. Boring and G. Lindzey, eds.), Appleton–Century–Crofts, New York, pp. 147–166.

Goldstein, K., and Sheerer, M., 1941, Abstract and concrete behavior: An experimental study with special tests, *Psychol. Monogr.* **53**(2, whole number 239):1–151.

Grafman, J., Salazar, A., Weingartner, H., Vance, S., and Amin, D., 1986, The relationship of brain-tissue loss volume and lesion location to cognitive deficit, *J. Neurosci.* **6**:301–307.

Head, H., 1926, *Aphasia and Kindred Disorders of Speech*, Volume 1, Macmillan, New York.

Humphrey, G. W., and Riddoch, M. J., 1987, On telling your fruit from your vegetables: A consideration of category-specific deficits after brain damage, *Trends Neurosci.* **10**:145–148.

Kashiwaga, A., Kashiwaga, T., and Hagesawa, T., 1987, Improvement of deficits in mnemonic rhyme for multiplication in Japanese aphasics, *Neuropsychologia* **25**:443–447.

Kinsbourne, M., 1981, Cognitive defect and the unity of brain organization: Goldstein's perspective updated, *J. Commun. Disord.* **14**:181–194.

Kohn, H., 1986, Qualitative neuropsychological assessment: Kurt Goldstein revisited, in: *Clinical Neuropsychology of Intervention* (B. P. Uzzell and Y. Gross, eds.), Martinus Nijhoff, Boston, pp. 51–58.

Luria, A. R., 1966, Kurt Goldstein and neuropsychology, *Neuropsychologia* **4**:311–313.

Maslow, A. H., 1954, *Motivation and Personality*, Harper Brothers, New York.

Meiers, J., 1968, Bibliography of the published works of Kurt Goldstein, in: *The Reach of Mind* (M. L. Simmel, ed.), Springer, New York, pp. 271–195.

Melzack, R., and Wall, P. D., 1981, *Challenge of Pain*, Basic Books, New York.

Milner, B., and Teuber, H.-L., 1967, Alterations of perception and memory in man: Reflections on methods, in: *Analysis of Behavioral Change* (L. Weiskrantz, ed.), Harper & Row, New York, pp. 268–375.

Phillips, C. G., Zeki, S., and Barlow, H. B., 1984, Localization of function in the cerebral cortex, *Brain* **107**:327–361.

Reise, W., 1960, Dynamics of brain lesions, *J. Nerv. Ment. Dis.* **131**:291–301.

Smith, A., 1962, Ambiguities in concepts and studies of 'brain damage' and 'organicity,' *J. Nerv. Ment. Dis.* **135**:311–326.

Smith, A., 1975, Neuropsychological testing in neurological disorders, in: *Current Reviews of Higher Nervous Dysfunction. Advances in Neurology*, Volume 7 (W. J. Friedlander, ed.), Raven Press, New York, pp. 49–110.

Smith, A., 1979, Practices and principles of clinical neuropsychology, *Int. J. Neurosci.* **9**:233–238.

Smith, A., 1981a, Principles underlying human brain function in neuropsychological sequelae of different neuropathological processes, in: *Handbook of Clinical Neuropsychology* (S. B. Filskov and T. J. Boll, eds.), John Wiley & Sons, New York, pp. 175–226.

Smith, A., 1981b, On the organization, disorganization and reorganization of language and other brain function, in: *Lateralization of Language in the Child, Neurolinguistics*, Volume 10 (Y. Lebrun and O. Zangwill, eds.), Swets & Zeitlinger, Lisse, pp. 51–70.

Smith, A., 1983, Clinical psychological practice and principles of neuropsychological assessment, in: *Handbook of Clinical Psychology* (C. E. Walker, ed.), Dorsey, Homewood, IL, pp. 445–500.

Smith, A., 1984, Early and long-term recovery from brain damage in children and adults: Evolution of the concepts of localization, plasticity, and recovery, in: *Early Brain Damage*, Volume 1. *Research Orientations and Clinical Observations* (R. Almli and S. Finger, eds.), Academic Press, Orlando, pp. 299–324.

Smith, A., 1986, Brain-mind philosophy, *Inquiry* **29**:203–215.

Teitelbaum, P., Cheng, M. F., and Rozin, P., 1969, Development of feeding parallels its recovery after hypothalamic damage, *J. Comp. Physiol. Psychol.* **67**:430–441.

Teuber, H.-L., 1959, Some alterations in behavior after cerebral lesions in man, in: *The Evolution of Nervous Control from Primitive Organisms to Man*, American Association for the Advancement of Science, Washington, pp. 157–194.

Teuber, H.-L., 1966, Kurt Goldstein's role in the development of neuropsychology, *Neuropsychologia* **4**:299–310.

Valenstein, E., 1973, *Brain Control*, John Wiley & Sons, New York.

Valenstein, E., 1975, Brain stimulation and behavior control, in: *1974 Nebraska Symposium on Motivation* (J. K. Cole and T. B. Sonderegger, eds.), University of Nebraska Press, Lincoln, NE, pp. 251–292.

6

Assumptions about the Brain and Its Recovery from Damage

ROBERT L. ISAACSON

1. MECHANISMS OF BRAIN FUNCTION

Any consideration of how the brain recovers from damage must be linked to some idea, however vague, about how the brain operates. Frankly, we are a very long way from understanding how the brain as an entity, or its subdivisions, acts to produce the mental and behavioral characteristics of people. Yet without any firm foundation of this knowledge, as scientists, we must proceed in our research programs to generate an empirical basis for the understanding that we anticipate we will someday achieve.

One of the most fundamental distinctions that can be found between experimenters is whether they believe more in the importance of network type of organization based on the loose and apparently incoherent interconnections of neurons and dendrites that constitute the neuropil of the brain or whether they believe in the fundamental importance of a "hardwired" approach to brain function. I stand on the network side of brain function. Historically, this bias comes from the study of the work originating in the laboratory of C. Judson Herrick and his classic books, including *The Brain of the Tiger Salamander* (Herrick, 1948). This book is a good starting place for all who would try to understand the neuroanatomic organization of the brain.

Based on the study of the brains of amphibians and reptiles, it seems rather clear that much of the processing of information and the organization of behavior can be accomplished by means of diffuse systems existing in the neuropil. The neuropil constitutes most of the central nervous system, including the spinal

ROBERT L. ISAACSON • Department of Psychology and Center for Neurobehavioral Science, University Center at Binghamton, Binghamton, New York 13901.

cord, of these "lower" vertebrates. A careful study of these brains shows that
the few cells of which they are composed are gathered around the ventricular
spaces with processes from them filling most of the rest of the tissue. This fine
feltwork of the neuropil represents the vast majority of the nervous system.

Very few people have attempted to develop theories of how the brain could
function on the basis of loosely organized sets of interconnections as found in the
neuropil. I was profoundly affected by Ross Ashby's *Design for a Brain* (Ashby,
1960). In recent years little attention has been directed toward theories that are
founded, in part, on diffuse systems of neuritic processes. The assumption that
information storage occurs in the form of holograms has been urged by Pribram
(1971) and represents one such model.

2. MACLEAN AND THE TRIUNE BRAIN

I have also been influenced by the theories of Paul MacLean. A synopsis of
his ideas can be found in the last chapter of my book on the limbic system
(Isaacson, 1982) and in many publications by MacLean himself (e.g., Mac-
Lean, 1970). The essential ideas put forward by MacLean include the assumption of a hierarchical organization of the brain based on the notion that the basics
of the vertebrate brain (the protoreptilean brain as he calls it) can cope with the
basic life-sustaining processes required by the individual, including mating,
procreation, and agonist and antagonistic behaviors. Species-typical displays and
important vocalizations are also encoded or specified in some fashion in this
oldest portion of the nervous system of vertebrates. Although MacLean does not
emphasize it, this protoreptilean portion of the brain is almost completely made
up of neuropil in the amphibian brain. As has been shown by a number of
authors, animals with little more than brainstem mechanisms can demonstrate
remarkable abilities for learning environmental contingencies (e.g., Huston and
Borbely, 1973).

The neuropil-abundant protoreptilean brain has direct access, or at least
virtually direct access, to the effectors that activate the somatic and autonomic
mechanisms of the body. It should be noted that this direct access does not
regulate individual muscle contractions but rather organized sets of extensions
and contractions of muscles that in the long haul form species-typical behavioral
patterns.

In the opinion of MacLean, through the course of evolution the early protoreptilean brain comes to be regulated by a new form of neural tissue, the
paleomammalian brain, which originates in the early forms of mammalian life.
This portion of the brain corresponds roughly to the areas of more advanced
brains that are collectively called the limbic system. Although these portions of
the brain may become extensively enlarged in the mammalian line, the rudiments
of the system may be found in amphibia. In my opinion, it is hard to argue for

more than a quantitative difference in the development of the more advanced brains, since many of the limbic structures can be found in premammals: portions of hypothalamus, amygdala, septal area, and hippocampus. The evolutionary step that can be found within the mammalian line is likely to be in the advancement of the limbic brain into more complex structures or an enhancement of its influences on the protoreptilean complex.

The final evolutionary addition to the brain of mankind is the creation of a neomammalian brain. Again, it may be inaccurate to say that this is a specific mammalian development, since there are some cells in areas of "general cortex" of amphibia and reptiles that may be similar in form, if not function, to cells that make up the neocortex of the more complex brain.

In general, it can be suggested that the neomammalian brain modifies brainstem and spinal cord activities that are embedded in the protoreptilean brain. The more or less direct pathways to the cord and brainstem areas are not shared by the limbic brain. The influences of the limbic brain do not extend much below the telencephalon itself. Only few fibers reach to nuclear regions of the brainstem. It may be, however, that the neomammalian brain's most important role in the life of the animal is its regulation of the limbic brain.

As a trend in interpretations of the functional anatomy of the brain, I see increased recognition of the neocortex as providing input to the lower regions of the telencephalon, including providing a major input to the limbic and basal ganglia regions. For example, we are just beginning to appreciate how the neocortex provides a direct regulation of the brainstem nuclei that provide the biogenic amine projections to the entire forebrain. It could be argued that the neocortex largely directs and regulates the input to the entire brain both through its input into basal ganglia regions and to the thalamus as well as through its projections to the spinal cord. This is because most of the fibers of the corticospinal tracts end in the dorsal portions of the cord, and a minority end on the motoneurons or the more ventral aspects of the cord. In the rat almost all of the corticospinal projections end in the mediodorsal aspects of the cord, presumably exerting strong and chronic control of the sensory information reaching spinal cord systems and higher centers of the brain (Brown, 1971). The emphasis on a sensory-regulatory control rather than a direct control by the neocortex is supported by the fact that sectioning of the medullary pyramids produces few noticeable motor effects (Barron, 1934; Bucy, 1966).

There is a close tie between the development of the neocortical mantle of the brain and the parallel evolution of the nuclei of the dorsal thalamus. Once again, examination of the role of the majority of the nuclei of the dorsal thalamus reveals that they are concerned with the transfer of information from the environment to the neocortical tissue.

To this point, several important assumptions that underlie my work on recovery of function after brain damage have been identified. They include a hierarchical organization of a brain that has mechanisms for the redundant processing of information at different levels. Furthermore, it is clear that the phylogenetically newest portions of the brain, the neomammalian areas, have direct connections with specific and general regulatory mechanisms. They include the specific, anatomically discrete inputs into the basal ganglia that, at least in part, use excitatory amino acids as their transmitter substances. This same cortical tissue also sends projections to the biogenic-amine-containing cells of the brainstem. Unilateral lesions of the neocortical mantle can produce changes in the amounts and utilization of these biogenic amines on both the damaged and the contralateral side of the brain (e.g., Robinson and Coyle, 1980). Therefore, this important tissue is involved in the modulation of neural activity as well as in the conveying of high-information messages.

3. MULTIPLE FUNCTIONS OF NEURAL SYSTEMS

We are just beginning to understand the multiple roles played by most, if not all, projection systems of the brain. Even the monosynaptic projections from the neocortex to the basal ganglia that may be carrying specific information are modulating other systems at the same time (e.g., O'Donohue *et al.*, 1985; Bartfai, 1985). As an example, the basal ganglia sites of the caudate and accumbens nuclei project to the globus pallidus, including the rather diffuse regions that contain cells that send the massive cholinergic projections to the neocortex. Therefore, there is a mixing of inputs of both a modulatory and specific nature in these basal ganglia regions, whose convergence may produce both modulatory and specific effects once again. Overall, it is clear that there is a mixing of modulatory transmitter types converging on the cells of the basal ganglia, and these mix with more and less direct specific inputs. It can be imagined that particular cell clusters of the basal ganglia are receiving the rapidly acting excitatory amino acids while also receiving the slower-acting modulatory influences such as those mediated by muscarinic receptors and by one or more of the biogenic amines.

This is far from the end of the story, since many transmitter systems contain multiple transmitters or modulators within each cell. These multiple systems may be arranged so that one substance modifies the release of the other or the action of the other at the postsynaptic receptor. In addition, the effects of neurohormones and neuropeptides must be considered. These substances influence the overall responsiveness of the cells in widespread regions of the brain, although not necessarily the same functions in each area. They are also involved in the maintenance of appropriate conditions for the integrity of the cell and its continued conduction of neural impulses. In other words, they may exhibit activational and trophic influences on cells in certain regions and, perhaps, in all of the central nervous system.

In addition, the influence of the long-studied hormones cannot be ignored in understanding brain functions. It is well known that the sex hormones can influence the activity of certain transmitters and receptors in some regions of the brain (e.g., Sar and Stumpf, 1977; Euvarrd *et al.*, 1980; Rainbow *et al.*, 1980). It is possible that these same sorts of effects will be found for the corticosteroids as well.

4. THE EFFECTS OF DAMAGE

Whether it be in a laboratory animal or at the human level, the job of the researcher in the field of recovery from brain damage is to determine the nature of the changes that have occurred after the insult and the kinds of therapeutic interventions that may be of help to the individual. It is obvious that lesions have both transitory and permanent effects. It has been commonly assumed that the permanent changes, at least those that are permanent in that they extend beyond the time limits of an experiment, result from disruption of hard-wired, specific connections of the brain. These hard-wired connections are ones thought to be indispensable for the executions of certain types of behavior.

However, this assumption about why some changes are permanent does not necessarily fit with the overview of brain activities as described above. Indeed, I do not think that common explanations for permanent and temporary effects of brain damage are correct.

If one accepts the principles I have outlined concerning brain function, this leads to certain predictions about what happens after brain damage that may seem unusual and, at first, contradictory to common experience. However, it is possible that what appears to contradict experience may not actually do so.

4.1. Are Any Changes "Absolute"?

An important generalization arising from my assumptions about brain function is that no physically restricted lesion of the brain should ever lead to an absolute impairment of any particular specific mental or behavioral capacity. This is because every function is multiply represented in the brain at different hierarchical levels and within each of the several levels. Even damage that destroys a "complete" sensory region (as the term is currently used) of the neocortex should not, in my view, lead to a permanent disturbance in the use of a particular capacity. If the damage were to occur to the "visual" neocortical surface, for example, the animal need not be blind, basically because visual information can be processed in many brain regions and not just those that receive highly organized projections from the lateral geniculate nucleus. The same can be said for the regions that receive "auditory" or "somatosensory" inputs.

An examination of the literature indicates that deficits in visual abilities seem slight, if even detectable, after complete destruction of the "visual" cortex in rodents and in some other nonprimate species. When impairments are observed, they usually are found in the performance of certain learned responses to visual stimuli and the performance of some species-typical acts usually occasioned by events in the visual world. Perhaps the most striking example of this residual capacity of the visual system can be found in the cat. Following extirpation of the entire visual cortex, this animal can show virtually perfect performance on most visual tasks and in everyday life. They do not show, however, a visual placing response. That is, when held suspended over the edge of a flat surface, they do not look at this surface and then extend their paws to make contact with it. However, if the back of a paw is put in contact with the table top, the animals will place the paw on the flat surface. This is called the tactile placing response. On the basis of this description it might be thought that lesions of the cat's visual cortex destroy the cat's mechanisms for the use of certain kinds of visual information for producing or placing response. However, if the animal is given amphetamine before testing, the animal can exhibit a normal visually guided placing response for as long as the drug is active (Meyer, 1963). Later (the following day for instance), the brain-damaged cat will again fail to exhibit the visual placing response. Therefore, the lesion does not destroy the cat's ability to perform the response but rather its inclination or motivation to do so. For a review of research in this area, see Meyer and Meyer (1982).

4.2. Motivational Changes following Brain Damage

Ten to 12 years ago at the University of Florida, two undergraduates and I undertook the study of the effects of serial destruction of the anterior neocortex of the rat. We hoped that the simple testing device designed by Castro (1972) for his doctoral research at that institution might prove useful in evaluating the effects of the serial versus all-at-once removal of the anterior telencephalic tissue. Castro's device consisted of a set of small trays that could be placed at the front of the rat's home cage. In each tray was a small food pellet. The trick was that there was a gap between the tray and the cage so that the animals had to carry the pellet over the gap in order to eat it. If they simply tried to scoop the pellet into their cage, as is their first inclination, the pellet would fall into the gap. Therefore, they had to learn to expand their digits, place their extended paw over the pellet, and then contract the digits, thus grasping the food pellet. This allowed them to carry the food pellet over the gap. All of the animals were trained for several weeks before the unilateral or bilateral anterior neocortical lesions were made.

I learned a great deal from this experiment. The first finding was that rats should not be trained to use their forepaws. Once the animals learned to use them, they found they could reach out to grasp the lever that secured the door of their cage, and after a few days of training when we went into the colony room the animals were running about all over the room. When we placed them back in their cages, they would reach through the grillwork, turn the lever, and open their doors again. To keep them confined I had to go buy 60 padlocks. As far as I know this ability of rats to use their forepaws as hands as a consequence of training has never been reported in the literature.

The postlesion behavioral results were difficult to analyze in the usual fashion for a "scientific publication" (which is the reason it was not submitted for publication). In both the serial and "all-at-once" lesion groups, some animals tried to get the pellets but because of awkwardness of their forepaws were only partially successful. Many times the pellets slipped from their grasps and fell into the gap. For a substantial number of these animals, these few failures were sufficient to induce a permanent reluctance ever again to try to obtain a pellet, even when food deprivation was increased, and even though they would readily grasp the pellets when they were held in front of them by forceps. What had happened was that the partially successful (as opposed to the formerly uniformly successful) responses of the brain-damaged animals induced a reluctance even to try the task again. The failure to keep trying to get the pellets was not a consequence of a long series of failures. Often, only two or three failures would be enough to induce an apparent reluctance to pursue the task. In fact, those few animals that did continue to try to get the pellets became almost as proficient as they had been before the lesion.

This seems to be another example of a change in the willingness of the animal to undertake a response: a change in motivation, a reluctance to act, but not an irreversible loss of motoric abilities. Unfortunately, we did not try to overcome this reluctance to respond with amphetamine, an oversight in retrospect.

A careful reading of Kurt Goldstein's (1942) accounts of the effects of traumatic brain damage to German soldiers of World War I indicates similar types of reactions when the patients attempted new tasks. One of the most prominent consequences of brain damage was the fear of experiencing failure. A corresponding fear was of an uncertain environment. According to Goldstein, the majority of actions of these brain-damaged men were directed by the fear of a disorganized world that they could no longer control. Therefore, great effort was expended to achieve a well-ordered "everything in its proper place and time" world, and there was an associated reluctance to try things that might produce "failure." The behavioral effects were observed following damage to almost any portion of the telencephalon in Goldstein's patients.

In the previous discussion, I have often placed quotation marks around the terms "sensory" and "motor" regions of the brain, usually referring to particular regions of the neocortex. The reason for these marks is to emphasize the difficulties of defining sensory and motor neocortical areas. Sensory areas as defined by the anatomic projections of thalamic nuclei (even making the untenable assumption that any thalamic nucleus acts as a pure sensory relay) do not always match up with the areas of responsiveness mapped electrophysically following peripheral stimulation. This means that there is no unambiguous way to define such areas. The boundaries of a sensory region, defined by evoked sensory potentials, are variable and depend on the level and type of drug used in the experiment. In unanesthetized animals, electrical responses to almost every type of sensory input can be found across the entire surface of the brain, and stimulation of all portions of the brain induce observable responses (Lilly, 1958).

In my opinion, there also is no convincing evidence of permanent, specific sensory or motor losses subsequent to focal brain lesions in nonprimates. Certainly brain damage can produce behavioral effects, some of which are of a profound nature. These can best be explained, I believe, by the assumption of damage-induced alterations in the remaining neural systems. The nature of the deficit changes over time and probably differs with the age and genetic endowments of the animals (Isaacson, 1975; Donovick and Burright, 1984).

5. SECONDARY EFFECTS OF BRAIN DAMAGE

As examples, consider the changes in behavior of cats and rats after the lesions I have described. Both seem to be explicable on the basis of secondary effects occurring at some distance from the site of the cortical destruction, including those involving the biogenic amines. There is evidence that the telencephalon influences the activities of such systems and that artificial pharmacological activation of these systems can offset at least some of the deficits (i.e., the amphetamine studies of Meyer *et al.*, 1963). In short, there is more compelling evidence for the effects of lesions in the sensory or motor regions arising from changes in the secondary, diffusely projecting systems than there is for specific impairments of sensory or motor capacities. Whether or not different principles apply to the human and nonhuman primates is another issue.

Probably the two examples of specificity of behavioral changes that best appear to argue for a loss of specific capacities are the effects of presumed localized damage to either the occipital or the left fronto–temporal–parietal lobes in people.

6. RESIDUAL VISUAL ABILITIES

With occipital cortical lesions the damage at first seems to produce a loss of visual abilities, sometimes even producing a ''cortical blindness.'' In the case of left hemispheric lesions, interference with the use of spoken or written language often appears. In actual patients, however, the damage caused by trauma never really confines itself to the neocortex *per se* but also typically involves the interruption of underlying white matter and subcortical structures. Even if the wound could be confined to the gray matter of the brain's surface by some rare accident, the effects would be widespread because of the interruption of normal neuronal input and trophic influences that affect many other regions.

In the visual domain, the most common consequence of occipital lobe destruction is the presence of scotomas. Generally, this word refers to a failure to be aware of visual information coming from certain portions of the visual field. However, the extent of the scotoma depends on how the visual deficit is measured. The area of the deficit is different depending on whether bright or dark "spots" are presented, the ambient brightness, and whether or not the target is moving, flickering, or stationary. These and many other features of the testing situation determine, at least in part, how large an area is affected. The point is that there is no absolute criterion for defining an area of scotoma but only ones closely tied to how it is measured. Furthermore, with the increased awareness of a phenomenon known as "blind sight," it is questionable whether or not there is a true failure to process visual input in any region of the scotoma or whether it is an inability to report, or to be consciously aware of, the sensory input.

"Blind sight" refers to the ability of an individual to use visual information despite a failure to report the presence of the stimuli after extensive occipital damage (see Weiskrantz, 1980). This has been shown in a few human cases in which the patients have even been able to "guess" the presence or the correct color of a stimulus despite an inability to "see" it. At least in people, a distinction needs to be made between being aware of sensory events and the processing and use of the information in guiding behavior.

Remarkable recoveries from cortical blindness have been achieved both by strenuous training in a testing situation and by frequent exposure to more natural environments after the lesion in nonhuman primates (e.g., Humphrey and Weiskrantz, 1967; Zihl, 1980). Commonly, after complete extirpation of the visual cortex, the nonhuman primate seems to be blind, bumping into objects and apparently unable to detect stimuli presented at various points in its visual field. However, if the stimulus is associated with a noise or is made to flicker, two procedures enhancing its attentional value, the animal may begin to respond correctly. It was found that when such animals were taken out of doors and allowed to explore the real world of trees, natural objects, and people, they came to demonstrate remarkable visuomotor skills, even learning to swing from one tree limb to another (Weiskrantz, 1980).

It would be of great interest to learn whether or not amphetamine could improve the performance of brain-damaged people with scotomas in their visual fields. However, such tests have not been carried out because of the moderately high doses of the drug that would be needed and the lack of medical justification for such treatment.

In regard to the loss of language abilities found in patients with left hemisphere damage, it has become popular to describe the changes in terms of alterations of certain specific language areas. Once again, however, accidents

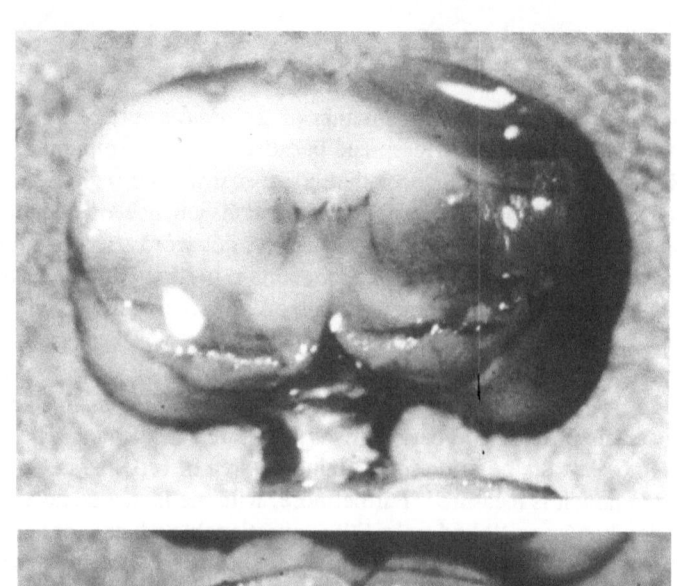

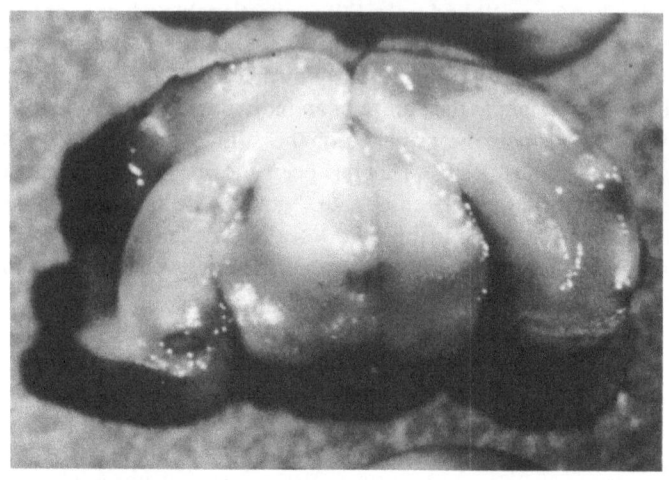

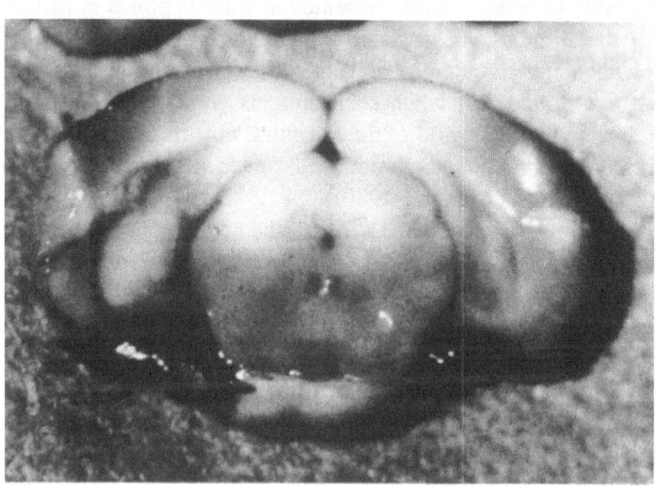

seldom if ever produce damage restricted to the neocortex and consequently to one cortical language center or to the fibers interconnecting them. In fact, in the opinion of some observers, language deficits are never seen unless there is some involvement of subcortical regions. It is also the case that subcortical damage by itself can induce language dysfunctions of various kinds (Fedio and Van Buren, 1975; Riklan and Cooper, 1975; Ojemann, 1976). Furthermore, the actual locations of specific brain sites associated with specific types of deficits vary to some degree among patients (see Lennenberg, 1967; Benson, 1979). In evaluating brain damage related to pathology, it must be noted that the lesions are usually large, irregular, and hardly the type to be reproducible. Furthermore, it must always be remembered that localizing the region from which a symptom arises is not the same as localizing the functional neural system(s) related to the impaired capacity (Jackson, 1884).

Another consideration concerning the loss or alterations in verbal abilities after brain damage comes from the evaluation of stroke patients. The effects of strokes, usually in one of the larger branches of the middle cerebral artery, are not restricted to cortical areas, and thus the extent of the damage cannot be fully appreciated by the evaluation of obviously necrotic regions at some extended time after the neural insult. Strokes can alter the metabolic activities of wide areas of the brain without necessarily producing the usual histological signs of necrosis. The areas affected by the stroke may be metabolically inactive for some period of time but subsequently recover to show a partial or complete return to normal metabolic activities.

7. THE EXTENT OF STROKE-INDUCED DAMAGE

The areas affected metabolically by a stroke are likely to be far more widespread than previously imagined, and knowledge about this in people has really only been possible with the advent of positron emission tomography (PET) and magnetic resonance scans, costly procedures not available to most patients.

Surprisingly, very little has been done to determine the metabolic sequelae

---

Figure 1. Photographs of three thick sections from a rat brain in which the right middle cerebral artery had been electrocoagulated 24 hr earlier. The coagulation was made on the left side of the sections. The sections were incubated with a nitro-BT solution with succinate added to indicate the presence of succinate dehydrogenase by a formazan deposit. Therefore, the darker the brain tissue, the greater the presumed metabolic activity. It should be noted in the top section that the area of reduced metabolic activity is found bilaterally in the cortex, mediodorsal caudate, and the nucleus of the diagonal band. In the middle section, reduced metabolic activity can be seen in the superior colliculus and the hippocampal formation. A reduction in the metabolic activity of the central gray can be seen in the middle and bottom sections, although it is less of a reduction than found in the superior colliculi above it. The extremely white "spots" found on some of the sections are artifacts caused by the glare of the lights used to photograph the sections.

of brain damage in animals either. This is in part because of the difficulties inherent in most of the common procedures used to evaluate regional brain metabolism. In using the 2-deoxy-D-glucose method, for example, the period in which time evaluations can be made is quite restricted: it is that period of time after the injection of the bolus of the radioactive label and extending to the time the animal is sacrificed. In this time frame, the animal must be restrained and subjected to periodic blood sampling—highly stressful procedures.

Another, although less easily quantified, approach to the evaluation of regional metabolic activities is the histochemical staining of enzymes of importance for aerobic metabolism in the brain. Figure 1 shows three large sections of brain tissue that have been stained to produce a formazan reaction product indicative of the presence of succinic acid dehydrogenase. These were from an animal that had suffered the electrocoagulation of the left middle cerebral artery 24 hr earlier. As can be seen, the areas that show greatly reduced oxidative metabolic activity are extraordinarily large and found on both sides of the brain. This extreme reaction is found soon after the lesion and gradually decreases over time. The regions that are lightly stained in this picture show higher levels of metabolic activity and stain more deeply in rats sacrificed at longer times after the lesion, although it is not certain whether they ever will return to normal levels. From what is known about the distribution of blood supplies to the rat brain, it is not likely that the unilateral coagulation of this one major vessel directly produces such widespread effects. Some of these effects must be caused by secondary, neurally mediated alterations in regional brain activities.

Since these metabolic changes are very widespread soon after the lesion but shrink fairly rapidly over time, they could represent the neural basis of "diaschisis." In any case, it seems likely that recovery from stroke or other forms of damage could entail the restoration of metabolic activities at some distance from the direct site of brain damage. It would be of great interest to study the return of metabolic activities in various brain areas and correlate these changes with alterations in behavior. Further, since localized brain damage can influence metabolic changes in distal brain regions, it might be useful to look for therapeutic interventions that can rapidly restore metabolic activities.

One aspect of my view of how the brain operates and its response to damage is that it provides hope for those who have the misfortune to suffer stroke or other forms of damage. According to these ideas, it should be possible to provide exogenous nonspecific modulators of affected systems in such a way as to allow some of them to function in a near-normal fashion. In the last few years we have been rather successful in some attempts to do this. Our work has included the diffusion of dopaminergic agonists into the basal ganglia, the systemic administration of a precursor to the formation of acetylcholine, and the reduction of circulating corticosterone in rats after bilateral hippocampal damage (for a review see Isaacson et al., 1986). However, each of these approaches has its limitations. None is a "magic bullet" that will cure all of the effects of the hippocampal destruction. On the other hand, the fact that animals with such large amounts of brain damage can have some nearly normal behavioral patterns restored is remarkable, and we are now able selectively to improve some lesion-induced effects without altering others. With sufficient data and wisdom, it may be possible to improve or remedy at least some of the damage-induced behavioral changes that are most debilitating to people. At least such benefits are conceptually possible.

REFERENCES

Ashby, W. R., 1960, *Design for a Brain. The Origin of Adaptive Behaviour*, second ed., Chapman and Hall, London.

Barron, D. H., 1934, The results of unilateral pyramidal section in the rat, *J. Comp. Neurol.* **60**:45–55.

Bartfai, T., 1985, Presynaptic aspects of the coexistence of classical neurotransmitters and peptides, *Trends Pharmacol. Sci.* **6**:331–334.

Benson, D. F., 1979, Aphasia, in: *Clinical Neuropsychology* (K. M. Heilman and E. Valentein, eds.), Oxford University Press, New York, pp. 22–58.

Brown, L. T., 1971, Projections and termination of the corticospinal tract in rodents, *Exp. Brain Res.* **13**:432–450.

Bucy, P. C., 1961, The delusion of the obvious, *Perspect. Biol. Med.* **9**:358–375.

Bucy, P. C., 1966, The delusion of the obvious, *Perspec. Biol. Med.,* **9**:358–368.

Castro, A. J., 1972, The effects of cortical ablations on digital usage in the rat, *Brain Res.* **37**:173–185.

Donovick, P. J., and Burright, R. G., 1984, Roots to the future: Gene–environment coaction and individual vulnerability to neural insult, in: *Early Brain Damage*, Volume 2 (S. Finger and C. R. Almli, eds.), Academic Press, New York, pp. 291–312.

Euvrard, C., Oberlander, C., and Boissier, J. R., 1980, Antidopaminergic effect of estrogens at the striatal level, *J. Pharmacol. Exp. Ther.* **214**:179–185.

Fedio, P., and Van Buren, J. M., 1975, Memory and perceptual deficits during electrical stimulation in the left and right thalamus and parietal subcortex, *Brain Language* **2**:78–100.

Goldstein, K., 1942, *After Effects of Brain Injuries in War. Their Evaluation and Treatment,* Grune & Stratton, New York.

Herrick, C. J., 1948, *The Brain of the Tiger Salamander, Annbystoma Trigrinum,* University of Chicago Press, Chicago.

Humphry, N. K., and Weiskrantz, L., 1967, Vision in monkeys after removal of the striate cortex, *Nature* **215**:595–597.

Huston, J. P., and Borbely, A. A., 1973, Operant conditioning in forebrain ablated rats by use of rewarding hypothalamic stimulation, *Brain Res.* **50**:467–472.

Isaacson, R. L., 1975, The myth of recovery from early brain damage, in: *Aberrant Development in Infancy: Human and Animal Studies* (N. R. Ellis, ed.), Lawrence Erlbaum, Potomac, MD, pp. 1–16.

Isaacson, R. L., 1982, *The Limbic system,* second ed., Plenum Press, New York.

Isaacson, R. L., Springer, J. E., and Ryan, J. P., 1986, Cholinergic and catecholaminergic modification of the hippocampal lesion syndrome, in: *The Hippocampus*, Volume 4 (R. L. Isaacson and K. H. Pribram, eds.), Plenum Press, New York, pp. 127–158.

Jackson, J. H., 1984, The evolution and dissolution of the nervous system, *Br. Med. J.* **1**(591):660–703.

Lennenberg, E. H., 1967, *Biological Foundations of Language,* John Wiley & Sons, New York.

Lilly, J. C., 1958, Neurophysiological activity and short term behavior, in: *Biological and Biochemical Basis of Behavior* (H. F. Harlow and C. N. Woolsy, eds.), University of Wisconsin Press, Madison, pp. 83–100.

MacLean, P. D., 1970, The triune brain, emotion, and scientific bias, in: *The Neurosciences. Second Study Program* (F. O. Schmidt, ed.), Rockefeller University Press, New York, pp. 336–349.

Meyer, P. M., 1963, Analysis of visual behavior in cats with extensive neocortical ablations, *J. Comp. Physiol. Psychol.* **56**:397–401.

Meyer, P. M., and Meyer, D. R., 1982, Memory, remembering, and amnesia, in: *The Expression of Knowledge* (R. L. Isaacson and N. E. Spear, eds.), Plenum Press, New York, pp. 179–212.

Meyer, P. M., Hovel, J. A., and Meyer, D. R., 1963, Effects of dl-amphetamine upon placing responses in neodecorticate cats, *J. Comp. Physiol. Psychol.* **56**:402–405.

O'Donohue, T. L., Millington, W. R., Handelman, G., Contreras, P. C., and Chronwall, B. M., 1985, On the 50th anniversary of Dale's law: Multiple transmitter neurons, *Trends Pharmacol. Sci.* **6**:305–308.

Ojemann, G. A., 1976, Subcortical language mechanisms, in: *Studies in Neurolinguistics*, Volume 1 (H. Whitaker and H. A. Whitaker, eds.), Academic Press, New York, pp. 103–108.

Pribram, K. H., 1971, *Languages of the Brain*, Prentice-Hall, Englewood Cliffs, NJ.

Rainbow, T. C., Degroff, V., Luine, V. N., and McEwen, B. S., 1980, Estradiol-17β increases the number of muscarinic receptors in hypothalamic nuclei, *Brain Res.* **198**:239–243.

Riklan, M., and Cooper, I., 1975, Psychometric studies of verbal functions following thalamic lesions in humans, *Brain and Language*, **2**:45–64.

Robinson, R. G., and Coyle, J. T., 1980, The different effects of right versus left hemisphere infarcton on catecholamines and behavior in the rat, *Brain Res.* **188**:63–78.

Sar, M., and Stumpf, W. E., 1977, Androgen concentration in motor neurons of cranial nerves and spinal cord, *Science* **197**:77–79.

Weiskrantz, L., 1980, Varieties of residual experience, *Q. J. Exp. Psychol.* **32**:365–386.

Zihl, J., 1980, Blindsight: Improvement of visually guided eye movements by systematic practice in patients with cerebral blindness, *Neuropsychologia* **18**:71–77.

7

Mass Action and Equipotentiality Reconsidered

BRYAN KOLB and IAN Q. WHISHAW

1. INTRODUCTION AND HISTORICAL ROOTS

Two conceptual views of cortical function have evolved and been used a number of times over the past 150 years to explain puzzling effects of brain damage: mass action and equipotentiality. The mass action hypothesis asserts that the entire cortex participates in every behavior. Thus, removal of any cortical tissue produces a behavioral change that is proportional to the amount of tissue removed. The equipotentiality hypothesis states that each portion of any given area is able to encode or produce the behavior normally controlled by the entire area. Thus, incomplete damage within a zone is compensated for by the remaining area. These concepts were debated extensively for the first half of this century and now are still invoked periodically as explanations for recovery of function. We revisit the concepts by briefly looking at the history before considering their current forms. We then examine the question of whether they are useful concepts to consider as explanations of recovery of function.

In the 1600s, Descartes, faced with potential conflicts with the church over his views on brain function, argued that the mind was indivisible and distinct from the body. This view became established and was still prominent 200 years later when Gall and Spurtzheim proposed that different psychological functions could be localized in different cortical areas. The problem their view posed was that if the mind was indivisible, theories that divided cortical function could not be correct. It was this paradox that led Flourens to the first formulation of the concept of equipotentiality. Flourens reported that ablation of various parts of the

BRYAN KOLB and IAN Q. WHISHAW • Department of Psychology, The University of Lethbridge, Lethbridge, Alberta T1K 3M4, Canada.

brain of a variety of vertebrates lead to a general loss in behaviors such as feeding, walking, wing flapping, and the like. In addition, he noted that, provided the damage was not too severe, the behavioral loss showed almost complete recovery. Flourens concluded that individual functions were not localized in discrete brain areas. He also observed that restitution of function occurred through the activity of the remaining brain. Thus, Flourens concluded that any surviving remnant of cortex could fulfill all cortical functions.

It was not until about 1880 that Goltz proposed a relationship between cerebral mass and behavior. Goltz, who was opposed to localization of function, had been intrigued by the results of Fritsch and Hitzig, who had shown that stimulation of a restricted portion of the neocortex resulted in specific movements. He reasoned that if a specific portion of the neocortex contained a particular function, then removal of the cortex should lead to a loss of that function. Goltz removed the neocortex, most of the basal ganglia, and parts of the midbrain of three dogs, one of which he studied for 18 months. Goltz found that his dogs were capable of many behaviors. For example, they walked, could dig for food, oriented to pinches, showed sleep–waking cycles and dream sleep, avoided obstacles, and were startled by sudden noises. Similarly, they were not deprived of the senses of smell or taste, for they would eat food soaked in milk and reject meat soaked in bitter quinine. What was most important, however, was that their higher functions seemed completely gone. Goltz concluded that Flourens was in error in arguing that any surviving remnant of cortex could fulfill all functions. Rather, in view of the obvious behavioral loss, he concluded that the cortex is the organ of "higher psychic functions" and that damage to the cortex produced a loss in intelligence that is proportional to the extent of damage. "In my opinion the most important failing observed after the removal of the cerebrum is the failure of all manifestations from which we might draw conclusions regarding the animal's understanding, memory, reasoning power, and intelligence" (Goltz, 1892, p. 607).

The relationship between cerebral mass and behavioral change was most clearly formulated by Lashley between 1920 and 1950. Lashley studied the behavior of rats and later monkeys on a variety of learning tasks. He trained rats in a maze and then removed portions of the cortex. If a small amount of tissue was destroyed, say 5–10%, the performance decrement was hardly detectable. In contrast, if large amounts of tissue were destroyed, say 50% or more, the habit was severely disrupted, and relearning required more sessions than did original learning. When Lashley correlated the errors in maze learning with the extent of brain injury, he found that the loss in maze performance was proportional to the amount of cortex destroyed. More important, the amount of loss from a given extent of cortical destruction was about the same no matter what part of the cerebral hemisphere was destroyed, provided that the destruction was roughly similar in both hemispheres.

These results suggested to Lashley that the efficiency of maze performance was governed by the quantity of tissue available, and it provided the strongest support for his concept of mass action. In order to understand what Lashley believed this concept meant, it is necessary to return to the concept of equipotentiality.

In addition to his studies of maze learning, Lashley did extensive studies on the ability of rats and monkeys to learn sensory discriminations after various cortical lesions. Lashley's results were unequivocal: the ability to discriminate visual stimuli survived the destruction of any part of the cerebral cortex except the primary visual projection area. Similarly, the ability to make auditory or tactile discriminations was dependent on the auditory or somatosensory cortex, respectively. But what of localization within these sensory areas? Lashley made small lesions within the primary visual cortex and found that so long as some part of the primary visual projection zone was intact, the ability to make fine visual discriminations was intact. In a now classic study, he counted undegenerated cells in the lateral geniculate nucleus to determine the smallest amount of visual cortex that allows a rat to make fine visual discriminations (Lashley, 1929). Discrimination could be successfully learned with only 2% of the visual cortex remaining. Therefore, Lashley concluded that no sensory subregion is more important than any other area and that all parts of the cortex that subserve a given habit must contribute equally to its efficiency. This led to the concept of equipotentiality, which he defined as "the capacity of any intact part of a functional area to carry out, with or without reduction in efficiency, the functions that are lost by the destruction" of other parts (Lashley, 1929, p. 25). This property was not absolute, however, for it was subject to the law of mass action.

Lashley continued to refine his views until his death in 1958, so that today there are two versions of mass action and equipotentiality. The general version of mass action postulates that the entire cortex participates in every function. In this sense the cortex is seen to have some nonspecific function, particularly in complex behavior. The specific version of mass action recognizes functional specialization but asserts that within a cortical region all portions of the region participate in all of the region's functions. The general version of equipotentiality states that each portion of the cortex is able to encode or produce the behavior normally controlled by the entire area. The specific version states that this rule applies only to restricted cortical zones. With the development of the specific versions of the concepts, new problems arose. The most important of these concern the definition of a functional zone. As we shall see, new anatomic information has greatly changed the way that we view the cortex, and it is now no longer possible to think of cortical zones in the way that Lashley viewed them.

2. DO WE NEED MASS ACTION AND EQUIPOTENTIALITY?

We have seen that the concepts of mass action and equipotentiality derived from a philosophy of mind that stressed the indivisibility of mental processes. Do these concepts retain any value? After all, our knowledge of cerebral activity is now rather different than it was when Lashley's conceptualizations were most influential (*ca.* 1930–1950). Although concepts like mass action and equipotentiality still seem to have intuitive appeal in discussions of recovery of function (e.g., Finger and Stein, 1982), we must ask whether they really have either neurological support or predictive value. Stated differently, do we gain understanding of recovery of function by using the concepts? To consider this question we must first examine the major changes in our knowledge of cortical organization, and of recovery of function, since Lashley.

2.1. Principles of Cortical Function

There are at least four principles of cortical function that were essentially unknown to Lashley that are important for assessing the role of mass action and equipotentiality in recovery.

2.1.1. Specific versus Nonspecific Afferents

The neocortex receives two types of subcortical inputs: specific and nonspecific. Specific afferents, which arise largely in the thalamus, project to specific cortical targets. For example, the lateral geniculate nucleus projects to primary visual cortex, the ventrobasal complex projects to primary somatosensory cortex, and so on. Nonspecific afferents are more recent discoveries. They are neocortical afferents that arise from cell populations in the basal forebrain, thalamus, or brainstem and project to all cortical areas. The best examples are the cholinergic projections from the nucleus basalis, the noradrenergic projections of the locus coeruleus, and the serotoninergic projections of the raphé. The function of these systems remains unknown, but they appear to have some role in activating the cortex (Vanderwolf *et al.*, 1985). These systems are an obvious candidate for a nonspecific general function of cortical tissue, yet it is very difficult to demonstrate a functional loss following their partial elimination.

2.1.2. Cortical Organization

It has long been known that the neocortex is composed of several types of neurons organized into approximately six layers. It was also well known that there were distinct cytoarchitectonic areas that were connected with one another. The last decade has shown that cortical organization is far more complex than it once appeared. If it is assumed that this organization has a functional purpose, then concepts like mass action and equipotentiality have a significant logical challenge. Consider some of the evidence.

First, it has now become clear that different cytoarchitectonic areas are neurochemically distinct (e.g., Rakic, 1984). Thus, it seems likely that equipotentiality would be restricted to neurochemically similar regions.

Second, corticocortical connections are now known to have important functional significance. Although the concept of disconnection syndromes was old (dating at least to Wernicke and Leipmann in the last century), it was Geschwind's (1965) description of behavioral syndromes that provided a modern view of the importance of cortical connections. Thus, Geschwind, and subsequently others such as Mishkin, showed that simply cutting the connections between two cortical regions could produce significant behavioral deficits (e.g., agnosias, apraxia) even though there is no change to the cortical mass. Disconnection effects clearly demonstrate the importance of sequential cortical processing in behavior. Furthermore, we are unaware of a single case of recovery of function following a complete disconnection.

2.1.3. Multiple Representations

When Ferrier, Sherrington, Penfield, and others (see Doty, 1969) stimulated the motor and somatosensory cortex of laboratory animals and humans, they found it was possible to identify regions of the pre- and postcentral gyri that appeared to represent localized body parts such as the leg, hand, and face. To summarize their findings they constructed homunculi in which different body parts were represented in discrete regions of the cortex. Similar representations of the visual and auditory fields were also constructed, presenting points in the visual field or tones. These sensory and motor representations were each surrounded by "association cortex" that was assumed to have a function in further associational analysis. Lashley's equipotentiality views predicted that this cortex could reorganize the homunculi. To demonstrate this he electrically stimulated the precentral gyrus of rhesus monkeys on four separate occasions over a period of 18 days. Each time he stimulated he was able to identify the arm, leg, and hand areas, etc. Over a given session the points were stable and yielded a consistent movement. However, on different test days stimulation at particular points resulted in widely different movements, and at different times the same movement was obtained from separate and shifting areas. These results suggested to Lashley that "within the segmental areas the various parts of the cortex may be equipotential for the production of all of the movements of that area" (Lashley, 1929, p. 154).

Lashley's experiments had methodological difficulties (see Doty, 1969) but their important point was that sensory maps are equipotential. It is now clear that they may be (see below), but there is a complication. Recent experiments using microelectrode recording and stimulation techniques have shown that each of the sensory representations is actually composed of multiple representations, each coding a different aspect of sensation. Thus, in somatosensory cortex there are separate representations for touch, pressure, and pain. In the owl monkey visual cortex there may be as many as ten independently organized fields. The entire visual field is systematically represented in each field, although the representations are not mere copies of one another: the size of each field is different, and the properties and connections of the neurons in each field differ. Again, it is reasonable to suppose that each field has a different contribution to the overall sensory experience, a conclusion that again presents a challenge for mass action and equipotentiality.

2.1.4. Neural Plasticity

It is only recently that there has been compelling evidence of plastic changes in the brain that might account for phenomena such as recovery of function. Consider four examples. First, recent work by Merzenich and his colleagues has reopened the concept of stability of points. Merzenich *et al.* (1984) recorded from the postcentral gyrus of owl monkeys and mapped the somatosensory field of the hand. One unexpected finding was that the map was not static: it could apparently change spontaneously in response to experience and as a result of peripheral or central injury. For example, if the peripheral nerve running to one finger was severed, the somatopic map showed a distinct and orderly shift. When the input from one finger is missing, the neighboring fingers expand their maps to occupy the now vacant space. Similarly, if the postcentral gyrus is damaged, the map again shifts to allow recovery of a tactile map of the entire surface, a map that now uses fewer neurons and allows less tactile acuity than before. This result implies that sensory points may not be stable and that a mechanism rather similar to Lashley's equipotentiality may account for recovery.

Second, connections in the cortex are not static. For example, it appears that synaptic connections are affected by experience. Rearing laboratory animals in specific environments or teaching them various habits alters the number of synapses in different cortical layers (e.g., Greenough and Volkmar, 1973). These changes can be surprisingly rapid and provide a potential mechanism not only for learning but also for some recovery phenomena. Further, it appears that more is not necessarily better. Thus, during development the brain appears to make more connections than it will actually need, and during childhood and adolescence cells and synapses are lost. Hence, in the human frontal cortex the number of synapses peaks at about 2 years of age, only to decline dramatically until about 16 years of age.

Third, although it may only be true of certain cell populations and in certain species, it appears that neurogenesis can continue throughout the adult life of both birds (e.g., Goldman and Nottebohm, 1983) and mammals (e.g., Bayer *et al.*, 1982). For example, neurons in certain forebrain nuclei of some songbirds appear to grow in the spring, support song, then die in the autumn, only to be regrown the following spring.

Fourth, there is now abundant evidence that brain injury can produce regrowth, or sprouting, of connections (e.g., Cotman, 1985). Further, these connections can be functional, although this does not always convey a behavioral advantage (e.g., Schneider, 1973; 1979). Sprouting has been used as an explanation for many types of behavioral recovery and for certain procedures that enhance recovery such as serial lesion phenomena.

Taken together, contemporary knowledge of neural plasticity permits many explanations of the processes of normal and recovery phenomena that seemingly once required the concepts of mass action and equipotentiality.

2.2. Principles of Behavior

Lashley and his contemporaries knew very little about the analysis of behavior. As a result, they largely studied the behavior of laboratory animals in a series of maze tasks that were in reality analogues of human environments and then generalized about cortical function from these tasks. It is obvious that the rodent brain did not evolve to solve these tasks and that even simple maze tasks are complex. It is now recognized that learning a particular maze task requires guiding movements in space on the basis of an enormous amount of sensory information.

To date, no one has devised a learning test that is affected by only one brain lesion. This does not imply, however, that functions are not localized in the cortex. There have been significant advances in techniques of behavioral observation and the development of a new technology of behavioral analysis (e.g., Robinson, 1983). Thus, there has been a trend away from studying complex behaviors that Lashley was most interested in, e.g., attention and memory, in favor of studying what animals actually do, such as limb use, tongue use, nest building, etc. These studies show that functions can be localized. For example, removal of the tongue area of the cortex abolishes tongue extension, whereas removal of the entire cortex except the tongue area has no effect on tongue extension (e.g., Whishaw and Kolb, 1983). Similarly, it appears that whereas performance of maze tests shows good recovery of function after various cortical lesions, performance of species-typical behaviors in the same animals fails to show any recovery at all (e.g., Kolb and Whishaw, 1981).

3. EVIDENCE OF RECOVERY AND NONRECOVERY

3.1. Distinguishing between Getting Better and Recovery

Before we can look at the role of mass action and equipotentiality in recovery, we must ask what we will take as evidence of recovery. Discussions of recovery processes do not normally address this question, but we believe that this may be the central question to discuss. We would like to propose three criteria for evaluating putative recovery from cortical damage.

1. The behavior in question must be gone when the cortex is gone. Thus, if a maze or other behavioral task can be solved, even if only slowly, by

the decorticate preparation, then that recovery cannot be shown to be a uniquely cortical phenomena. It could be entirely subcortical. Thus, a great deal of the recovery on learning tasks falls into this category because decorticate rats are capable of learning a surprising array of discriminations and mazes and even of forming learning sets (e.g., Oakley, 1979). There are tasks, however, that they cannot learn, even if the decortication is performed at birth (e.g., Whishaw and Kolb, 1984), and these tasks can quite clearly be regarded as cortically dependent.

2. It must be shown that the recovered behavior is the same behavior that was lost. It is not sufficient to demonstrate that an animal can relearn some task. For example, if a cortical lesion abolishes orienting by disrupting eye movement but orienting recovers by substituting a head movement, then that is not recovery (Steele-Russell, 1982).

3. If a treatment is to be shown to produce recovery, it must be shown that recovery would not have occurred without it. Treatments that merely speed up the end point are interesting clinically but have a different theoretical implication than do treatments that produce recovery that would not have occurred normally. For example, if a treatment speeds up the learning of a maze task, but animals without the treatment would have solved the task eventually, then unless it can be demonstrated that there is a qualitative difference between the effects of the treatments on behavior, it is difficult to argue that there is a fundamental difference produced by the treatment. Stated differently, quantitative differences in behavior cannot be taken as evidence of qualitative differences in the brain.

3.2. Examining the Evidence for "Recovery of Function"

Studies of recovery of function have emphasized the role of several factors that are alleged to promote recovery, including (1) subtotal lesions, (2) serial lesions, (3) pharmacological manipulations including drugs, hormones, and trophic factors, (4) environmental treatments, (5) grafts, and (6) early lesions. We have searched the literature for evidence of recovery (as we have defined it), and good examples are very hard to find. We reach three conclusions: (1) most behaviors fail to recover following cortical injury; (2) only subtotal lesions and early lesions provide any evidence of recovery; and (3) recovery is more likely to occur in young than in older people or laboratory animals, but recovery is not complete and only occurs under certain circumstances. Consider the following examples.

3.2.1. Subtotal Lesions

It appears to be a common observation that recovery is most likely to occur if lesions are subtotal. Bucy et al. (1964) studied a man with a pyramidal tract surgically sectioned in the lower brainstem as a treatment for involuntary movements. During the first 24 hr after surgery he had complete flaccid hemiplegia, followed by slight return of voluntary movements in his extremities. By the tenth day he was able to stand alone and walk with assistance. By the 24th day he could walk unaided. Within 7 months maximum recovery seemed to have been reached, and he could move his feet, hands, fingers, and toes with only slight impairment. At autopsy $2\frac{1}{2}$ years later, about 17% of his pyramidal tract fibers were found to be intact. The recovery in his ability to move his toes and fingers seem attributable to that remaining 17% doing the job previously done by the entire tract. This suggest that the system displays substantial equipotentiality. Lawrence and Kuypers' (1968) experiments with pyramidal tract lesion in rhesus monkeys might be interpreted as supporting this conclusion. They found that a few remaining pyramidal tract fibers could support relatively independent finger movements, apparently irrespective of the location of the fibers. If all the fibers were removed, there was no recovery of finger movement. Finally, Glees and Cole (1950) defined the thumb area of the motor cortex in monkeys by electrical stimulation. They then removed this area and followed the course of recovery of thumb movements from the flaccid paralysis caused by the lesion. After complete recovery the brain was reexposed, and they found that the borders of the lesion now produced thumb movements although they had not during the first stimulation session. This area was then removed, and flaccid paralysis occurred, followed again by recovery.

All of these experiments seem to demonstrate recovery, but there are several problems. First, the studies of recovered movements looked at only one aspect of movement. The Bucy study was a clinical description and not a careful study of movement. The Lawrence and Kuypers experiment looked at independent finger movements. It is now known, however, that the motor cortex has other functions in movement such as controlling rotation of the wrist and so on (e.g., Passingham *et al.*, 1983). Second, partial lesions outside the motor and somatosensory systems have not been shown to be associated with recovery. For example, primary visual cortex lesions virtually never allow recovery of fine pattern vision, frontal cortex lesions are always associated with personality change, and temporal lobe patients virtually always have some form of memory impairment (for a review see Kolb and Whishaw, 1985a).

3.2.2. Early Lesions

There is substantial evidence that under the right circumstances there can be impressive recovery of function following brain damage in early childhood. The best example comes from studies of language. Damage to the cortical language zones, especially the posterior zone, in adults produces chronic language impairments such as aphasia. Damage to the language zones in the young infant can result in the development of one or both language zones in the opposite hemisphere. Similar damage in early childhood allows the development of new language zones, but rather than shifting to the opposite side, the new language region is in the damaged hemisphere. Damage in adolescence or later fails to produce either (e.g., Rasmussen and Milner, 1977). This is a clear case of recovery, as we have defined it. There may be a price for such recovery, however, as other functions may be sacrificed for language (e.g., Teuber, 1975; Woods, 1981). Another prime example of recovery comes from the studies of childhood hemidecortication: removal of an entire hemisphere allows recovery of many functions that are clearly cortical. Like the studies of language reorganization, studies of hemispherectomized patients normally show a substantial price for this recovery (e.g., St. James-Roberts, 1981).

Studies of recovery following cortical ablations in infant laboratory animals have shown that although some recovery can occur, it occurs only at certain developmental ages and is never complete. Consider the following example. We gave rats frontal or parietal lesions at different ages in early infancy and subsequently tested them as adults on a broad test battery (e.g., Kolb, 1987; Kolb and Whishaw, 1981, 1985b; Whishaw and Kolb, 1984b). Several results surprised us. First, on the basis of previous work by ourselves and others, we expected to find recovery of function, with the greatest recovery occurring in the youngest operates. This was not the case. Rather, we found that lesions around 7–10 days of age allowed significant recovery, whereas lesions earlier actually produced a greater behavioral effect than similar removals in adulthood. A parallel result has also been seen in humans, as damage in the first year of life has been shown to produce greater cognitive effects than lesions later in life. Second, even the animals that showed recovery failed to show complete recovery, and when tested on tasks at which adult operates performed normally, these animals showed a deficit. Thus, although the animals showed recovery at some tests, they showed new deficits, a result parallel to the effects of lesions in the language zones of human infants. Third, recovery was most frequently observed on learning tests; tests of species-typical behaviors seldom showed recovery.

3.2.3. Other Treatments

We failed to find any examples of recovery in studies of trophic factors including drugs and hormones, experience, or early lesions. To be sure, recovery was often claimed, but we are unaware of a single instance in which the demonstration satisfied our criteria for recovery. Furthermore, many claims of recovery fail to control adequately factors other than the treatment that could produce enhanced performance. Consider an example. Two groups have shown that aphagia and adipsia following lateral hypothalamic lesions can be attenuated or prevented by serial lesions. In one study Glick and Greenstein (1972) reported the remarkable finding that if rats were given frontal cortex lesions 30 days prior to lateral hypothalamic lesions, they did not exhibit the typical aphagia and adipsia of the lateral hypothalamic syndrome. They proposed that the prophylactic effect of the frontal lesions was related to the development of denervation supersensitivity in catecholaminergic neural systems that support feeding behavior. Unfortunately, Glick and Greenstein failed to consider that the first lesion might have reduced body weight, and it was the reduced body weight that blocked the aphagia and adipsia. This hypothesis is reasonable since body weight is known to interact with the effects of hypothalamic lesions (e.g., Powley and Keesey, 1970). We repeated Glick and Greenstein's experiment but held body weight constant with special feeding, only to find that the serial lesion effect evaporated (Kolb et al., 1978)!

In sum, we are not yet convinced by the evidence that recovery is common, nor is it every complete.

3.3. Recovery, Mass Action, and Equipotentiality

Finally, we can address the question of whether concepts like mass action and equipotentiality have any role in explaining recovery. The specific versions of equipotentiality and mass action have clear implications for recovery of function. If a cortical region displays equipotentiality, then it follows that if the damage to the area is subtotal there will be recovery of function. In contrast, if the damage is complete in an equipotential area, there will be no recovery. Further, when recovery occurs, it will be dependent on the total mass lost in an area: the greater the amount lost, the less the recovery. It further follows that for regions that exhibit equipotentiality and mass action, recovery cannot be complete since it is the mass of tissue that is crucial to function.

We have seen that although true recovery of function is probably not very common, it seems to occur if there is sparing of tissue in a functional area or if the lesion is acquired at particular developmental ages. Does this recovery conform to the concepts of mass action and recovery? On the whole, the evidence from subtotal lesions appears to be at least partially consistent with mass action and equipotentiality. Recovery can occur if the damage is subtotal, but recovery is never complete. We know of no direct evidence relating recovery to amount of tissue remaining, but in principle this is not unreasonable. One difficulty, however, is that recovery does not occur everywhere in the cortex. For example, we know of absolutely no evidence of recovery of fine pattern vision following primary visual cortex lesions. This seems to be a major problem, since we need to account for why the theory works only sometimes and at some places.

Recovery from early lesions is more difficult to incorporate at all into mass action or equipotentiality, since the theory works only under very limited circumstances. Thus, it is hard to explain why the theory holds if the lesions are performed at 10 days of age but at no other time (see above). Lesion size was held constant in the different age groups, so it is difficult to understand why mass action or equipotentiality should apply in only one specific instance.

4. CONCLUSIONS

We conclude by stating that we find little use for the concepts of mass action or equipotentiality either as explanations for the function of the normal cortex or as explanations for changes in ability following brain damage. The concepts were derived from studies of maze learning that were designed to locate memory traces. Today we recognize that memory traces cannot be separated from innate predisposition to behave in a certain way and that animals have an array of predispositions available to help them solve even the simplest mazes. Furthermore, the cortex is involved in many other kinds of behaviors, including the seemingly simple acts of licking and chewing, and the neural control of these behaviors seems surprisingly well localized. Even as explanations for apparent recovery of function, the concepts are wanting. A great deal of seeming recovery occurs through continued activity in parallel systems, many of which are located subcortically. Even when limited recovery does occur, it can usually be explained by plastic changes in brain organization, and often this recovery is accompanied by impairments in other systems. Finally, when the effects of brain damage acquired at different ages are considered, the frequently encountered new or wide-ranging impairments are neither predicted nor explained by any version of either concept. Thus, it seems to us that, given our new knowledge about the organization of the cortex and our insights into the many anatomic, electrophysiological, and behavioral changes that occur both during normal behavior and following brain damage, the concepts of mass action and equipotentiality are no longer useful.

REFERENCES

Bayer, S. A., Yackel, J. W., and Puri, P. S., 1982, Neurons in the rat dentate gyrus granular layer substantially increase during juvenile and adult life, *Science* **216**:890–892.

Bucy, P. C., Keplinger, J. E., and Siquerira, E. B., 1964, Destruction of the ''pyramidal tract'' in man, *J. Neurosurg.* **21**:385–398.

Cotman, C. W., 1985, *Synaptic Plasticity*, Guilford Press, New York.

Doty, R. W., 1969, Electrical stimulation of the brain in behavioral context, *Annu. Rev. Psychol.* **20**:289–320.

Finger, S., and Stein, D. G., 1982, *Brain Damage and Recovery: Research and Clinical Perspectives*, Academic Press, New York.

Geschwind, N., 1965, Disconnexion syndromes in animals and man, *Brain* **88**:237–294.

Glees, P., and Cole, J., 1950, Recovery of skilled motor functions after small repeated lesions of motor cortex in macaque, *J. Neurophysiol.* **13**:137–148.

Glick, S. D., and Greenstein, S., 1972, Facilitation of recovery after lateral hypothalamic damage by prior ablation of frontal cortex, *Nature (New Biol.)* **239**:187–188.

Goldman, S. A., and Nottebohm, F., 1983, Neuronal production, migration, and differentiation in a vocal control nucleus of the adult female canary brain, *Proc. Natl. Acad. Sci. U.S.A.* **80**:2390–2394.

Goltz, F., 1892, Der Hund ohne Grosshirn. Siebente Abhandlung uber die Verrichtungen des Grosshirns, *Pflugers Arch.* **51**:570–614.

Greenough, W. T., and Volkmar, R. F., 1973, Pattern of dendritic branching in occipital cortex of rats reared in complex environments, *Exp. Neurol.* **40**:491–504.

Kolb, B., 1987, Factors affecting recovery from early cortical damage in rats. 1. Differential behavioral and anatomical effects of frontal lesions at different ages of neural maturation, *Behav. Brain Res.* **25**:205–220.

Kolb, B., and Whishaw, I. Q., 1981, Neonatal frontal lesions in the rat: Sparing of learned but not species-typical behavior in the presence of reduced brain weight and cortical thickness, *J. Comp. Physiol. Psychol.* **95**:863–879.

Kolb, B., and Whishaw, I. Q., 1985a, *Fundamentals of Human Neuropsychology*, 2nd ed., Freeman, New York.

Kolb, B., and Whishaw, I. Q., 1985b, Earlier is not always better: Behavioral dysfunction and abnormal cerebral morphogenesis following neonatal cortical lesion in the rat, *Behav. Brain Res.* **17**:25–43.

Kolb, B., Nonneman, A. J., and Whishaw, I. Q., 1978, Influence of frontal neocortex lesions and body weight manipulation on the severity of lateral hypothalamic aphagia, *Physiol. Behav.* **21**:541–547.

Lashley, K. S., 1929, *Brain Mechanisms and Intelligence*, University of Chicago Press, Chicago.

Lawrence, D. G., and Kuypers, H. G. J. M., 1968, The functional organization of the motor system in the monkey. 1. The effects of bilateral pyramidal lesions, *Brain* **91**:1–14.

Merzenich, M. M., Nelson, R. J., Stryker, M. P., Cynader, M. S., Schoppmann, A., and Zook, J. M., 1984, Somatosensory cortical map changes following digit amputation in adult monkeys, *J. Comp. Neurol.* **224**:591–605.

Oakley, D. A., 1979, Cerebral cortex and adaptive behavior, in: *Brain, Behavior and Evolution* (D. A. Oakley and H. C. Plotkin, eds.), Methuen, London, pp. 154–188.

Passingham, R. E., Perry, V. H., and Wilkinson, F., 1983, The long-term effects of removal of sensorimotor cortex in infant and adult rhesus monkeys, *Brain* **106**:675–705.

Powley, T. L. and Keesey, R. E., 1970, Relationship of body weight to the lateral hypothalamic feeding syndrome. *J. Comp. Physiol. Psychol.* **70**:25–36.

Rakic, P., 1984, Defective cell-to-cell interactions as causes of brain malformations, in: *Malformations of Development* (E. S. Gollin, ed.), Academic Press, New York, pp. 239–285.

Rasmussen, T., and Milner, B., 1977, The role of early left-brain injury in determining lateralization of cerebral speech functions, *Ann. N.Y. Acad. Sci.* **299**:355–369.

Robinson, T. E. (ed.), 1983, *Behavioral Approaches to Brain Research*, Oxford University Press, New York.

St. James-Roberts, I., 1981, A reinterpretation of hemispherectomy data without functional plasticity of the brain, *Brain Lang.* **13**:31–53.

Schneider, G. E., 1973, Early lesions of the superior colliculus: Factors affecting the formation of abnormal retinal projections, *Brain, Behavior and Evolution* **8**:73–109.

Schneider, G. E., 1979, Is it really better to have your brain lesion early? Revision of the Kennard principle, *Neuropsychologia* **17**:557–584.

Steele-Russell, I. S., 1982, Some observations on the problem of recovery of function following brain damage, *Hum. Neurobiol.* **1**:68–72.

Teuber, H.-L., 1975, Recovery of function after brain injury in man, *Ciba Found. Symp.* **34**:159–186.

Vanderwolf, C. H., Leung, L. W. S., and Stewart, D. J., 1985, Two afferent pathways mediating hippocampal rhythmical slow activity, in: *Electrical Activity of the Archicortex* (G. Buzsaki and C. H. Vanderwolf, eds.), Akademiai Kiado, Tokyo, pp. 47–66.

Whishaw, I. Q., and Kolb, B., 1983, "Stick out your tongue": Tongue protrusion in neocortex and hypothalamic damaged rats, *Physiol. Behav.* **30**:471–480.

Whishaw, I. Q., and Kolb, B., 1984a, Decortication abolishes place but not cue learning in rats, *Behav. Brain Res.* **11:**123–134.

Whishaw, I. Q., and Kolb, B., 1984b, Behavioral and anatomical studies of rats with complete or partial decortication in infancy, in: *Early Brain Damage* (C. R. Almli and S. Finger, eds.), Academic Press, New York, pp. 117–137.

Woods, B. T., 1981, The restricted effects of right-hemisphere lesions after age one: Wechsler test data, *Neuropsychologia* **18:**65–70.

8

Margaret Kennard and Her "Principle" in Historical Perspective

STANLEY FINGER and C. ROBERT ALMLI

1. INTRODUCTION

The present chapter centers around the idea that the behavioral effects of brain damage may be less deleterious if sustained early in life, when the central nervous system is in an immature state, and more severe if endured after the period of rapid brain growth has ended. This belief has been called the "Kennard principle" by some writers (see Isaacson, 1975), and it has been the subject of a number of critiques and reviews (e.g., Finger and Stein, 1982; Johnson and Almli, 1978).

Because Margaret Kennard's papers on early versus later brain damage first appeared approximately 50 years ago, and because her name has been associated with the possibility that the signs and symptoms that follow brain damage may vary with developmental status at the time of insult, it seems fitting to reexamine and reevaluate her contributions at this time. This is especially important because a careful examination of Kennard's writings reveals that her papers are not always interpreted accurately; that contributions attributed to her sometimes had earlier antecedents; that she has not received adequate credit for some of her observations and conclusions; and because previous reviewers have not approached Kennard's early brain damage studies with reference to her own objectives. In short, the present chapter looks historically at Margaret Kennard, the so-called "Kennard principle," and some of her related scientific contributions in an attempt to put each into better perspective.

STANLEY FINGER • Psychology Department and Neural Sciences Program, Washington University, St. Louis, Missouri 63130. C. ROBERT ALMLI • Programs in Occupational Therapy and Neural Sciences, Departments of Anatomy and Neurobiology, Preventive Medicine and Psychology, Washington University School of Medicine, St. Louis, Missouri 63110.

2. EDUCATION AND BACKGROUND

Margaret Alice Kennard (Fig. 1) was born on September 25, 1899 in Brookline, Massachusetts. She received her undergraduate degree from Bryn Mawr in 1922 and graduated from Cornell University Medical School in 1930. During this time she conducted research on proteins and plasma and served as a coauthor of two scientific papers, the earlier of which was published in 1924 (Fremont-Smith *et al.*, 1924; Allison *et al.*, 1926). Kennard did her internship at Strong Memorial Hospital in Rochester, New York and passed her specialty boards in Neurology and Psychiatry in 1942.

Kennard was associated with the Physiology Department of the Yale University School of Medicine from 1931 to 1943. Her appointments were as Research Fellow, Instructor, and, beginning in 1933, Assistant Professor. From 1933 to 1943 she also held an appointment at New Haven Hospital. Margaret Kennard was the recipient of a Rockefeller Traveling Fellowship for study in Europe from 1934 to 1936 and visited The Netherlands, Germany, and England during this time.

Kennard's research program at Yale was concerned with the effects of lesions of the motor pathways and cortex in primates. These studies were conducted largely with Fulton and Jacobsen, two very well-known scientists who were interested not only in motor systems but in the effects of frontal association cortex damage on emotionality and problem-solving performance in monkeys and apes—investigations that stimulated the use of frontal lobotomies to treat patients with behavioral disorders (Fulton, 1951, 1952; Fuster, 1980).

Figure 1. Margaret Kennard. Portrait courtesy of Lydia Kennard and Margaret Burt.

Kennard's first motor study appeared in 1932 and was written with both Fulton and Jacobsen (Fulton *et al.*, 1932). This study showed that lesions of the motor cortex (area 4) in mature monkeys resulted in flaccid paralysis and hemiplegia and that damage to the "premotor" cortex (area 6) led to forced grasping, spasticity, and impaired skilled movements. Lesions of the frontal association cortex, in contrast, were found to have no effect on motor performance.

In 1933, Kennard and Fulton went on to show that the simple Babinski response of extension of the toes to stroking of the plantar surface was associated with area 4 lesions in mature monkeys and that one could dissociate this response from the fanning and lateral deviation of the toes, which resulted from area 6 involvement. In related studies conducted at about the same time, she demonstrated that ephedrine could increase reflexes in spinal monkeys (Jacobsen and Kennard, 1933); that cutting the corpus callosum before or after lesions of area 4 or 6 did not affect symptomatology, recovery, or the response to electrical stimulation of the premotor cortex (Kennard and Watts, 1934); and that symptoms comparable to those seen in higher primates could be observed in a human patient with a tumor confined to area 6 (Kennard *et al.*, 1934). Changes in skin temperature on the side of the body opposite the tumor in the latter case suggested that damage to the premotor cortex could also affect autonomic nervous system functioning, and this was the subject of an article that appeared in *Science* in 1934.

The afore-mentioned publications, and related articles on motor systems (Kennard, 1935a; Fulton and Kennard, 1934), show that Kennard's interest in examining the effects of brain damage in subjects of different ages did not stem from a background or an orientation in developmental neurology or physiology, but rather came about as an extension of her efforts and those of senior scientists at Yale to learn more about the motor circuitry of the brain by utilizing a variety of techniques and preparations. This may be surmised from the fact that prior to publishing her first study on the effects of motor and premotor cortex lesions in "young" monkeys in 1936, Kennard was involved in research assessing post-traumatic movement responsivity to pharmacological agents (Jacobsen and Kennard, 1933), analyzing movements after brain lesions in adult monkeys (Fulton and Kennard, 1933, 1934), examining the effects of brain pathology and injury on motor function in mature humans (Kennard *et al.*, 1934), electrically stimulating motor areas of the brain (Kennard and Watts, 1934), and anatomically tracing efferent pathways (Kennard, 1935b)

3. LESION DEVELOPMENT AND MOTOR FUNCTION

Kennard's first developmentally oriented brain lesion study appeared in the *American Journal of Physiology* in 1936. This report only involved two monkeys with lesions sustained early in life and cited data previously collected on older monkeys that were operated on by Fulton. At 10 days of age, one of the infants had the motor and premotor areas of the left hemisphere ablated, and the other infant had the entire hemisphere removed. Kennard described the first subject as follows:

> The immediate recovery after the operation was surprising. Within 24 hours the animal walked about, using all four extremities, with only a slight lag in those of the right side. In purposeful movement, as grasping or picking up an object, the right fingers and toes were used less frequently and a trifle more awkwardly than the left, but even this disability disappeared within ten days. Forced grasping, after the first day when the right hand and foot showed weakness, was at all times equal on the two sides. It disappeared gradually during the second month of life and simultaneously on the two sides of the body. The animal then developed at a normal rate for a healthy infant and showed no motor deficit (p. 142).

Kennard also reported only limited lesion effects in the infant monkey with the entire hemisphere removed. She commented that the animal walked and climbed 1 week later and showed no forced grasping and only a bit of awkwardness 4 months after surgery. She went on to remove the opposite motor and premotor cortices in the first subject when the animal was 5 months old and again described surprising sparing and recovery of motor functions, surprising because adults with bilateral lesions of these areas often are unable to sit up, stand, or readily feed themselves. In contrast, the young subjects even showed movements of prehension.

Over the next few years, Kennard (1938, 1939, 1940a, 1942, 1944a) expanded her subject sample to include primates of other ages and, in some instances, with different lesion configurations. This work led her to emphasize that motor deficits observed in subjects sustaining damage at different ages only gradually changed to the adult pattern of symptoms. Although newborn subjects showed the most sparing and recovery of function, differences were found between monkeys that received lesions even at 1 versus 2 years of age. All monkeys older than 3 years at the time of surgery showed severe spasticity, reflex grasping, and no progression from the "thalamic" posture, except for one animal who could right itself from one side.

As noted, Kennard (1939) also recognized that there were other variables that could interact with the age factor to affect relative performance after early brain damage. One variable was whether the lesions were made in stages (Section 7), and another was the species studied. Regarding the latter factor, Kennard observed that chimpanzees showed even greater age-related differences in sparing and recovery of function than baboons or macaque monkeys. She attributed this to the fact that motor deficits are even more marked in adult apes than they are in mature monkeys. She also related her findings on infrahuman primates to children by examining records from the Neurologic Clinic of the Children's Hospital of Boston (Kennard, 1940a). Paresis of the face was encountered much more frequently in children with injuries occurring well after the time of birth than at the time of birth. In addition, she noted that spasticity (increased resistance to passive manipulation) was rare in cases of motor cortex injury in young children and more common in older children. Even with older children, however, the observed spasticity was not as great as that associated with damage sustained during adulthood.

Careful examination of the series of publications on early brain damage that began in 1936 and ended in 1944 reveals a number of facts that are sometimes but not always recognized about Kennard or her contributions. One is that Kennard acknowledged a number of times that she was not the first to report age-related differences in response to brain damage. A second is that she recognized from her own work that the infant sparing effect was not universal or absolute and also noted that some symptoms could worsen or become manifest with maturation or the passage of time. A third is that Kennard presented and evaluated various theories to explain the age effects and generated hypotheses, which she then systematically tested and modified. A fourth is that she discussed how making lesions in stages could also affect performance and recovery in these studies. And a fifth fact, rarely noted by individuals interested only in recovery of function, is that these developmental lesion studies only represented a fraction of her output at the time and well under 10% of her lifetime publications. Each of these points merits elaboration.

4. HISTORICAL ANTECEDENTS

Kennard's celebrated 1936 journal article did not cite previous studies on early brain damage, but beginning with the next paper in the series (Kennard, 1938), references were made to earlier findings by other scientists. Here, Kennard cited Vulpian (1866), who advised the use of young animals in his hemidecortication experiments because they withstood the procedure better than mature subjects, and Soltmann (1876), who noted that hemidecorticated puppies still developed highly coordinated movements with little difference between the limbs on the two sides of the body.

A number of citations also were made to studies of decerebrate rigidity with very young animals (Brown, 1915; Langworthy, 1924; Weed, 1917). These experiments showed that fetal or newborn cats and rabbits (but not guinea pigs) may not show extensor stiffness of the forelegs and resistance of the hindlegs to passive flexion after removal of the cerebral hemispheres and basal ganglia but that such symptoms may be seen when subjects of more than 1 or 2 weeks of age were operated on. Kennard also cited Langworthy's 1926 study in which decerebrate rigidity was correlated with central nervous system maturation at the time of insult, and specifically with myelination of the spinal cord and long motor pathways.

In addition, a number of earlier clinical studies with children were mentioned to show that age-related lesion effects were not limited to laboratory animals. In the clinical sphere, reference was made to Little (1862), McNutt (1885), Freud and Rie (1891), Freud (1897), Sachs (1895), Osler (1899), and others.

It is interesting that Kennard did not make reference to a classic early paper by Broca (1865). Broca was presented with the case of an epileptic woman at the Salpetriere who, upon autopsy, showed a complete absence of the left hemispheric region that included Broca's area. This was thought to have resulted from a congenital absence of the Sylvian artery. The woman had not shown an aphasia, thus forcing Broca to consider the issue of plasticity of the nervous system early in life. It was in this context that Broca remarked almost casually: "I am convinced that a lesion of the left third frontal convolution, apt to produce lasting aphemia (aphasia) in an adult, will not prevent a small child from learning to talk" (see Schiller, 1979, Chapter 10).

In any case, Kennard's own citations show very clearly that she recognized that she was not the first to study the effects of brain damage in young laboratory animals or children and, more importantly, that she was predated in showing that early brain damage may be less deleterious than brain damage acquired later in life. In this context, Kennard's contribution may be that her lesions were more circumscribed and better defined than those of her predecessors, that she measured behavior over a longer time period, and that she used a primate model. Further, one cannot take from her clarity of thought or exemplary writing style. But, because her most important findings had well-defined antecedents and were consistent with the earlier literature, one could argue that that the so-called "Kennard principle" clearly preceded Kennard.

5. DEFICITS FOLLOWING EARLY LESIONS

It is important to note that Kennard never claimed that her subjects with early motor cortex lesions were indistinguishable from unoperated, control subjects. For instance, she (Kennard, 1936) stated that some motor problems remained in the monkey that had the motor areas removed on one side of the brain at 10 days of age and on the opposite side in the fifth month. These included movements that were slower than normal, galloping and hopping motions when locomoting, high stepping and hyperextension of the extremities, awkward climbing, and abnormal clinging. Further, in 1938, Kennard mentioned that the motor abnormalities became more pronounced as her young brain-damaged monkeys matured. For example, one 3-year-old monkey with bilateral motor and premotor area ablations (operations at 3 weeks and 6 months of age) now displayed increased spasticity, less adequate and less discrete finger movements in feeding, and walking with the feet everted and fingers and toes spread like a "stalking cat" with slightly bent forelegs.

Kennard (1939, 1940a) recognized that the paresis exhibited by the brain-damaged monkeys typically emerged at the time when coordinated skilled movements would have appeared in control subjects and that spasticity occurred somewhat later. The following description was provided in her presentation to the American Neurological Association in 1939:

> When areas 4 and 6 are removed bilaterally from a 3-weeks infant monkey, there follows only a very slight alteration in motor performance. Reflex grasp continues, and there is no spasticity. During the next few months as the performance of this animal becomes more complex, the differences from normal begin to appear. At the age of 3 or 4 months such an infant, although it can run, climb, and feed itself, shows a more immature type of behavior than does a normal infant of the same age; the fingers are not well used for feeding, the animal tends to climb as does the infant, and reflex grasping persists. At this stage, motor performance is as complex as it will ever become, but although volitional prehension is present, indicating cortical function, there is no increase in resistance to passive manipulation. Later tendon jerks become hyperactive, and during the second half of the first year of life some spasticity develops but is not apparently correlated with any advance in type of motor behavior. Gradually these animals become more rigid, and the posture of the extremities more extended, but the extreme spasticity which appears after the damage to the motor regions of the adult is never present (pp. 58–59).

It is important to realize that the aforementioned idea of "growing into deficits" following brain injury also preceded these publications. In her 1942 review with Fulton, the authors opened the paper with the following quotation from Sachs in 1895:

> . . . It is a matter of fact that the symptoms of such intrauterine cerebral defects (motor paresis) are not always manifest at birth, and indeed a number of months may pass before it becomes evident to the physician that the child's cerebral condition is not a normal one. . . . Premature delivery is responsible for many of the cerebral palsies; but the symptoms may not be fully developed until months after birth (p. 594).

In the same paper, Kennard and Fulton made another important but often overlooked point. It is that marked deficits that resemble the adult pattern may be present immediately after some early lesions and that some of these deficits may not diminish appreciably over time. They noted that infant monkeys show severe, adultlike paresis of conjugate deviation of the eyes after damage to area 8, even at ages when most motor functions are minimally affected by area 4 and 6 lesions. Placing and hopping responses also seem to be permanently abolished in infant monkeys with area 4 and 6 lesions. Kennard and Fulton (1942) also made reference to marked deficits with early lesions outside the motor areas. They pointed out that vision did not return after large ablations of the occipital lobes in infant or adult monkeys and that bilateral removal of frontal association cortex produced a loss of immediate memory and hyperactivity in primates of all ages.

These examples show that Kennard did not view the absence of deficits or the more mild deficits sometimes observed following early brain damage as an inevitable response, a rule, or a principle. Indeed, Kennard had found many exceptions to sparing and recovery of function in her own early lesion studies and in observations by her colleagues, and she carefully worded her conclusions to reflect this point. Kennard further recognized that dynamic changes could take place in symptomatology over time and that initial impressions of sparing or recovery sometimes may be associated with the later emergence of behavioral dysfunction.

6. THEORETICAL FORMULATIONS

Kennard devoted almost all of the space in her first paper (Kennard, 1936) to detailed descriptions of her experimental subjects. Her later manuscripts, however, had more space allocated to discussing theoretical mechanisms that might account for the early sparing and recovery of motor function seen under *some* circumstances.

In her 1936 and 1938 articles, Kennard hypothesized that motor function was mediated subcortically in newborn primates. This assumption was based on her finding that even after extensive bilateral motor and premotor cortex lesions, newborn monkeys typically show few immediate motor deficits compared to control animals of the same age.

In contrast, Kennard thought that other cortical areas could have mediated the more complex motor functions that emerged after the first few months of postnatal life in her brain-damaged monkeys. Previous studies conducted by other researchers were cited to support the idea that the opposite hemisphere could play a role in recovery of function after an early unilateral lesion. In addition, Kennard found that right motor and premotor cortex ablations, placed months after neonatal left hemispheric lesions, affected motor functions on both sides of the body in contrast to unilateral lesions, which only seemed to affect the contralateral side. Nevertheless, she also stressed the idea that surviving cortical regions on the same side as the damage could mediate performance in the infants, especially after bilateral insult. Kennard (1940a, p. 388) argued that "this is shown by the appearance of additional motor deficit on removal of the post-central areas together with area 8, which in the adult have no such influence on motor status." Kennard referred specifically to the Betz cells in this regard. The Betz cells are large pyramidal neurons that are heavily concentrated in the pre-central gyrus but are also found in fair number in the postcentral gyrus and in frontal areas 9 and 12. She raised the possibility that Betz cells outside the classic motor areas might enable some sort of reintegration of motor activity, thus implicating both frontal association cortex and the parietal lobes in the recovery process.

In a theoretical review published in 1942, Kennard and Fulton considered three ways to account for the apparent recovery of function sometimes seen following early brain damage. These were compensation without real recovery, vicariation, and dynamic reorganization within a partially damaged system (see Jacobsen *et al.*, 1936). Kennard and Fulton readily dismissed compensation. They postulated that the frontal and parietal areas formed a single functional unit and theorized that sparing or recovery did not represent a takeover of new, unusual functions by unrelated parts of the brain (vicariation). Instead, they favored the idea that recovery occurred through "reorganization" within the partially damaged "frontal–parietal system." Kennard and Fulton emphasized that such reorganization was most likely to occur at a time before cortical growth was complete and equated maturation with greater specificity of function within areas such as the sensorimotor cortex.

Kennard obtained indirect support for this reorganizational theory in an experiment that she published with McCulloch in 1943. Monkeys with cortical areas 4 and 6 removed in infancy were subjected to electrical stimulation of the postcentral gyrus. Unlike adult-operated or control monkeys, which displayed only limited movements in response to stimulation of the postcentral gyrus, the infant-operated monkeys exhibited vigorous motor responses when these areas were stimulated. No areas were found to be highly excitable that were not excitable to at least some degree in the normal monkey, suggesting that behavior is most likely mediated within remaining parts of the broadly defined system.

It is interesting to note that Kennard also postulated that dendritic growth could contribute to within-system functional reorganization. She (1942, p. 239) hypothesized that if some synapses were removed, "others would be formed in less usual combinations" and that "it is possible that factors which facilitate cortical organization in the normal young are the same by which reorganization is accomplished in the imperfect cortex after injury." It is worth remembering that when Kennard advanced these possibilities approximately one-half century ago, she did so without the benefit of the currently available arsenal of research techniques that allow for detailed descriptions of reactive events such as axonal sprouting and growing axons being rerouted to different locations following injury.

The relationship between normal growth and the reactive neural growth seen after injury, the idea that remaining elements within the same neural "system" are those most likely to account for good performance, and the belief that the capacity for functional reorganization diminishes as elements within a system undergo maturation and become "committed" to specific functions remain important theoretical possibilities. These issues continue to demand the attention of contemporary investigators interested in understanding the relationship between brain damage and recovery of function (Almli and Finger, 1984; Finger and Almli, 1984, 1986; Finger and Stein, 1982; Goldman, 1974; Simons and Finger, 1984).

7. SERIAL LESIONS

Kennard made her lesions in stages in many of her experimental subjects, often spacing the surgeries many months apart. This allowed her to collect data on the effects of unilateral as well as bilateral lesions or control operations without having to double or triple the number of animals in a study, and it permitted her to ask whether recovery or sparing after a unilateral lesion could at least in part by mediated by areas damaged in another operation.

As early as 1936, Kennard wrote that the deficits found after successive unilateral lesions often were not as severe as those observed with the same brain areas damaged all at once. She pointed out that this, however, was dependent on the amount of time that was permitted between the surgeries. For example, she noted that Bieber and Fulton (1938) found that adolescent monkeys who were given two-stage lesions of areas 4 and 6 with 2–3 weeks between surgeries were just as impaired as subjects sustaining simultaneous lesions of the same cortical areas. In contrast, she found that a greater degree of voluntary power could return to adult monkeys if comparable two-stage lesions were separated by at least 1 month.

After analyzing data from 16 monkeys, Kennard discussed the staged-lesion phenomenon in greater detail. In 1942, she wrote the following about the effects of one-stage and multistage lesions of the motor and premotor areas:

> That reorganization is greater when ablations from the two hemispheres are performed *seriatim* is well demonstrated in this series. Experiments 1 and 2 were both carried out on young infants; yet the recovery in the case of monkey 1 was greater than that of monkey 2 because there was an interval of 5 months between operations in the first case. In the higher age groups, the same can be shown. . . . Recovery was greater in the case of monkey 9 than in that of monkey 8 because the former had an interval of 8 months between operations and the latter none (pp. 237–238).

The term ''serial-lesion effect'' is now used to signify greater sparing or more recovery following successive lesions than simultaneous lesions (Finger *et al.*, 1973), and the more contemporary literature on this phenomenon is by no means small (see Finger, 1978; Finger and Stein, 1982). The serial-lesion effect has been demonstrated many times over with different lesions, tasks, and species, although, as with other brain lesion effects, examples also exist of no behavioral differences relating to this procedural manipulation.

There are many historical parallels between Kennard's recognition of differences stemming from multiple-stage versus single-stage surgeries, even in adult animals, and her recognition that there could be age-related differences in performance after motor and premotor cortex lesions. As was the case with age factors in recovery, the record shows that she was not the first to describe lesion momentum effects (see Finger and Stein, 1982; Finger *et al.*, 1973). However, she again provided an excellent primate model of an important phenomenon— one that attracted much attention and stimulated new research, and one that confirmed that even with lesions placed in just two stages, as opposed to one, there could be behavioral differences.

This was an important contribution for at least three reasons. First, it provided rigorous experimental support for the observations of such clinical investigators as John Hughlings Jackson (1873), who held that slow-growing lesions (e.g., tumors), unlike strokes or acute brain injuries, can fool the physician by not being associated with deficits for a long time, even after so-called critical parts of the brain were damaged. Secondly, it showed how dangerous it was to make assumptions about recovery or structure–function relationships in the brain solely on the basis of acute lesion data. And third, it made the study of sparing and recovery just as legitimate an endeavor in adults as it was in immature organisms.

8. OTHER PURSUITS AND LATER CONTRIBUTIONS

The research that Margaret Kennard conducted on age factors and recovery represented only a small part of her output while she was associated with Yale. During these years the age variable was examined in just a few of her manuscripts, although she published almost 50 articles. The majority of her papers at the time evaluated adult animals with various brain lesions. These included the basal ganglia (Kennard, 1940b, 1944b), the parietal areas (Kennard and Kessler, 1940), and the frontal association areas (Kennard et al., 1941) in addition to lesions of the motor and premotor parts of the brain (Kennard and Welch, 1944; Welch and Kennard, 1944). At the same time, she conducted a series of studies on the results of routine autopsies on large numbers of monkeys and apes (Kennard, 1941; Kennard and Willner, 1941a,b,c) and began to use the EEG to monitor development and to assess the effects of various lesions in her experimental animals (Kennard, 1943; Kennard and Nims, 1941, 1942a,b). Also notable was a study to see if drugs could improve performance in monkeys with damage to the motor areas. Doryl (carbaminol choline) was found to enhance performance for at least 6 months after injury and 3 months after the medication ended (Ward and Kennard, 1942). Watson and Kennard (1945) later showed that some sedatives (e.g., phenobarbital sodium) could have the opposite effect.

After leaving Yale, Kennard was affiliated with New York University from 1943 until 1948, during which time she was also associated with Bellevue Hospital where she worked with Loretta Bender. This was followed by an appointment at the University of British Columbia Medical School and Provincial Mental Hospital (1951–1956) and a position as director of the Washington State Mental Health Research Institute and an affiliation with the University of Washington (1956–1960).

It is important to emphasize that Kennard remained very productive in New York and on the West Coast. She published numerous articles during this period but presented no new papers on early or staged lesions and showed little inclination to continue to study motor systems in monkeys. Most of her new publications were on humans, not monkeys and apes; her predominant experimental groups changed from cases of focal lesions to behaviorally disordered populations; and indices of motor performance gave way to analyses of EEG patterns.

Kennard looked at such things as the role of inheritance on the EEG (Kennard, 1947) and changes in EEG resulting from shock therapy (Kennard and Willner, 1948). She also evaluated EEG correlates of delirium tremens (Kennard et al., 1945), learning disabilities (Kennard et al., 1952), and psychosocial, neurotic, and psychotic behavior (Kennard and Levy, 1952; Levy and Kennard, 1953a,b). A few studies also were conducted on pain (e.g., Kennard and Haugen, 1955). These various achievements contributed to her being elected President of the Society for Biological Psychiatry (1956–1957) and Vice-President of the American Neurological Society (1958–1959).

Kennard's very active research career essentially ended in 1960 when she returned to New England. She settled in New Hampshire, where she took a position at the Elliot Hospital in Bedford and helped to organize and direct the Community Guidance Center in Manchester. Kennard presented an occasional paper (e.g., Kennard, 1969) before succumbing to amyotrophic lateral sclerosis in 1976. A tribute to her appeared a year later in Biological Psychiatry (Himwich, 1977).

9. CONCLUSIONS

Margaret Kennard was a very prominent scientist during her lifetime, one who published well over 100 scientific papers and made many contributions to both clinical and experimental fields. Unfortunately, her contributions have not always been put into proper perspective, especially in the brain damage literature. On the one hand, Kennard is sometimes cited as if she "discovered" new brain lesion effects (e.g., age factor, staged lesions), with little regard for the earlier contributions of her predecessors. But on the other hand, there are many instances where she simply is given inadequate credit for what she did do—her experimental methods, observations, thoughts, and ideas.

As was pointed out, Kennard was not the first to notice that the effects of early brain damage may not be as deleterious as those of damage sustained later in life, at least under some conditions. To call this the "Kennard effect" is not entirely accurate. Kennard herself cited previous work on the early brain damage issue, such as that of Brown and Langworthy earlier in the 20th century and of Soltmann and Vulpian in the 19th century. Nor was Kennard the first to look at staged lesions and recovery, or even drugs and recovery. What she did in each of these cases was to follow up on earlier ideas as a result of her basic interest in motor systems. The fact that her name is strongly associated with some of these contributions probably stems from her use of primates and a well-studied model system and from the fact that she exhibited an experimental rigor and methodological thoroughness that did not characterize the efforts of many of her predecessors. It is also to her credit that she wrote well and published in respected journals at a time when people seemed more eager to listen than in the past. There can be no question about the fact that Kennard's papers on age factors (and staged lesions), unlike those of her predecessors, immediately stimulated a wealth of new experimentation to see whether similar phenomena could be confirmed after lesions in different systems, with other species, and on a variety of other tasks (Finger and Stein, 1982). It is in this respect that much credit is due.

If there is one place in particular where Kennard seems to be given inadequate recognition, it is in understanding the limits of the phenomenon that she observed after early brain damage. She never claimed that lesions sustained early in life were always associated with sparing or recovery of function, and she never discussed her motor cortex findings as if they represented an invariant rule that could generalize across neural systems and experimental conditions. Kennard pointed out that deficits may emerge well after the time of insult, that early brain-damaged subjects were not identical to control animals on all measures of motor integrity, and that with lesions in other parts of the brain, such as area 8 or occipital cortex, young animals could in fact show marked deficits that persisted. In short, Kennard was aware of the complexity of the effects she was studying and attempted to account for different results by looking at factors such as maturation of parts of the nervous system at the time of insult and at interacting variables such as time between surgeries.

Kennard's place in the history of the neural sciences in general, and recovery of function in particular, is still being debated. Her efforts have not always been viewed in proper context, and secondary sources appear to be increasingly consulted or cited although her original articles remain readily accessible. Her papers are filled with a wealth of "forgotten" material that deserves to be rediscovered by contemporary investigators and by students of the history of science. These papers merit special attention on more than just the 50th anniversary of her first publication on sparing after lesions of the infant motor cortex.

REFERENCES

Allison, N., Fremont-Smith, F., Dailey, M., and Kennard, M. A., 1926, Comparative studies between synovial fluid and plasma, *J. Bone Joint Surg.* **8**:758–764.

Almli, C. F., and Finger, S., 1984, *Early Brain Damage*, Volume 1, *Research Orientations and Clinical Observations*, Academic Press, New York.

Bieber, I., and Fulton, J. F., 1938, Relation of the cerebral cortex to the grasp reflex and to postural and righting reflexes, *Arch Neurol. Psychiatry* **39**:433–454.

Broca, P., 1865, Siège de la faculté de langage articulé dans l'hemisphere gauche du cerveau, *Bull. Soc. Anthropol.* **6**:377–393.

Brown, T. G., 1915, On the activities of the central nervous system of the unborn foetus of the cat; with a discussion of the question whether progression (walking, etc.) is a "learnt" complex, *J. Physiol. (Lond.)* **49**:208–215.

Finger, S., 1978, Lesion momentum and behavior, in: *Recovery from Brain Damage: Research and Theory* (S. Finger, ed.), Plenum Press, New York, pp. 135–164.

Finger, S., and Almli, C. R., 1984, *Early Brain Damage*, Volume 2, *Neurobiology and Behavior*, Academic Press, New York.

Finger, S., and Almli, C. R., 1985, Brain damage and neuroplasticity: Mechanisms of recovery or development? *Brain Res. Rev.* **10**:177–186.

Finger, S., and Stein, D. G., 1982, *Brain Damage and Recovery: Research and Clinical Perspectives*, Academic Press, New York.

Finger, S., Walbran, B., and Stein, D. G., 1973, Brain damage and behavioral recovery: Serial lesion phenomena, *Brain Res.* **63**:1–18.

Fremont-Smith, F., Ayer, J. B., Kennard, M. A., and Dailey. M., 1924, The normal and abnormal quantitative protein content, *J. Nerv. Ment. Dis.* **4**:100–103.

Freud, S., 1897, *Die Cereballahmung*, A. Holder, Vienna.

Freud, S., and Rie, O., 1891, *Klinische Studie uber dis halbeitige Cerebrallahmung der Kinder*, Moritz Perles, Vienna.

Fulton, J. F., 1951, *Frontal lobotomy and Affective Behavior: A Neurophysiological Analysis*, W. W. Norton & Co., New York.

Fulton, J. F., 1952, *The Frontal Lobes and Human Behavior*, Charles C. Thomas, Springfield, IL.

Fulton, J. F., and Kennard, M. A., 1933, The relationship of the motor and premotor areas of the cortex to spasticity and postural reflexes in primates, *Am. J. Physiol.* **105**:35.

Fulton, J. F., and Kennard, M. A., 1934, A study of the flaccid and spastic paralyses produced by lesions of the cerebral cortex in primates, *Proc. Assoc. Res. Nerv. Ment. Dis.* **13**:158–210.

Fulton. J. F., Jacobsen, C. F., and Kennard, M. A., 1932, A note concerning relation of frontal lobes to posture and forced grasping in monkeys, *Brain* **55**:524–536.

Fuster, J. M., 1980, *The Profrontal Cortex*, Raven Press, New York.

Goldman, P., 1974, An alternative to developmental plasticity: Heterology of CNS structures in infants and adults, in: *Plasticity and Recovery of Function in the Central Nervous System* (D. G. Stein, J. J. Rosen, and N. Butters, eds.), Academic Press, New York, pp. 149–174.

Himwich, W. A., 1977, In memoriam: Margaret Kennard (1899–1976), *Biol. Psychiatry* **12**:603–605.

Isaacson, R. L., 1975, The myth of recovery from early brain damage, in: *Aberrant Development in Infancy* (N. R. Ellis, ed.), Erlbaum, Potomac, MD, pp. 1–25.

Jackson, J. H., 1873, Lectures on the diagnoses of tumors of the brain, *Med. Times Gaz.* **2**:139–140.

Jacobsen, C. F., and Kennard, M. A., 1933, Influence of ephedrine sulphate on reflexes of spinal monkeys, *J. Pharmacol. Exp. Ther.* **49**:366–374.

Jacobsen, C. F., Taylor, F. V., and Haslerud, G. M., 1936, Restitution of function after cortical injury in monkeys, *Am. J. Physiol.* **116**:85–86.

Johnson, D. A., and Almli, C. R., 1978, Age, brain damage and performance, in: *Recovery from Brain Damage: Research and Theory* (S. Finger, ed.), Plenum Press, New York, pp. 115–134.

Kennard, M. A., 1934, Vasomotor representation in the cerebral cortex, *Science* **79**:348–349.

Kennard, M. A., 1935a, Vasomotor disturbances resulting from cortical lesions, *Arch. Neurol. Psychiatry* **33**:537–545.

Kennard, M. A., 1935b, Corticospinal fibers arising in the premotor area of the monkey as demonstrated by the Marchi method, *Arch. Neurol. Psychiatry* **33**:698–711.

Kennard, M. A., 1936, Age and other factors in motor recovery from precentral lesions in monkeys, *Am. J. Physiol.* **115**:138–146.

Kennard, M. A., 1938, Reorganization of motor function in the cerebral cortex of monkeys deprived of motor and premotor areas in infancy, *J. Neurophysiol.* **1**:477–496.

Kennard, M. A., 1939, Relation of "spasticity" to age in young monkeys and chimpanzees, *Trans. Am. Neurol. Assoc.* **65**:58–62.

Kennard, M. A., 1940a, Relation of age to motor impairment in man and in subhuman primates, *Arch. Neurol. Psychiatry* **44**:377–397.

Kennard, M. A., 1940b, Function of basal ganglia in monkeys, *Trans. Am. Neurol. Assoc.* **66**:131–130.

Kennard, M. A., 1941, Abnormal findings in 246 consecutive autopsies on monkeys, *Yale J. Biol. Med.* **13**:701–712.

Kennard, M. A., 1942, Cortical reorganization of motor function, *Arch. Neurol.* **48**:227–240.

Kennard, M. A., 1943, Effects on EEG of chronic lesions of basal ganglia, thalamus and hypothalamus of monkeys, *J. Neurophysiol.* **6**:405–416.

Kennard, M. A., 1944a, Reactions of monkeys of various ages to partial and complete decortication, *J. Neuropathol. Exp. Neurol* **3**:289–310.

Kennard, M. A., 1944b, Experimental analysis of functions of basal ganglia in monkeys and chimpanzees, *J. Neurophysiol.* **7**:127–148.

Kennard, M. A., 1947, Inheritance of electroencephalogram patterns in families of children with behavior disorders, *Trans. Am. Neurol. Assoc.* **72**:177–179.

Kennard, M. A., 1969, EEG abnormality in first grade children with "soft" neurological signs, *Electroencephalogr. Clin. Neurophysiol.* **27**:544.

Kennard, M. A., and Fulton, J. F., 1933, The localising significance of spasticity, reflex grasping and the signs of Babinski and Rossolimo, *Brain* **56**:213–225.

Kennard, M. A., and Fulton, J. F., 1942, Age and reorganization of central nervous system, *J. Mt. Sinai Hosp.* **9**:594–606.

Kennard, M. A., and Haugen, F. P., 1955, The relationship of subcutaneous focal sensitivity to referred pain of cardiac origin, *Anesthesiology* **16**:297–311.

Kennard, M. A., and Kessler, M. M., 1940, Studies of motor performance after parietal ablations in monkeys, *J. Neurophysiol.* **3**:248–257.

Kennard, M. A., and Levy, S., 1952, The meaning of the abnormal electroencephalogram in schizophrenia, *J. Nerv. Ment. Dis.* **116**:413–423.

Kennard, M. A., and McCulloch, W. S., 1943, Motor response to stimulation of cerebral cortex in absence of areas 4 and 6, *J. Neurophysiol.* **6**:181–189.

Kennard, M. A., and Nims, L. F., 1941, Changes in electroencephalograms appearing coincident with growth in infant monkeys, *Am. J. Physiol.* **143**:349.

Kennard, M. A., and Nims, L. F., 1942a, Changes in normal electroencephalogram of *Macaca mulatta* with growth, *J. Neurophysiol.* **5**:325–334.

Kennard, M. A., and Nims, L. F., 1942b, Effect on electroencephalogram of lesions of cerebral cortex and basal ganglia in *Macaca mulatta*, *J. Neurophysiol.* **5**:335–348.

Kennard, M. A., and Watts, J. W., 1934, The effect of section of corpus callosum on the motor performance of monkeys, *J. Nerv. Ment. Dis.* **79**:159–169.

Kennard, M. A., and Welch, W. K., 1944, The relation of the cerebral cortex to spasticity and flaccidity, *Trans. Am. Neurol. Assoc.* **70**:158–164.

Kennard, M. A., and Willner, M. D., 1941a, Findings at autopsies of seventy anthropoid apes, *Endocrinology* **28**:967–976.

Kennard, M. A., and Willner, M. S., 1941b, Weights of brains and organis of 132 New and Old World monkeys, *Endocrinology* **28**:977–984.

Kennard, M. A., and Willner, M. D., 1941c, Findings in 126 routine autopsies of *Macaca mulatta*, *Endocrinology* **28**:955–966.

Kennard, M. A., and Willner. M. D., 1948, Significance of changes in electroencephalogram which result from shock therapy, *Am. J. Psychiatry* **105**:40–45.

Kennard, M. A., Viets, H. R., and Fulton, J. F., 1934, Syndrome of premotor cortex in man: Impairment of skilled movements, forced grasping, spasticity, and vasomotor disturbance, *Brain* **57**:69–84.

Kennard, M. A., Spencer, S., and Fountain, G., 1941, Hyperactivity in monkeys following lesions of frontal lobes, *J. Neurophysiol.* **4**:512–524.

Kennard, M. A., Bueding, E., and Wortis, S. B., 1945, Some biochemical and electroen-
cephalographic changes in delirium tremens, *Q. J. Stud. Alcohol* **6**:4–14.
Kennard, M. A., Rabinovitch, R. D., and Wexler, D., 1952, Abnormal electroencephalogram as
related to reading disability in children with disorders of behaviour, *Can. Med. Assoc. J.*
67:330–333.
Langworthy, O. R., 1924, A physiological study of the reactions of young decerebrate animals, *Am.
J. Physiol.* **69**:254–264.
Langworthy, O. R., 1926, Relation of onset of decerebrate rigidity to the time of myelination of
tracts in the brain-stem and spinal cord of young animals, *Contrib. Endocrinol.* **17**:127–140.
Levy, S., and Kennard, M. A., 1953a, A study of the electroencephalogram as related to personality
structure in a group of inmates of a state penitentiary, *Am. J. Psychiatry* **109**:832–839.
Levy, S., and Kennard, M. A., 1953b, The EEG pattern of patients with psychiatric disorders of
various ages, *J. Nerv. Ment. Dis.* **118**:416–428.
Little, W. J., 1862, On the influence of abnormal parturition, difficult labours, premature birth and
asphyxia neonaterum on the mental and physical condition of the child, especially in relation to
deformities, *Trans. Obstet. Soc. Lond.* **3**:293–344.
McNutt, S., 1885, Double infantile spastic hemiplegia, with report of a case, *Am. J. Med. Sci.*
89:58–79.
Osler, W., 1899, *The Cerebral Palsies of Children*, H. K. Lewis, London.
Sachs, B., 1895, *A Treatise on the Nervous Diseases of Children for Physicians and Students*,
William Wood & Co., New York.
Schiller, F., 1979, *Paul Broca*, University of California Press, Berkeley.
Simons, D., and Finger, S., 1984, Some factors affecting recovery after brain damage early in life,
in: *Early Brain Damage*, Volume 2, *Neurobiology and Behavior* (S. Finger and C. R. Almli,
eds.), Academic Press, New York, pp. 327–347.
Soltmann, O., 1876, Experimentalle Studien uber die Functionen des Grosshirns der Neugeborenen,
Jahrb. Kinderheilkd. **9**:106–148.
Vulpian, A., 1866, *Leçons dur la Physiologie Générale et Comparée du Système Nerveux*, Balliére,
Paris.
Ward, A. A., Jr., and Kennard, M. A., 1942, Effect of cholinergic drugs on recovery of function
following lesions of central nervous system in monkeys, *Yale J. Biol. Med.* **15**:189–288.
Watson, C. W., and Kennard, M. A., 1945, Effect of anticonvulsant drugs on recovery of function
following cerebral cortical lesions, *J. Neurophysiol.* **8**:221–231.
Weed, L. H., 1917, The reactions of kittens after decortication, *Am. J. Physiol.* **43**:131–157.
Welch, W. K., and Kennard, M. A., 1944, Paralysis in flexion and tremor in monkey following
cortical ablations, *J. Neurosurg.* **1**:258–264.

9

Infant Brain Injury
The Benefit of Relocation and the Cost of Crowding

N. DAVIS LeVERE, SUSAN GRAY-SILVA,
and T. E. LeVERE

1. INTRODUCTION

The present chapter is concerned with the consequences of brain pathology when it occurs very early in life, which is to say, ''how much recovery may be expected following injury sustained early in life as compared to brain injury sustained later in life?'' Although the answers ''less recovery,'' ''about the same recovery,'' or ''more recovery'' would seem simple enough, establishing one alternative over the others has proven difficult (see Finger and Almli, 1984; Isaacson, 1975; Schneider, 1979). This is partly because our experiments have sometimes been flawed and thus difficult to interpret. It is more so, we think, because our research has been dominated by the supposition that recovery of function following early brain injury occurs because of relocation of functions. We do not wish to quarrel with, or attempt to establish the veracity of, the so-called ''Kennard principle'' (Schneider, 1979). Nor do we wish to quarrel with, or attempt to establish the veracity of, the notion that the developing nervous system possesses a significant degree of plasticity (Lund, 1978). Rather, it is our purpose to determine if relocation of function is a general consequence of early brain injury and if it can account for the alleviation of behavioral dysfunctions or whether it is, in fact, part of the problem. For example, the same developmental plasticity that can form the basis of functional relocation can also produce aberrant behavior patterns in adults (Schneider, 1979), and, from the Syrian ham-

N. DAVIS LeVERE • Dorothea Dix Hospital, Raleigh, North Carolina 27611. SUSAN GRAY-SILVA and T. E. LeVERE • Behavioral Neuropsychology Laboratory, Department of Psychology, North Carolina State University, Raleigh, North Carolina 27695-7801.

ster's point of view, a unilateral superior colliculus lesion very early in life is not at all necessarily better than a similar lesion in adulthood.

At this time, and to our knowledge, recovery of function as mediated by the relocation of functions following early brain injury has only been convincingly established in humans and then with respect to but a single behavioral capacity, speech. So it is here that we begin, and we start by reviewing the human literature, which defines the parameters that are important in the relocation of speech. We then consider a series of experiments that suggest that there may be a similar consequence of early brain injury in animals. And finally, we discuss the relevance of these data to the general question of recovery of function following brain injury.

2. THE RELOCATION OF SPEECH

2.1. *The Phenomenon*

Since the early writings of Bouilland and Broca (see Schiller, 1979) there has been little doubt that speech is a highly localized human capacity that is restricted to an anterior speech area in the inferior frontal region (Broca's area) and a posterior speech area in the temporoparietal region (Wernicke's area). There is also a third speech area in the dorsal medial supplemental motor area that can be defined by electrical stimulation (Penfield and Roberts, 1959), although it is apparently not critical to maintaining overt speech functions. Moreover, Milner and her colleagues (Milner, 1974; Milner *et al.*, 1966; Rasmussen and Milner, 1977) have now established that speech functions are typically restricted to a single hemisphere, the sometimes inappropriately labeled dominant hemisphere, which is also related to handedness.

The procedure used by Milner and her group to demonstrate this hemispheric asymmetry and specialization was initially developed by the Japanese neuroscientist Juhn Wada nearly 40 years ago. It involves the intracarotid injection of sodium amobarbital (Amytal) into a patient who is lying down with knees drawn up, arms extended, fingers moving, and counting aloud. Within a few seconds following the injection, there is a contralateral flaccid paralysis. If the injection is to the hemisphere mediating speech, then the patient also becomes aphasic. If the injection is to the hemisphere that does not mediate speech, then normal verbal behavior is maintained.

Using the Wada procedure with over 450 patients, Rasmussen and Milner (1977) report that in patients without evidence of early left hemisphere pathology, speech was localized in the left hemisphere in 96% of right-handed patients, with the remaining showing right hemisphere speech localization. In left-handed or ambidextrous patients, speech was localized in the left hemisphere in 70%; 15% had speech localized in the right hemisphere, and the remainder showed bilateral speech representation.

The above picture changes quite dramatically, however, in patients with clinical evidence of early left hemisphere injury. In this group, 81% of the right-handed patients had speech localized in the left hemisphere, and in 12% and 7% speech was localized, respectively, in the right hemisphere or bilaterally. In the left-handed or ambidextrous patients, speech was localized in the left hemisphere in only 28% of the cases, whereas the percentages for right and bilateral speech localization were, respectively, 53% and 19%. Clearly, speech is usually localized in one hemisphere, typically the left, but in response to early left hemisphere injury, it may become relocated to the right hemisphere.

It is necessary to keep in mind, however, that notwithstanding the relocation of speech following early unilateral brain injury, recovery of speech functions is not complete. The data relative to this are provided by Dennis and Whitaker (1976), who extensively studied the language capabilities of three 9- to 10-year-old children, two with only a right hemisphere and one with only a left hemisphere. In all cases the hemispherectomy was performed to correct severe unilateral seizures and was completed prior to the fifth month of age. On tests of phonemic and semantic ability, Dennis and Whitaker found that all the children were quite comparable irrespective of which hemisphere they possessed. However, without the capacities of the left hemisphere, there were significant deficiencies in syntactic competence. For example, with only a right hemisphere, the children were unable to do such things as (1) understand auditory language when meaning was conveyed by syntactic diversity, (2) integrate semantic meaning and syntax to replace missing pronouns, and (3) detect and correct errors of surface syntactic structure. Sentences like "I paid the money by the man" were judged acceptable or judged incorrect for the wrong reasons ("You shouldn't pay money to a stranger.") by the children possessing only a right hemisphere. Similar sentences were judged unacceptable and correctly corrected by the left-hemisphere child. These results thus suggest a fundamental hemispheric asymmetry in humans that favors the left hemisphere for speech functions even though speech may be relocated to the right hemisphere following early left hemisphere injury. A similar asymmetry favoring the right hemisphere for nonspeech functions has been suggested by Milner (1974).

2.2. Two Necessary Conditions for Speech Relocation

There appear to be two conditions that are necessary for the relocation of speech following unilateral injury. One of these is, of course, age at the onset of the unilateral pathology. The other is that mass action does not seem to apply, and it is the precise location of the lesion that plays the critical role.

Although most of the data indicate that it is not necessary that the lesion occur prior to the establishment of language, Milner (1974) has suggested that speech will not be relocated in the noninjured hemisphere if the unilateral injury occurs after the age of 6. This does not mean, however, that prior to the age of 6 the relocation of speech following left hemisphere injury will always occur and will always be equivalent.

For example, Lansdell (1969) studied some 15 adult patients who had had early left-hemisphere injury and showed right-hemisphere speech localization on the basis of intracarotid amobarbital tests. These individuals were tested on four verbal and five performance subtests of the Wechsler–Bellevue battery and compared to a large group of brain-injured subjects of comparable IQ but with speech localized in the left-hemisphere. Factor and correlation analysis showed that for the right-hemisphere speech group, as compared to the left hemisphere speech group, verbal capabilities were better the earlier their symptoms of brain damage appeared. On the other hand, their performance scores were better the later the onset of the neurological disorder. In short, verbal performance was negatively correlated with the onset of neural pathology, and nonverbal performance was positively correlated with the onset of neural pathology in those patients in whom speech was localized in the right-hemisphere. Thus, it appears that the earlier the left-hemisphere lesion, the more effective will be the relocation of speech, but there will also be a greater disruption of nonverbal functions (see below). Identical findings have recently been reported by Vargha-Khadem *et al.* (1985), who also suggest that lesion size is of relatively little importance in determining the relocation of speech.

The question of lesion size versus lesion location was specifically addressed some time ago by Rasmussen and Milner (1977), who demonstrated that early left-hemisphere pathology, by itself, is not sufficient to establish speech functions in the right hemisphere. One hundred thirty-four patients with evidence of early left-hemisphere damage were studied with the amobarbital procedure prior to surgery for the relief of focal epilepsy. One-third of these patients were right-handed, and in 81% of these speech was localized in the left hemisphere. In the left-handed patients, only about 33% tested left-hemisphere dominant for speech, and 50% showed speech localized in the right hemisphere; the remainder had bilateral speech representation. At the time of surgery, when it was possible to determine the actual location of the early left-hemisphere lesion, it became clear that a significant involvement of either the anterior or posterior speech area was required before speech would become lateralized in the right hemisphere irrespective of the actual size of the early lesion. Moreover, involvement of one speech area did not necessarily predispose the relocation of all speech functions in the right hemisphere. Rather, only those functions mediated by the involved speech area were relocated in the opposite hemisphere, which would account for the relatively high percentage of bilateral language representation in this group of patients. This also attests to the risky nature of assuming right dominance for speech in left-handed individuals, even in those with early left-hemisphere lesions. Speech may indeed be lateralized in the right hemisphere in these individuals, but only if the early lesion involved the speech areas.

2.3. The Cost of Relocation: Crowding of Functions

Although speech may be relocated following unilateral injury to the hemisphere where it would normally reside, other unilateral functions do not appear to enjoy the same resiliency. For example, Woods and Teuber (1973) studied some 50 cases of early postnatal or congenital hemipareses. When tested as young adults, those individuals with left-hemisphere involvement showed some language dysfunctions but rather severe deficits on visuospatial tasks. A similar result was found with those individuals having early right-hemisphere involvement. In other words, left-hemisphere speech functions will transfer to the right following left-side lesions, but right-hemisphere visuospatial functions will not transfer to the left following right-side lesions.

As is usually the case, Hans-Lukas Teuber (1974) summarized the issue most succinctly:

> All in all, these findings suggest a definite hemisphere specialization at birth, with a curiously greater vulnerability to early lesions for those capacities that depend, in the adult, on the right hemisphere—as if speech were relatively more resilient or simply earlier in getting established. Yet this resiliency is purchased at the expense of non-speech functions as if one had to admit a factor of competition in the developing brain for terminal space, with consequent crowding when one hemisphere tries to do more than it had originally been meant to do (p. 73).

2.4. Conclusions

On the basis of the human literature, it appears that the relocation of functions following brain injury early in life requires a particular set of circumstances. First, only certain behaviors are apparently capable of being relocated. Second, these behaviors are highly localized and restricted to a single hemisphere. Third, the relocation is to the opposite hemisphere, which implies a considerable degree of hemispheric specialization and asymmetry. And finally, when relocation does occur, there is always a cost associated with it inasmuch as those behaviors normally mediated by hemisphere where the relocation occurs are compromised.

3. RELOCATION OF FUNCTIONS AND CROWDING IN ANIMALS

3.1. Hemispheric Specialization and Asymmetry

All in all, the above conditions represent a relatively restricted constellation, which suggests that functional relocation, and crowding, may not be a general consequence of early brain injury, particularly with regard to early brain injury in animals. On the other hand, it is also possible that the experimental procedures used with animals have not been designed to detect the phenomenon, a possibility to which we now turn.

Perhaps the most difficult condition to meet with respect to the relocation of functions in animals, at least as defined by the relocation of human speech, is that speech is localized within a single hemisphere and, when relocated, is relocated to the opposite hemisphere. This necessarily means that the human brain possesses a degree of hemispheric specialization and asymmetry that has not been clearly documented in subhuman species. The situation was described by Levy (1974) as follows:

> The abrupt evolutionary change from the functionally symmetric hemispheres of the ape to the profoundly asymmetric hemispheres of man is correlated with the discontinuity from mute to speaking animals. Those ancient ape-men who possessed both brain bisymmetry and language, but did not possess the ability to see a hungry lion embedded like a hidden figure in the tall savannah grass, paid for their speech with their lives. Their cousins, also possessing language and, in addition, a mute Gestalt-synthesizing, figure-ground-separating hemisphere, saw the lions, escaped, and, for good or ill, fathered the race of man (p. 180).

Levy's assertion that subhuman species have symmetrical brains is supported by several lines of evidence (see reviews by Walker, 1980; Warren, 1977). For example, although animals may exhibit a hand preference on certain tasks, this is not consistent across a particular species or even over different tasks for the same animal. Moreover, visual discriminations are learned equally well by either hemisphere in monkeys with unilateral lesions or split brains (Hamilton, 1977). On the other hand, singing is apparently lateralized, and to the left, in certain songbirds (Nottebohm, 1970, 1971, 1977). Additionally, Dewson (1977) reports that a delayed conditional visual discrimination, where the correct visual cue depends on a previously heard sound, is somewhat more disrupted by left superior temporal lesions than by similar lesions in the right hemisphere. Yet, on the whole, after his extensive review, Walker (1980) concludes:

> On the basis of the reports of hemispheric asymmetries in nonhuman species reviewed here, it is difficult to reject the null hypothesis that the vertebrate nervous system is an entirely symmetrical device, with the possible exceptions of the brains of humans and canaries. In only these two species is there strong evidence that damage to one side of the brain has behavioral effects different from those which result when the other side suffers similar injuries (p. 361).

We concur with Walker on this. Moreover, should functional relocation be restricted to lateralized behaviors such as speech in humans, then indeed its occurrence may be so limited that it would add little to our general understanding of brain pathology. Accordingly, we now turn to the consequences of bilateral brain injuries and to a possible demonstration of functional relocation that is not unilaterally left to right but bilaterally back to front.

3.2. *Bilateral Brain Injury and Relocation*

This particular empirical trek began with a report by Don and Pat Meyer and their group at The Ohio State University (Howarth *et al.*, 1979). The experiment was concerned with the effects of bilateral anterior neocortex injury on the performance of a learned brightness discrimination in adult rats that had been previously prepared with bilateral posterior visual neocortex lesions either as infants or as adolescents. In adult rats, bilateral anterior neocortical injury has about the same consequences whether or not the animal has also sustained a posterior visual lesion as an adult. Moreover, these consequences are quite trivial, and the animal is able to reverse any dysfunctions within 10 to 12 training trials. However, if the rat had sustained its bilateral posterior injury as an infant, then the consequences of an anterior lesion in adulthood were quite different.

Specifically, Howarth *et al.* bilaterally removed the posterior visual neocortex of a group of infant (7-day-old) and a group of adolescent (63-day-old) rats and then, after they had reached 90 days of age, trained both groups on a two-choice brightness discrimination. Both of the lesion groups, as well as two matched control groups, mastered the brightness discrimination at about the same rate and averaged roughly 30 trials to attain a criterion of nine first-choice-correct responses out of ten trials, a result that does not support the assertion that it is better to have your brain injury early. In fact, in this situation it is much worse, as shown by the consequences of a subsequent bilateral anterior lesion. Here the rats that had sustained their posterior injury as adolescents, as well as the control groups, regained criterion performance in approximately ten trials. On the other hand, the rats that had received their posterior injuries as infants were quite impaired and required an additional 30 to 40 training trials before again reaching criterion. In short, an anterior neocortical injury in a rat that has lost its visual cortex as an infant has about the same consequences for a learned brightness discrimination as bilateral visual injury in an adult.

Howarth *et al.* concluded:

> . . . a perinatal visual cortical ablation induces a reorganization of the functions of the extravisual cortex, and that effect is not producible by injuries to the visual cortex in adulthood (p. 163).

We became abruptly aware of, and involved in, this conclusion at a meeting of the Psychonomic Society when Don Meyer presented a preliminary report of the experiment some time prior to its publication. At the end of this presentation, Meyer suggested that "perhaps Professor LeVere would care to comment on these findings with respect to his position that brain injury does not induce neural reorganization" (LeVere, 1975). "Does not induce neural reorganization in the adult" was the only response I could muster on such short notice. Don's reply, "I see," was not terribly convincing to either myself or the others in the audience, and some collaborative research was obviously in order regarding this question of reorganization and relocation of functions.

This collaboration was an electrophysiological investigation stemming from an earlier report by Rebillard et al. (1977). It was the findings of these investigators that visual evoked potentials could be recorded in the primary auditory cortex of white cats that have a hereditary syndrome that causes the cochlea to degenerate within the first few weeks of life. Since it was impossible to record visual potentials in other sensory areas, say the somatic area (SI), the early peripheral degeneration and loss of auditory function appeared to be the critical factors related to this possible relocation of visual functions. Our question was whether infant injuries to visual neocortex could induce similar changes in anterior neocortex that would be detectable with electrophysiological techniques and thus provide an explanation of the Howarth et al. (1979) findings.

The experiment was relatively simple and involved obtaining the averaged visual evoked response to 25 light flashes (one flash per second) from a number of different anterior–posterior locations in adult normal rats, adult rats prepared with visual lesions as adults, and adult rats prepared with visual lesions at 7 days of age (Gray and LeVere, 1980). The infant surgeries were accomplished at the Ohio State University by Thackery Gray, and the adult lesions were performed at North Carolina State University, where all electrophysiological recordings were also completed.

The data from a normal rat are shown in Fig. 1, and it can be seen that the electrophysiological response to the visual stimulation corresponds nicely to the classically defined primary visual neocortex in rat (see LeVere, 1978). Figure 2 presents a similar set of recordings from an adult injured rat, and it is clear that the visual neocortical injury completely abolishes the visual evoked response. Moreover, no visual evoked responses were obtained from any of the anterior electrode locations. With the protocol thus verified, similar recordings were taken from the rats that had sustained their visual lesions as infants. As shown in Fig. 3, and to our surprise given the severe behavioral consequences reported by Howarth et al., these recordings from the infant injured rats were the same as those obtained from the adult injured rats. In fact, when our search was expanded to cover virtually the entire anterior dorsal neocortex of the infant injured rat, we were still unable to find even the slightest evidence of visually driven electrical activity (see Fig. 4).

Clearly, these data suggest that visual functions in rat, at least those detectable by visual evoked responses, are not relocated in anterior neocortex following infant bilateral posterior neocortical lesions. The data also suggest that there may be significant differences between the consequences of early neocortical injury and the consequences of the more peripheral neural degeneration investigated by Rebillard et al. (1977), which is complementary to a recent report by Hyvarinen et al. (1981) concerning unit responses in area A7 following binocular deprivation early in life. It thus appears that functional relocation and crowding may not follow early bilateral neocortical injuries in the symmetrical brains of animals. But then, our journey had not quite finished.

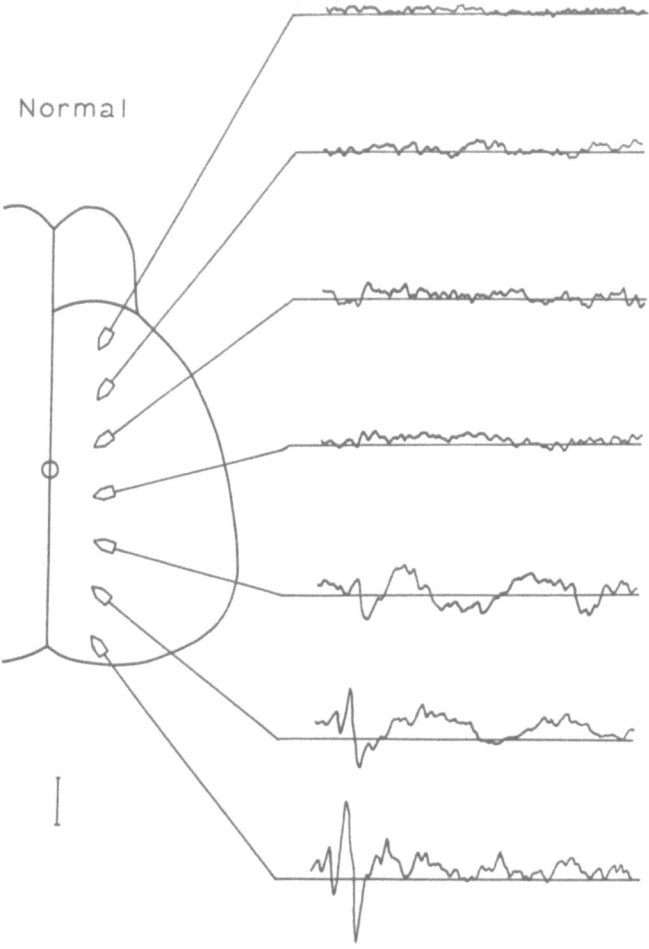

Figure 1. Average visual evoked responses from an adult normal rat. The linear electrode array was positioned 2 mm from the midline, and the interelectrode distance was 2 mm. The small circle represents the bregma. The vertical line in the lower left is an average of a 50-μV calibration signal.

3.3. Infant Lesions and Compound Cue Discriminations

Some years after our conclusion that infant posterior visual neocortex lesions do not induce reorganization of anterior neocortex in rats, we returned to the problem. We did so for two reasons. First, we continued to be perplexed about why we were unable to detect any physiological correlates of the severe visual deficits produced by bilateral anterior lesions in adult animals that had sustained visual neocortical injury as infants. Second, a model of recovery of function that had proved useful in our work with adult rats suggested that we

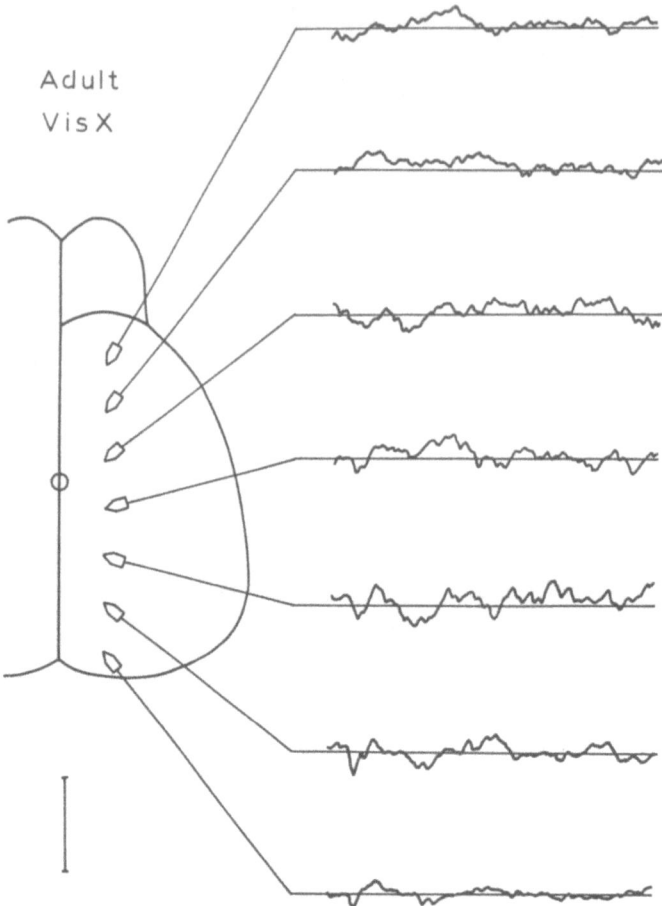

Figure 2. Average visual evoked responses from an adult rat prepared with a visual neocortex lesion at 90 days of age. See Fig. 1 for details.

might have our cake and eat it too. That is, we thought it possible to explain the severe visual consequences of adult anterior cortex lesions without necessarily postulating the relocation of visual functions following infant visual cortex injury.

The model that we considered appropriate holds that significant damage to one functional neural system disrupts the normal balance among all functional systems to the extent that the other, noninjured systems dominate behavioral control (LeVere, 1980; see also Chapter 2 for details). The result of this system imbalance is that a behavior may not be expressed not because the neural mechanisms responsible for it are lost but rather because the functional system with which these mechanisms are associated is not utilized to control behavior.

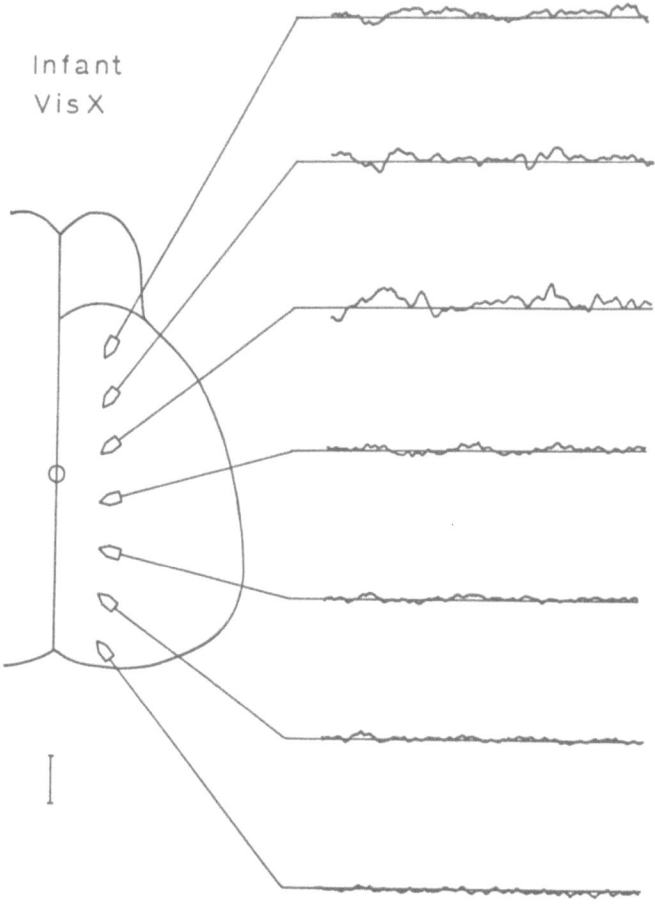

Figure 3. Average visual evoked responses from an adult rat prepared with a visual neocortex lesion at 7 days of age. See Fig. 1 for details.

We tested this position by training normal and visual decorticate adult rats on a compound-cue discrimination involving visual and haptic cues and then testing the contribution of each component cue to the animal's initial learning by further training with only the visual cues or only the haptic cues (LeVere and LeVere, 1982). The results of these experiments showed that whereas the normal rat solved the initial compound-cue discrimination by responding equally to both the visual cues and the haptic cues, the visual decorticate rat responded to only the haptic cues of the visual–haptic compound cue. In fact, the numbers were quite impressive inasmuch as the visual decorticate rats learned the compound-cue discrimination significantly faster than the normal rats even though the

Infant
VisX

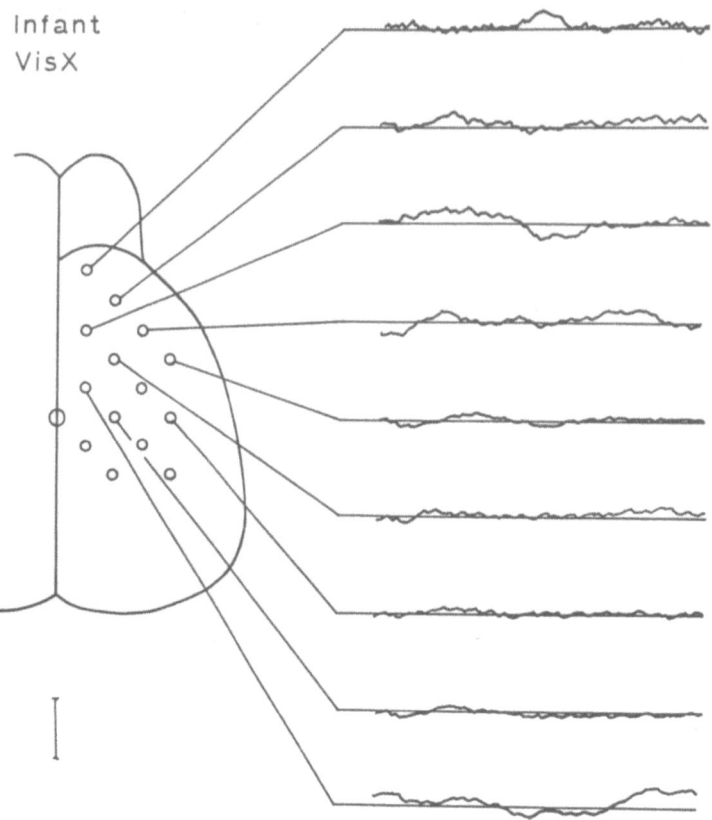

Figure 4. Average evoked responses from the matrix electrode array used with an animal prepared with visual neocortex lesion at 7 days of age. See Fig. 1 for details.

transfer training showed that they were totally naive about the visual cues. These data are presented in the topmost panel of Fig. 5, which shows the mean errors to criterion (nine correct choices in ten trials) on the compound-cue discrimination (leftmost pair of bars), retraining with just the visual cues (middle pair of bars), and retraining with just the haptic cues (rightmost pair of bars).

On the basis of this model and its supporting data, we were enticed to reinvestigate the consequences of early brain injury. Our argument was that since adult visual neocortex lesions seemed to shift behavioral control to somatosensory systems, then removal of the visual neocortex in infancy might be expected to amplify this shift on the basis of time alone. Thus, the Howarth *et al.* result might not reflect a relocation of functions but rather an extraordinary utilization of anterior cortex and dependence on the spatial aspects of the the two-choice discrimination problem. Accordingly, we replicated the compound-cue experi-

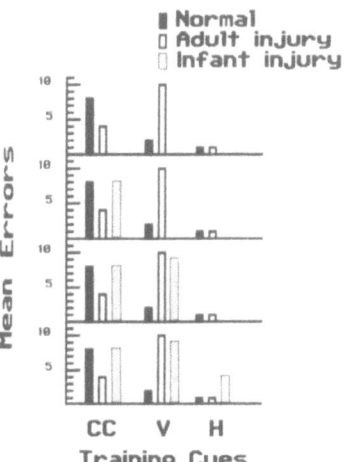

Figure 5. Compound-cue training (leftmost set of bars) and subsequent transfer training to only the visual component cues (middle set of bars) and to only the haptic component cues (rightmost set of bars) for normal rats (black bars), rats prepared with visual lesions and tested as adults (open bars), and rats prepared with visual lesions as infants and trained as adults (stippled bars). Mean errors are mean errors to criterion: nine first-choice-correct responses within a day's training of ten trials. Abbreviations: CC, compound visual–haptic cue training; V, subsequent transfer training with only the visual cues; H, subsequent transfer training with only the haptic cues.

ment with adult rats that had sustained their visual injuries in infancy. Our prediction was that these animals should learn the compound cue discrimination even faster than the 3.8 mean error score shown by the animals who had sustained their visual neocortex injury as adults. This prediction was far from the mark, but the results were nonetheless of some interest to us.

The second from the top panel in Fig. 5 presents the performance of the infant-injured rats (stippled bar) on the initial compound-cue discrimination. Not only did the infant-visual-decorticate rats not perform better than the adult-visual-decorticate rats, they performed significantly worse and were, in fact, no different from normal rats. Another demonstration of the Kennard principle that it is better to have your brain injury early? Perhaps.

The "perhaps" comes from the results of subsequent training with only the visual cues or the haptic cues, which are shown in the lower two panels of Fig. 5. Consider first the third from the top panel, which adds to the picture the performance of the infant-injured rats when subsequently trained with only the visual cues. Here it is apparent that it is not better to have your brain injury early in life, since the infant-injured animals performed just exactly like the rats that had sustained their visual decortication as adults. A refutation of the Kennard principle that it is better to have your brain injury early? Perhaps.

This time the "perhaps" is associated with the infant-injured rats' performance when trained with only the haptic cues (bottom panel of Fig. 5). Here the infant-injured rat's performance is not like that of either the normal rat or the adult-injured rat but is significantly worse than both.

Thus, notwithstanding the apparent relocation of some visual functions to anterior cortex following infant posterior cortex lesions (Howarth et al., 1979), the present data, like those reported by Woods and Teuber (1973), suggest that this relocation, if it occurs, is purchased at a nontrivial cost. This cost is a compromise of the functions normally mediated by the brain area where the relocation supposedly takes place. In the present instance, this was a disruption of somatosensory abilities as evidenced by the infant-injured rats' performance on both the compound discrimination problem, which was no better than that of normal rats, and their subsequent transfer training with only the haptic cues, which was worse than both normal rats and rats prepared with posterior lesions as adults.

However, because of our rejection of the notion of functional relocation (Gray and LeVere, 1980), we first attempted to explain these data by suggesting that the infant visual lesions encroached on somatosensory cortex. Two things forced us to reject this hypothesis. First, if there were significant involvement of somatosensory cortex, then one would expect the infant-injured rats to be impaired, relative to the normal rats, on the initial compound-cue discrimination, but they were not. Second, histological analysis revealed that the lesions of the animals operated on as adults and those operated on as infants were virtually identical in terms of both cortical involvement and subcortical consequences. Other similarly unsuccessful attempts to explain the results began to make it clear that perhaps we had been a bit hasty in our initial conclusion concerning reorganization and relocation of functions following infant brain injury.

In point of fact, we now feel that we were quite incorrect and believe that damage to the developing brain can induce relocation of functions. Moreover, the results suggest that this can occur following bilateral injury in the symmetrical brains of subhuman species. Although the present example of functional relocation is not nearly as complete as speech relocation is in humans, nor is it presently clear what visual capacities are relocated, the cost of the relocation, in terms of crowding of functions in the anterior cortex, is clearly evident from the infant-injured rats' poor performance on the initial compound-cue problem and their subsequent training with only the haptic component cues. It thus appears that the rat, like the human, has only a limited amount of cortical capacity, and when an infant lesion induces the relocation of some function, then certain compromises must necessarily occur in that brain area where the relocation takes place. Whether there are other similarities between relocation of functions in human and subhuman species, such as the hierarchical capacity for relocation described by Woods and Teuber (1973), is a question of some importance that we are presently asking.

4. SOME FINAL COMMENTS

We believe that it is necessary to mention two final points. The first concerns the physiological mechanisms that might account for our animal results and may be in part responsible for the relocation and crowding of functions in humans. The second concerns whether reorganization and relocation of functions provide a viable explanation of recovery of function following brain injury.

Concerning a possible mechanism, it is necessary to point out that although we found rather strong evidence of crowding following infant posterior lesions, we did not observe the clear relocation of functions that has been reported for speech in humans. However, it is also important to remember that neocortex is relatively less important to the rat than to the human. For example, visual neocortex lesions produce blindness in humans but only a loss of pattern vision in rat (Lavond *et al.*, 1978). Accordingly, one should perhaps not expect the animal results to parallel exactly the relocation of speech functions in humans. Moreover, the neocortical dependence and highly localized nature of speech in humans may also attribute to the completeness with which this function may be relocated. Other functions, such as visuospatial abilities, which are less cortically dependent and less localized in humans, are also much less easily relocated.

Nonetheless, the anterior neocortex in rats sustaining posterior injury in infancy does appear to assume visual functions and at the expense of the somatosensory functions normally mediated by that part of the brain. Since our electrophysiological data did not indicate that vision *per se* is relocated in anterior cortex, it is worth speculating on what these visual functions might be. It is tempting, in this regard, to think that the visual role of anterior cortex following infant posterior lesions is one of corticofugal facilitation of subcortical visual centers. If this were so, then one might not necessarily detect visual evoked electrical activity in anterior cortex even though the area could serve an important visual function. This is, of course, similar to the facilitation afforded the superior colliculus by visual cortex in cats (Sprague, 1966; Goodale, 1973).

Moreover, our understanding of developmental neurobiology provides a framework for how this might come about. During development, in both subhuman species (Oppenheim, 1985) and humans (Huttenlocher, 1984), and in many neural systems including human cerebral neocortex (Huttenlocher, 1984), there is an initial imbalance between afferent neurons and their target structures. This imbalance, which favors the afferent neurons by as much as 50% in some cases, is ultimately corrected by naturally occurring cell death. Although the exact mechanism of "the harsh judgment of epigenesis" (Oppenheim, 1985) has yet to be established, one not unreasonable suggestion is that cell survival may depend on some postsynaptic trophic factor of which there is but a limited quantity. In the normal brain, it seems likely that an overproduction of visual corticofugal projections would preempt other neocortical areas from establishing and maintaining connections with subcortical visual centers. However, this would not be the case with a very early and complete visual neocortex lesion, and the same overproduction of neocortical afferents from other areas might establish synaptic linkage with subcortical visual centers and survive the period of normally occurring cell death. Moreover, if the target area can impart what Jacobson (1978) has termed "functional validation" to the afferent neuron, then it would be quite possible that anterior neocortical neurons could very well develop significant visual attributes.

There is, however, one particularly nasty fly in this theoretical ointment. This is that it is based on the proposition that it is the excess or "overproduced" anterior neocortical neurons that subsume visual capacities. And if this is so, then anterior cortical functions should not be compromised by crowding since there should be ample neurons to cover both anterior and posterior neocortical functions. It is possible, of course, to suggest that the mere existence of neurons serving visual functions could degrade the normal somatosensory functions of anterior cortex independent of whether or not there are sufficient neurons to accomplish both. This, however, adds a layer of theoretical complexity that is less than completely satisfying.

Finally, we would like to close with a comment concerning neural plasticity, neural reorganization, and relocation of functions as viable explanations of recovery of function following brain injury. The data we have reviewed do not change our opinion that reorganization and relocation cannot account for recovery of function. That is, when it occurs, even under the most favorable conditions of infant brain injury, it appears to be of little benefit to the individual's total behavioral capacities. In fact, if the correct observations are completed, then it is possible to demonstrate that it is detrimental. Notwithstanding the reported occurrences of one brain area "taking over" the functions of some injured brain area, we believe that the present data emphasize a logical point often overlooked in the interpretation of these experiments. This point is: what happens to the functions normally served by the area vicariously assuming the new duties? The available data indicate that these functions most probably will be severely compromised. In short, we suggest that when we speak of relocation and "taking over functions," it might be closer to the mark to speak of "exchanging deficits."

REFERENCES

Dennis, M., and Whitaker, H. A., 1976, Language acquisition following hemidecortication: Linguistic superiority of the left over the right hemisphere, *Brain Lang.* 3:404–433.

Dewson, J. H., 1977, Preliminary evidence of hemispheric asymmetry of auditory function in monkeys, in: *Laterization in the Nervous System* (S. Harnad, R. W. Doty, L. Goldstein, J. Jaynes, and G. Krauthamer, eds.), Academic Press, New York, pp. 63–71.

Finger, S., and Almli, C. R., 1984, *Early Brain Damage,* Volume 2, *Neurobiology and Behavior,* Academic Press, New York.

Goodale, M. A., 1973, Cortico-tectal and intertectal modulation of visual responses in the rat's superior colliculus, *Exp. Brain Res.* 17:75–86.

Gray, T., and LeVere, T. E., 1980, Infant posterior neocortical lesions do not induce visual responses in spared anterior neocortex, *Physiol. Psychol.* 8:487–492.

Hamilton, C. R., 1977, Investigations of perceptual and mnemonic lateralization in monkeys, in: *Lateralization in the Nervous System* (S. Harnad, R. W. Doty, L. Goldstein, J. Jaynes, and G. Krauthamer, eds.), Academic Press, New York, pp. 45–62.

Howarth, H., Meyer, D. R., and Meyer, P. M., 1979, Perinatal injuries to the visual cortex enhance the significance of extravisual cortex for performance of a visual habit, *Physiol. Psychol.* **7:**163–166.

Huttenlocher, P. R., 1984, Synapse elimination and plasticity in developing human cerebral cortex, *Am. J. Ment. Defic.* **88:**488–496.

Hyvarinen, J., Hyvarinen, L., and Linnankoski, I., 1981, Modification of parietal association cortex and functional blindness after binocular deprivation in young monkeys, *Exp. Brain Res.* **42:**1–8.

Isaacson, R. L., 1975, The myth of recovery from early brain damage, in: *Aberrant Development in Infancy* (N. R. Ellis, ed.), Erlbaum, Hillsdale, NJ, pp. 1–25.

Jacobson, M., 1978, *Developmental Neurobiology*, 2nd ed., Plenum Press, New York.

Lansdell, H., 1969, Verbal and nonverbal factors in right-hemisphere speech, *J. Comp. Physiol. Psychol.* **69:**734–738.

Lavond, D., Hata, M. G., Gray, T. S., Geckler, C. L., Meyer, P. M., and Meyer, D. R., 1978, Visual form perception is a function of the visual cortex, *Physiol. Psychol.* **6:**471–477.

LeVere, T. E., 1975, Neural stability, sparing, and behavioral recovery following brain damage, *Psychol. Rev.* **82:**344–358.

LeVere, T. E., 1978, The primary visual system of the rat: A primer of its anatomy, *Physiol. Psychol.* **6:**142–169.

LeVere, T. E., 1980, Recovery of function after brain damage: A theory of the behavioral deficit, *Physiol. Psychol.* **8:**297–308.

LeVere, N. D., and LeVere, T. E., 1982, Recovery of function after brain damage: Support for the compensation theory of the behavioral deficit, *Physiol. Psychol.* **10:**165–174.

Levy, J., 1974, Cerebral asymmetries as manifested in split-brain man, in: *Hemispheric Disconnection and Cerebral Function* (M. Kinsbourne and W. L. Smith, eds.), Charles C. Thomas, Springfield, IL, pp. 165–183.

Lund, R. D., 1978, *Development and Plasticity of the Brain*, Oxford University Press, New York.

Milner, B., 1974, Hemispheric specialization: Scope and limits, in: *The Neurosciences, Third Study Program* (F. O. Schmitt and F. G. Worden, eds.), MIT Press, Cambridge.

Milner, B., Branch, C., and Rasmussen, T., 1966, Evidence for bilateral speech representation in some non-right-handers, *Trans. Am. Neurol. Assoc.* **91:**306–308.

Nottebohm, F., 1970, Ontogeny of bird song: Different strategies in vocal development are reflected in learning stages, critical periods, and lateralization, *Science* **167:**950–956.

Nottebohm, F., 1971, Neural lateralization of vocal control in a passerine bird. I. Song, *J. Exp. Zool.* **177:**229–262.

Nottebohm, F., 1977, Asymmetries in neural control of vocalization in the canary, in: *Laterization in the Nervous System* (S. Harnad, R. W. Doty, L. Goldstein, J. Jaynes, and G. Krauthamer, eds.), Academic Press, New York, pp. 23–44.

Oppenheim, R. W., 1985, Naturally occurring cell death during neural development, *Trends Neurosci.* **8:**487–493.

Penfield, W., and Roberts, L., 1959, *Speech and Brain Mechanisms*, Princeton University Press, Princeton.

Rasmussen, T., and Milner, B., 1977, The role of early left-brain injury in determining lateralization of cerebral speech functions, *Ann. N.Y. Acad. Sci.* **299:**328–354.

Rebillard, G., Carlier, E., Rebillard, M., and Pujol, R., 1977, Enhancement of visual responses on the primary auditory cortex of the cat after an early destruction of cochlear receptors, *Brain Res.* **129:**162–164.

Schiller, F., 1979, *Paul Broca*, University of California Press, Berkeley.

Schneider, G. E., 1979, Is it really better to have your brain lesion early? A revision of the "Kennard principle," *Neuropsychologia* **17**:557–583.

Sprague, J. M., 1966, Interaction of cortex and superior colliculus in mediation of visually guided behavior in the cat, *Science* **153**:1554–1547.

Teuber, H.-L., 1974, Why two brains? in: *The Neurosciences, Third Study Program* (F. O. Schmitt and F. G. Worden, eds.), MIT Press, Cambridge, pp. 71–74.

Vargha-Khadem, F., O'Gorman, A. M., and Watters, G. V., 1985, Aphasia and handedness in relation to hemispheric side, age at injury and severity of cerebral lesion during childhood, *Brain* **108**:677–696.

Walker, S. F., 1980, Lateralization of function in the vertebrate brain: A review, *Br. J. Psychiatry* **71**:329–361.

Warren, J. M., 1977, Functional lateralization of the brain, *Ann. N.Y. Acad. Sci.* **299**:273–280.

Woods, B. T., and Teuber, H. L., 1973, Early onset of complementary specialization of cerebral hemispheres in man, *Trans. Amer. Neurol. Assoc.* **98**:113–117.

10

Arguments against Redundant Brain Structures

ULF NORRSELL

1. INTRODUCTION

The term redundancy is sometimes used in papers that review the various effects of different lesions of the nervous system with regard to functional recovery (e.g., Rosner, 1970; Laurence and Stein, 1978; Freund and Bauer, 1982; Singer, 1982), although more rarely in the original reports. It is used for discussions of an unexpected lack of functional defects or surprisingly rapid recovery after some lesions. Its usage implies that the damaged structure was superfluous or contained an overabundance of the relevant elements. The idea is important if true but difficult to argue, since a lack of defect can only be ascertained if all of the possible functions are known and have been studied. Perhaps that is why the problem usually is avoided in research papers that are concerned with restricted themes. Nevertheless, the idea of overabundant neurons in the brain is an old one.

O'Leary (1962) cited a report by Piorry, a French pathologist of the early 19th century, who "discovered that the subject of an autopsy he was performing had been getting along with but one cerebral hemisphere. . . . The subject's intellect was said to be unimpaired." Wiener (1961) was under the impression that Pasteur had done "some of his best work" after having suffered extensive lesions of the parietal and temporal regions of the right cerebral hemisphere, which were revealed at autopsy. The two examples are sufficient to illustrate a widespread belief about neuronal redundancy, although it can be added that Wiener (1961) appears to have been misled. At least Dr. Patrick Sourander, a well-known neuropathologist who has been interested in the subject, gave me a

ULF NORRSELL • Departments of Physiology and Neurology, University of Göteborg, S-400 33 Göteborg, Sweden.

different version of Pasteur's illness. Pasteur did suffer a stroke in 1868, i.e., before some of his best work. The disorder has been sufficiently well documented to permit the assertion that the lesion must have been restricted. There is no mention of an autopsy in the ordinary biographies about Pasteur, but if it was performed, and the above-mentioned extensive lesions were observed, then it seems likely that they will have appeared towards the end of his life, when he was quite ill. That is not to claim, on the other hand, that one cannot overlook the effects of even large cerebral lesions. A dramatic example is the much-discussed "split-brain" group of symptoms, which appear in human patients after cerebral commissurotomy. It may be noted that originally more than two dozen such patients were examined without any of those symptoms being observed (Sperry *et al.*, 1969).

2. TOO MUCH BRAIN

Each cerebral hemisphere contains sensory and motor cortical areas that are essential for primary functions of the contralateral body half, and these are discussed in Sections 5 and 8. Other parts of the cerebral cortex (belonging to the so-called association areas) are essential for higher functions, which are not restricted to individual organs. The capacity and sometimes even the character of such higher functions are not predictable, and it is therefore almost hopeless to try to exclude their deterioration following brain damage without knowledge of the subject's total premorbid endowments. Nevertheless, an inkling about the possible lack of redundant cerebral structures may be obtained from the results of human hemispherectomies.

McFie (1961) reviewed 34 hemispherectomy cases from the National Hospital (Queen Square, London) who had undergone pre- and postoperative psychological testing. His patients (median age at operation 12, range 1–31 years) suffered from infantile hemiplegia, and the results were compared to those of other available reports including hemispherectomies of adult tumor (glioma) cases. The latter operations had been found to cause "a profound but not absolute impairment of speech" after removal of the left hemisphere and "at least a reduction in judgment and insight" after removal of the right hemisphere. In contrast, it was found that hemispherectomies in cases of infantile hemiplegia could produce an improvement of the functional performance. In cases with brain damage sustained before 1 year of age and a healthy second hemisphere (as judged by the EEG), the verbal and especially the nonverbal IQ improved after the operation. The results were found to support "Krynauw's suggestion that the remaining hemisphere is free to function normally when the abnormal influences of the damaged hemisphere are removed." The postoperative IQ remained below the normal average, however, and McFie (1961) concluded "that these results indicate a limit to the capacity of the remaining hemisphere, imposed by the mediation of the normal functions of both hemispheres by one hemisphere alone.''

Higher than average IQ values later have been obtained after unilateral hemispherectomy in at least two cases (Griffith and Davidson, 1966; Smith and Sugar, 1975). However, IQ values should be expected to vary between single individuals, and those two values appear to be statistically insignificant and so far do not contradict McFie's (1961) above-cited conclusion.

3. NEURONAL REDUNDANCY DURING INFANCY

The results of hemispherectomy evoke the supplementary question about possible infantile redundancy. The above-described cases of infantile hemiplegia had similar linguistic capacities independent of which hemisphere had been removed (McFie, 1961; cf. also Wilson, 1970). The same is not true for patients who have one hemisphere removed because of a pathological cerebral process that commences at a later stage in life. Such patients show an irreparable lateralization of speech function to one, usually the left, hemisphere (cf., e.g., Norrsell, 1983). The difference probably may be explained at least partially by a greater plasticity, i.e., a later-disappearing aptitude of the immature brain to rearrange its functional connections (Goldman and Galkin, 1978).

The immature brain, on the other hand, has been shown to contain superfluous neurons and axonal contacts (Purves and Lichtman, 1980). These redundant structures seem to disappear around birth, before the behavioral development that involves an adaptation to external influences, but the findings could indicate that an older infant's brain might contain unoccupied (i.e., redundant in the context of the present paper) neural circuits that could replace others that had become damaged.

Such a hypothesis appears to be at least partially refuted, however, by the observations of Teuber and Rudel (1962). These two authors were trying to find out whether early brain injuries were invariably more or less damaging than injuries sustained later in life. They found instead that the differential effects of early compared to later brain injuries were qualitative rather than quantitative. Three different tests involving localization during body tilt were used for normal and brain-damaged children and adults. The results of one test were found to deviate from the normal after brain damage in both children and adults. The results of another test were found to deviate from the normal after brain damage only in children, and the results of the third test were exactly opposite, i.e., brain-damaged adults, but not children, deviated from the normal. These findings are perhaps not surprising to the reader of modern textbooks of developmental psychology, who will have been told that the behavior of children may be categorically different from that of adults in certain respects. The findings do show, on the other hand, that the absence of certain expected symptoms after brain damage in a child does not prove that the lesion involved redundant structures, since the same lesion may cause other symptoms that are unexpected on the basis of previous knowledge of adult patients.

4. RESTATEMENT OF PROBLEM

It has been stated that a search for redundant nervous structures may be indecisive regarding regions that are devoted to individual functions that are based on personal experience. The effort could perhaps be more rewarding if directed towards structures involved in sensory or motor functions. Those functions have the advantage of being executed fairly uniformly by a given species and are predictable to a certain extent on the basis of neuroanatomic and neurophysiological data. Before turning to specific examples it may be worthwhile to consider the usage of the term redundancy. That word covers several meanings, and the two most common ones, according to contemporary dictionaries, are overabundance and superfluity. Therefore, when the term redundancy is applied to nervous or other structures, it may have ambiguous functional implications. The human right hand, for example, is overabundant for tooth brushing, which can be done with the left hand and therefore seems strictly superfluous in that context. It is, on the other hand, obviously not superfluous for tying the shoestrings. Consequently, it might be regarded to be more or less redundant by two different communities, one of which brushes its teeth and wears slippers, whereas the other is partial to brogues but indifferent to sparkling smiles.

5. THE MOTOR SYSTEM

Pribram *et al.* (1955–1956) used cortical ablations to evaluate the functional specialities of the precentral motor cortex. Monkeys were observed pre- and postoperatively with regard to spontaneous behavior, and several formal tests were done, including a latchbox-opening task. The lesions embraced the precentral agranular cortex either subtotally or extensively and were made uni- and/or bilaterally. Mainly contralateral postoperative deficits of varying severity were found, but sufficient retention and/or recovery was observed in all instances to permit the suggestion that the performance had been "impaired, but not abolished." The authors concluded among other things that the precentral motor cortex did not appear to be the "principal cerebral efferent or effector mechanism by which the brain expresses its activity." In the present context, the findings could perhaps be taken to indicate a certain redundancy of that cortical area for motor performance.

Penfield (1954), on the other hand, noted that removal of the precentral gyrus in man permanently removed the capacity for independent finger skills.

Pribram *et al.* (1955–1956) did not comment on the independent finger movements of their monkeys, and there appear to be no comparable later studies. After transection of the corticospinal fibers of the monkey's pyramidal tracts, however, the independent finger movements disappear permanently (Lawrence and Kuypers, 1968). In addition, Muir and Lemon (1983) demonstrated corticospinal neurons in the precentral gyrus of the monkey that were specifically active during the performance of a precision grip with individual finger movements, in contrast to a power grip in which all digits are used in concert. It therefore appears possible that the monkey's precentral motor cortex may in fact be a principal, and perhaps unique, cerebral effector mechanism for independent finger movements. Considering the importance of manual dexterity for survival, it therefore may be misleading to use the term redundancy in connection with the primate motor cortex under any circumstances.

6. THE VISUAL SYSTEM

Galambos *et al.* (1967) studied the visual capacity of cats with extensive bilateral lesions of the optic tract. Two of their cats had lesions that were estimated to have interrupted 98% and 98.5% of the fibers, respectively. These cats were found to resemble normal cats in gross tests of visual abilities: they performed a flux discrimination task without difficulty a few days postoperatively, and they were able to perform a skilled pattern discrimination. Lesions that were even larger disturbed the visual behavior severely (Norton *et al.*, 1967). Galambos *et al.* (1967) concluded that "brain matter which both hard facts and common sense tell us should be indispensable proves to be dispensable." Rosner (1970) suggested that these findings showed "a striking case of redundancy," and, depending on one's attitude to the above-presented toothbrush metaphor, perhaps it is.

The optic tracts contain the afferent axons of ganglion cells that serve restricted parts of the ipsilateral halves of the two retinas. Partial lesions of the optic tracts therefore could be expected to cause partial denervations of the retinal receptors, but since a cat may use any part of the retina for visual behavior, there is no necessity for such a lesion to impede its ability to perform a flux discrimination. The same would be true for the "gross visual tests" and the pattern discrimination task unless the lesion had caused a reduction of the amount of visual information below a critical level. Galambos *et al.* (1967) did not measure their cats' visual acuities. The pattern discrimination task, however, involved comparison of the figures 6 and 9, each 9 × 4 cm large and viewed at a distance of 115 cm. A symbolic human patient with a visual acuity of 0.1 (20/200), on the other hand, ought to be able to perform that task, which in this case seems to require less than 1% of the afferent fibers (Frisén and Quigley, 1984). It appears improbable that the patient would regard the missing 99% of his optic afferent pathway as redundant, and whether the above-mentioned cats did seems difficult to tell.

The eye is not a uniform receptive organ but provides different types of information for several, including neuroendocrine, functions. Therefore, it is necessary to establish the preservation of all of those functions before the conclusion is reached that a given severed fiber of its afferent system was superfluous. Such conclusions appear to be premature. There is convincing evidence, however, that the visual afferent pathway of man does not contain any superfluous fibers with regard to those functions that are invoked by testing visual acuity (Frisén, 1980).

7. THE AUDITORY SYSTEM

The auditory system, like the visual system, embraces several partially independent functions and therefore invokes similar restraints in the present context. An additional complication is present, however, in the spatial domain. In the visual system each of the cerebral hemispheres receives primary information from only one-half of the visual field, whereas primary information from all of the auditory space is projected to both hemispheres in the auditory system.

It seems to be generally accepted that large bilateral lesions of the cerebral auditory areas do not affect behavioral tone thresholds or the ability to discriminate changes of tone intensities (Aitkin et al., 1984). The finding perhaps is not surprising, or, to paraphrase Kaas et al. (1967), it is not likely that the cerebral cortex evolved simply to replace functions already present in the reptile tectum. Other auditory functions are disturbed by removal of the cortical projection areas, however, and these include the ability to localize sound in space. The cat has been a popular subject for this type of experiment, and Neff and Casseday (1977) found its ability for monaural sound localization to be severely disturbed or abolished by contralateral lesions but undisturbed by ipsilateral lesions. During binaural sound localization, on the other hand, additional cues become available, and the lesion effects are different. Strominger (1969) used a behavioral situation that was similar to that of Neff and Casseday (1977), but his cats were permitted to use both ears and were found to localize quite well after unilateral, but hardly at all after bilateral, lesions. When the data of Neff and Casseday (1977) and of Strominger (1969) are compared, one gets the definite impression that it is easier for a cat to perform the task after unilateral destruction of the cerebral auditory projection areas than after having been made monaural by unilateral cochlear destruction. Sufficient bilateral projection of binaural information therefore seems to be available for this type of auditory spatial localization to allow the suggestion of a possibly redundant organization on the basis of this evidence.

Whitfield et al. (1972), on the other hand, noted that experiments of the above type, which involve free-field localization of a single sound source in an anechoic chamber, are somewhat artificial. They continued: "In a more 'natural' environment any sound is usually accompanied by reflections from surrounding objects, although the echoes are not heard as such unless the reflector is quite distant." These authors also studied the cat's ability to localize a sound binaurally after unilateral removal of the auditory projection areas. They used a two-choice situation, however, and their sounds were compound stimuli originating from two sources. Their cats showed an irreparable and categorical deficit of the ability to localize sound, which was predictable depending on the side of the lesion, and they consequently postulated that "ablation of the auditory cortex on one side destroys the laterality of a compound stimulus normally localized on the contralateral side." In the present context, it thus seems that the suggested apparent redundancy of the auditory system may have been produced by an experimentally induced reduction of the stimulus properties.

The auditory cortex embraces several areas; i.e., to a certain extent there is parallel projection of primary afferent information to multiple cerebral targets. Multiple projections, of course, could mean redundant organization, but too little is known about the differential importance of these separate areas (Aitkin et al., 1984) to permit conclusions along these lines. A somewhat better opportunity is perhaps offered by the somatosensory system.

8. THE SOMATOSENSORY SYSTEM

It has been shown for a large number of different animals that primary somatosensory information is projected in parallel to several different areas of the cerebral cortex (Woolsey, 1984). There are still uncertainties about these areas, and their number seems to vary among different species (Kaas, 1984). For example, in the cat four areas (SI–IV) have been found so far, which are similar in that they all recieve primary tactile information from the contralateral body half but are different with regard to afferent and efferent connections (Clemo and Stein, 1983; Stein et al., 1983). The areas' different numbers denote the order of discovery, and they are arranged in such a way that SI and III are found medially and SII and IV are found laterally on the hemisphere. There may be greater or smaller numbers of areas for other species, but the intrahemispheric bimodal arrangement with regard to laterality seems to be general and has been investigated with lesion experiments. The findings are complex but permit the conclusion that the medial and the lateral projections are involved in different types of somatosensory function in primates, carnivores, and rats (Norrsell, 1978, 1980). The area traditionally called SI involves parallel projections in the monkey in contrast to other species (Kaas, 1984), but again, lesions restricted to one or another subdivision have been found to cause differential effects (Randolph and Semmes, 1974; Schwartz, 1983). Therefore, there are no indications so far that the parallel projections to the somatosensory cortex constitute a redundant organization.

Primary afferent information belonging to the somatosensory system not only reaches multiple cortical areas but is also transmitted via several independent ascending pathways in the spinal cord. For that reason, restricted lesions of the spinal cord almost invariably spare the somatosensory functions partially, which may give a false impression of redundancy (Norrsell, 1980). The impression may be of the type that is caused by placing overemphasis on a single functional aspect and can be illustrated by the effects of spinal lesions on thermosensitivity.

Symmetrical bilateral lesions have been made in the cervical spinal cords of different cats that together covered the spinal cord's entire transverse area. All of those cats were afterwards able to use their paws for temperature discriminations (Norrsell, 1979), and the findings thus might indicate the presence of several redundant thermosensory ascending pathways. A lesion involving the middle part of only one lateral funiculus, on the other hand, was found to cause a severe and permanent contralateral thermosensory deficiency (Norrsell, 1979, 1983). These apparently contradictory findings can probably be explained by the simple fact that the effects of the unilateral lesions for natural reasons were tested with a behavioural technique permitting assessment of the individual body halves, whereas the animals with bilateral lesions were permitted to use all of their paws at the same time for reasons of convenience. The lack of defects after the bilateral lesions indicates that there is more than one ascending thermosensory pathway for each body half. Each of these pathways has to transmit one or more types of information, e.g., the quality, quantity, and localization of the thermosensory stimuli, and it is likely that the different types are distributed unequally among the pathways. Thus, a unilateral lesion involving one of the pathways would cause a functional asymmetry that would be apparent to an animal who was tested for each body half separately and probably would cause a behavioral disturbance. A comparable bilateral lesion, on the other hand, would not cause any asymmetry and might pass unnoticed unless the behavioral test was designed to evaluate the importance of each type of thermosensory information quantitatively (Norrsell, 1983, 1984).

This reasoning may appear extravagant, but it is supported by some findings in human patients. After spinal lesions, thermal dysesthesias have been observed that involved loss of the subjective sensations warm and cool but not the ability to discriminate between warm and cool stimuli (Wilson, 1927; Davison and Schick, 1935). These findings are difficult to explain except by the presence of several ascending pathways serving different thermosensory properties. Furthermore, the same area that was involved in the unilateral lesions that caused contralateral thermosensory defects in the cat seems to be critical for the thermosensory defects that appear in human patients after spinal cordotomies (Norrsell, 1979). I have examined two such patients myself, who described both warm and cold water receptacles as feeling warm. When touched by both at the same time, however, and asked for the difference, they said that the cold but not the warm receptacle was "prickly" or "buzzed." Neither stimulus was anywhere close to the pain threshold, and thus the patients appeared to be able to discriminate the temperature differences despite suffering from a severe thermosensory defect. The term redundancy would of course be highly inappropriate in that particular situation, but perhaps not so much in a corresponding animal experiment with the means of communication restricted to recording of the results of a two-choice task.

9. AUTONOMIC FUNCTIONS

A physiologist will sometimes divide the nervous structures into "somatic" and "autonomic" categories, although in later years probably mainly for reasons of convenience supported by tradition. It may be argued that the foregoing discussion has dealt almost entirely with the consequences of lesions involving structures belonging to the somatic category and that lesions involving autonomic structures of the brainstem have different effects. These structures are paired but participate in unitary functions regarding the inner milieu (e.g., organ oxygenation or extracellular ionic balance) and certain essential, unitary behavioral functions (drives or motivation). The hypothalamus, for example, has been suggested to contain "feeding" and "satiety" centers, and the following citation is from a paper by Mayer and Barrnett (1955), who had studied those structures:

> Apparently lateral feeding centers have to be bilaterally destroyed in order that anorexia prevail . . . , whereas unilateral medial lesions may cause hyperphagia. . . . The removal of either one (of the medial centers) permits the development of about half the obesity caused by the destruction of both.

The observation could perhaps be cited as an example of redundancy, i.e., for the "lateral feeding centers."

The ideas about possibly unique feeding centers of the hypothalamus have been modified, however, during subsequent years (cf., e.g., Sahakian, 1982), and the anorexia or obesity following lateral or medial hypothalamic lesions now is interpreted differently. Marshall and co-workers (Marshall *et al.*, 1971; Marshall, 1975) found that a lesion of the relevant lateral hypothalamic site caused a behavioral neglect of the contralateral hemifield and that the medial lesions caused a corresponding increase of sensory responsiveness. Marshall (1975) has suggested that the altered responsiveness may be a major cause for the ensuing either aphagia or hyperphagia, and since both types of lesions now are known to have specific effects depending on laterality, any theories about redundancy seem unnecessary.

It is well established that unilateral electrical stimulation of the lateral hypothalamus of the awake cat can cause the animal to make a savage attack. Since an attack is a function involving the whole animal, the bilateral representation of the sensitive region might indicate a redundant organization. The hypothesis seems to be precluded, however, by certain observations. MacDonnell and Flynn (1966) searched the lateral hypothalamus of the awake cat with electrodes until they found points from which attacks could be elicited. They subsequently studied their cats' faces for sensory fields of biting reflexes and found that the electrical stimulation in addition established such fields on the contralateral side of the snout and that the size of the fields depended on the intensity of stimulation. The finding again seems to indicate a specific and detailed functional role of a minute brainstem region, which fits badly with theories of the presence of an overabundance of replaceable neurons.

10. COMMENT

Physiological studies have increasingly revealed the thrifty relevance of the functional organization of different organs. The possible overabundance or superfluity, i.e., redundancy, of nervous structures therefore appears to constitute a departure from the rule, which calls for an explanation. It may be noted, in addition, that a hypothesis about nervous redundancy carries the extra burden of being difficult to falsify, since the nervous systems of most species serve adaptable functions and therefore lack well-defined limits of performance. Finally, it is difficult to find controlled studies of the problem, although two reports (McFie, 1961; Teuber and Rudel, 1962) have been cited that can be interpreted in the opposite way, i.e., to deny the presence of nervous redundancy in man. Instead, it appears easier to adopt the reverse attitude, that the nervous system lacks redundancy. It is possible to show how individual observations can be erroneously interpreted to indicate the redundancy of a certain nervous structure, and several examples have been submitted. It should be noted, however, that those examples were chosen to illustrate the point and not for the sake of criticizing the original authors. The variables of an experiment are often reduced to a minimum in order to produce rapidly accountable results. A negative result, i.e., the lack of symptoms that might indicate redundancy, on the other hand, is relevant only for the prevailing conditions and carries less value as general evidence. Like the right hand for toothbrushing, a nervous structure can be redundant for a function in which it participates normally. Like the right hand with regard to the tying of shoestrings, it can be irreplaceable for something else at the same time. Thus, it may perhaps be concluded that the word redundancy is useful in many ways but without the merit of being a principle for nervous organization.

ACKNOWLEDGMENTS. Dr. Lars Frisén kindly read and advised on the text. My own work was supported by the Swedish Medical Research Council (project no. 2857).

REFERENCES

Aitkin, L. M., Irvine, D. R. F., and Webster, W. R., 1984, Central neural mechanisms of hearing, in: *Handbook of Physiology*, Volume 3 (I. Darian-Smith, ed.), Americal Physiological Society, Bethesda, pp. 675–737.

Clemo, H. R., and Stein, B. E., 1983. Organization of a fourth somatosensory area of cortex in cat, *J. Neurophysiol.* **50:**910–925.

Davison, C., and Schick, W., 1935, Spontaneous pain and other subjective sensory disturbances. A clinicopathologic study, *Arch. Neurol. Psychiatry* **34:**1204–1237.

Freund, H.-J., and Bauer, H. J., 1982, Regeneration and repair of the nervous system: Clinical aspects, in: *Dahlem Konferenzen, Life Sciences Research Reports*, Volume 24 (J. G. Nicholls, ed.), Springer-Verlag, Berlin, Heidelberg, New York, pp. 187–202.

Frisén, L., 1980, The neurology of visual acuity, *Brain* **103:**639–670.

Frisén, L., and Quigley, H. A., 1984, Visual acuity in optic atrophy: A quantitative clinicopathological analysis, *Graefes Arch. Clin. Exp. Ophthalmol.* **222:**71–74.

Galambos, R., Norton, T. T., and Frommer, G. P., 1967, Optic tract lesions sparing pattern vision in cats, *Exp. Neurol.* **18:**8–25.

Goldman, P. S., and Galkin, T. W., 1978, Prenatal removal of frontal association cortex in the fetal rhesus monkey: Anatomical and functional consequences in postnatal life, *Brain Res.* **152:**451–485.

Griffith, H., and Davidson, M., 1966, Long-term changes in intellect and behaviour after hemispherectomy, *J. Neurol. Neurosurg. Psychiatry* **29:**571–576.

Kaas, J. H., 1984, The organization of somatosensory cortex in primates and other mammals, in: *Wenner-Gren Center international studies*, Volume 41 (C. von Euler, O. Franzén, U. Lindblom, and D. Ottoson, eds.), Macmillan, London, pp. 51–59.

Kaas, J., Axelrod, S., and Diamond, I. T., 1967, An ablation study of the auditory cortex in the cat using binaural tonal patterns, *J. Neurophysiol.* **30:**710–724.

Laurence, S., and Stein, D. G., 1978, Recovery after brain damage and the concept of localization of function, in: *Recovery from Brain Damage* (S. Finger, ed.), Plenum Press, New York, pp. 369–407.

Lawrence, D. G., and Kuypers, H. G. J. M., 1968, The functional organization of the motor system in the monkey. I. The effects of bilateral pyramidal lesions, *Brain* **91:**1–14.

MacDonnell, M. F., and Flynn, J. P., 1966, Control of sensory fields by stimulation of hypothalamus, *Science* **152:**1406–1408.

Marshall, J. F., 1975, Increased orientation to sensory stimuli following medial hypothalamic damage in rats, *Brain Res.* **86:**373–387.

Marshall, J. F., Turner, B. H., and Teitelbaum, P., 1971, Sensory neglect produced by lateral hypothalamic damage, *Science* **174:**523–525.

Mayer, J., and Barrnett, R. J., 1955, Obesity following unilateral hypothalamic lesions in rats, *Science* **121:**599–600.

McFie, J., 1961, The effects of hemispherectomy on intellectual functioning in cases of infantile hemiplegia, *J. Neurol. Neurosurg. Psychiatry* **24:**240–249.

Muir, R. B., and Lemon, R. N., 1983, Corticospinal neurons with a special role in precision grip, *Brain Res.* **261:**312–316.

Neff, W. D., and Casseday, J. H., 1977, Effects of unilateral ablation of auditory cortex on monaural cat's ability to localize sound, *J. Neurophysiol.* **40**:44–52.

Norrsell, U., 1978, Sensory defects caused by lesions of the first (SI) and second (SII) somatosensory areas of the dog, *Exp. Brain Res.* **32**:181–195.

Norrsell, U., 1979, Thermosensory defects after cervical spinal cord lesions in the cat, *Exp. Brain Res.* **35**:479–494.

Norrsell, U., 1980, Behavioral studies of the somatosensory system, *Physiol. Rev.* **60**:327–354.

Norrsell, U., 1982, Comment on the partial roles of the cerebral hemispheres for speech, in: *Wenner-Gren Center International Symposium Series*, Volume 36 (S. Grillner, B. Lindblom, J. Lubker, and A. Persson, eds.), Pergamon Press, Oxford, pp. 67–73.

Norrsell, U., 1983, Unilateral behavioural thermosensitivity after transection of one lateral funiculus in the cervical spinal cord of the cat, *Exp. Brain Res.* **53**:71–80.

Norrsell, U., 1984, Possible redundance of spinal pathways for behavioural thermosensitivity, in: *Wenner-Gren Center International Symposium Series*, Volume 41 (C. von Euler, O. Franzén, U. Lindblom, and D. Ottoson, eds.), Macmillan, London, 273–283.

Norton, T. T., Galambos, R., and Frommer, G. P., 1967, Optic tract lesions destroying pattern vision in cats, *Exp. Neurol.* **18**:26–37.

O'Leary, J. L., 1962, Discussion first session, in: *Interhemispheric Relations and Cerebral Dominance* (V. B. Mountcastle, ed.), The Johns Hopkins University Press, Baltimore, pp. 39–41.

Penfield, W., 1954, Mechanisms of voluntary movement, *Brain* **77**:1–17.

Pribram, K. H., Kruger, L., Robinson, F., and Berman, A. J., 1955–1956, The effects of precentral lesions on the behavior of monkeys, *Yale J. Biol. Med.* **28**:428–443.

Purves, D., and Lichtman, J. W., 1980, Elimination of synapses in the developing nervous system, *Science* **210**:153–157.

Randolph, M., and Semmes, J., 1974, Behavioural consequences of selective subtotal ablations in the postcentral gyrus of *Macaca mulatta*, *Brain Res.* **70**:55–70.

Rosner, B. S., 1970, Brain functions, *Annu. Rev. Psychol.* **21**:555–594.

Sahakian, B. J., 1982, The interaction of psychological and metabolic factors in the control of eating and obesity, *Scand. J. Psychol. [Suppl.]* **1**:48–60.

Schwartz, A. S., 1983, Functional relationship between somatosensory cortex and specialized afferent pathways in the monkey, *Exp. Neurol.* **79**:316–328.

Singer, W., 1982, Recovery mechanisms in the mammalian brain, in: *Dahlem Konferenzen, Life Sciences Research Reports*, Volume 24 (J. G. Nicholls, ed.), Springer-Verlag, Berlin, Heidelberg, New York, pp. 203–226.

Smith, A., and Sugar, O., 1975, Development of above normal language and intelligence 21 years after left hemispherectomy, *Neurology* **25**:813–818.

Sperry, R. W., Gazzaniga, M. S., and Bogen, J. E., 1969, Interhemispheric relationships: The neocortical commissures; syndromes of hemisphere disconnection, in: *Handbook of Clinical Neurology*, Volume 4 (P. J. Vinken and G. W. Bruyn, eds.), North-Holland, Amsterdam, pp. 273–290.

Stein, B. E., Spencer, R. F., and Edwards, S. B., 1983, Corticotectal and corticothalamic efferent projections of SIV somatosensory cortex in cat, *J. Neurophysiol.* **50**:896–909.

Strominger, N. L., 1969, Localization of sound in space after unilateral and bilateral ablation of auditory cortex, *Exp. Neurol.* **25**:521–533.

Teuber, H.-L., and Rudel, R. G., 1962, Behaviour after cerebral lesions in children and adults, *Dev. Med. Child. Neurol.* **4**:3–20.

Whitfield, I. C., Cranford, J., Ravizza, R., and Diamond, I. T., 1972, Effects of unilateral ablation of auditory cortex in cat on complex sound localization, *J. Neurophysiol.* **35**:718–731.

Wiener, N., 1961, *Cybernetics or Control and Communication in the Animal and the Machine*, 2nd ed., MIT Press, Cambridge, p. 153.

Wilson, P. J. E., 1970, Cerebral hemispherectomy for infantile hemiplegia, a report of 50 cases, *Brain* **93**:147–180.

Wilson, S. A. K., 1927, Dysaesthesiae and their neural correlates, *Brain* **50**:428–462.

Woolsey, C. N., 1984, Comparative evoked potential studies on somatosensory cortex of mammals, in: *Wenner-Gren Center International Symposium Series,* Volume 41 (C. von Euler, O. Franzén, U. Lindblom, and D. Ottoson, eds.), Macmillan, London, pp. 19–49.

11

Another Look at Vicariation

MARY D. SLAVIN, SCOTT LAURENCE, and
DONALD G. STEIN

1. VICARIATION: RELATIONSHIP TO LOCALIZATION OF FUNCTION

Vicariation has long been suggested as a mechanism for recovery of function following brain damage. Essentially, this concept involves the ability of one part of the brain to substitute for the function of another. Vicariation has been the source of much debate in neuroscience and lies at the heart of the enduring controversy known as localization of function. The idea that another area of the brain can take over the function of a damaged area following brain injury is difficult to reconcile with the concept that specific functions are located in specific areas of the brain. If there exists an isomorphic relationship between the neurological and behavioral, how could the behavior return without a parallel regrowth of the neural tissue? And if other parts of the brain also have an isomorphic relationship with other sets of behaviors, how is it possible for one or more of these neurological areas to take on new behaviors?

As a result of this dilemma, proponents of localization of function propose that the recovery observed results from resumption of normal activity in un-damaged brain tissue. This view was proposed near the turn of this century by Constatin von Monakow (1914). He theorized that acute brain injury produces "shock" or depression of activity in normal tissue around the locus of damage or even at sites distant from the trauma. Loss of function was thought to be caused by "diaschisis," and recovery was seen as the result of a "disinhibition" of

MARY D. SLAVIN and SCOTT LAURENCE • The Brain Research Laboratory, Department of Psychology, Clark University, Worcester, Massachusetts 01610.   DONALD G. STEIN • Dean of the Graduate School and Associate Provost for Research, Rutgers University, Newark, New Jersey 07102.

activity in neural pathways. Today this basis for recovery is most often explained as the resolution of trauma to the intact tissue caused by edema, the release of disruptive chemicals (neurotoxins), transneuronal degeneration, or other post-traumatic events. Following this resolution, normal functions are resumed. and there is an apparent recovery without any underlying change in brain function. The concept that recovery occurs as a result of release from diaschisis, in its various forms, suggests that the remaining cortex continues to be used as it was premorbidly, and this view is consistent with the localization-of-function perspective.

At the other end of the localization-of-function controversy is the theory that the brain, at least the cerebral cortex, exhibits equipotentiality. According to this perspective, functions are equally distributed throughout the cortex, and recovery can occur through a reorganization of the remaining brain tissue. However, the theory of equipotentiality has been difficult to support in light of neurological observations that there are fairly specific and consistent relationships between lesion sites and the deficits that accompany them. In addition, it is true that recovery does not always occur, often requires prolonged and extensive training procedures, and is often incomplete. For these reasons and others, the theory of equipotentiality does not have many proponents.

Concepts of normal brain function and theories of recovery from brain damage have been greatly influenced by the polarization of ideas concerning localization of function, and the concept of vicariation must be considered in this context. In this chapter, we look at the differences and similarities between vicariation and some other explanations of recovery and at experiments designed to investigate these theories. We examine how new theories of brain function can bridge the gap between the extreme positions of discrete localization of function and equipotentiality, give new meaning to the concept of vicariation, and consequently provide more insight into recovery of function.

2. WHAT IS RECOVERY?

One of the enduring problems faced by anyone interested in recovery of function is the criteria involved in judging what constitutes recovered behavior. First, some would require that the recovered behaviors be identical to those that are lost. Clearly, if a rigorous and complete examination of the recovered behavior reveals that it is precisely as it was prior to the neurological damage, there would be a strong challenge to an isomorphic, localizationist theory of brain function. If, indeed, there are specific behaviors mediated by local neural areas, evidence of recovery could be challenged on the grounds that the ''recovered'' behavior is simply a series of clever ''tricks'' used by the organism to accomplish its ends. Thus, in a strict structure–function paradigm, the original behavior is permanently lost along with its underlying physiology, and what is left is a clever combination of behavioral and physiological substitutes with only surface similarity to the real thing. In this context, claims that pre- and postmorbid behaviors are identical are always prey to the criticism that the recovered behavior was not analyzed in sufficient detail, that the organism's pre- and posttrauma level of learning was too low to show the deficit, which would, under greater pressure, reveal itself. Even the smallest, most clearly defined behavior—the movement of a small muscle group, the utterance of a single word—might be challenged on theoretical grounds under the assumption that greater analysis would have revealed an underlying difference from the original behavior. Thus, there is a conceptual fortress within which one might find permanent refuge from the apparent challenges of the recovered brain.

3. VICARIATION AND OTHER THEORIES OF RECOVERY

Compensation is often used as an explanation for recovery, and in light of the discussion in the previous section, it is necessary to distinguish between vicariation and compensation. From our view, vicariation is true recovery—there is an adaptation to the loss of tissue, but the organism does not merely make external adjustments to such loss. For example, following a stroke a hemiplegic may learn to walk again in a fairly normal fashion with the use of a brace. This is quite clearly a case in which there has been compensation for the loss. Other patients may be able to regain the capacity to walk unassisted only when they are quite conscious of each step, and their capabilities may vary depending on their attention to the task of walking properly. Here, the goal of "normal" walking is accomplished, but not exactly as it was before the stroke. It is most likely that new brain areas are mobilized to accomplish the task and that there is a substantial difference in the quality of movement when compared to normal locomotion. Certainly, the distinction between compensation and vicariation is not as clear in most analyses of recovered behavior; often definitions require subtle shifts in emphasis. We see vicariation as a component of the dynamic, active, and incorporative nature of human behavior in the accomplishment of a goal. In contrast, compensation is more related to reactive, coping mechanisms that are less effective in improving overall performance after brain injury.

Another position sometimes put forth to explain recovery is that within a given brain area, there is considerable "redundancy," that is, an excess or "spare set" of neurons that mediate the same functions despite significant injury to the system itself. Support comes from the observation that, under certain circumstances, there can be a considerable loss of neural tissue without any apparent loss of function. For example, sectioning of most of the optic tract in cats produces only small impairments of performance of visually guided acts of discrimination (Galambos *et al.*, 1967). In addition, recent demonstrations of multiple representations in many sensory and motor systems (see R. W. Dykes and R. Metherate, Chapter 15, this volume) may be used to support redundancy as a mechanism for recovery of function. From a localizationist perspective, one would have to argue for numerous "back-up systems" that can, when injury occurs, "spring into action" to mediate functions in exactly the same way as the tissue that was lost or damaged. However, from an evolutionary perspective such redundancy may not have much significance (Finger and Almli, 1985). However, the inadequacy of redundancy as an explanation for recovery is evident in the following quote by P. D. Wall (1976):

> The observation that a dog can run on three legs is not usually followed by the statement that one leg is redundant, and yet this is exactly the logic of most statements about recovery from brain damage (p. 363).

Wall notes that it is important to remember that the dog running on three legs has dramatically changed his motor pattern formation.

Another concept that is often evoked along with redundancy to account for recovery of function is the theory of multiple control. According to this theory, functions are mediated by several different areas in the nervous system, and when one input has been lost, the remaining intact inputs can maintain the function. Although there is considerable evidence that there are multiple centers for control of certain functions (for example, vision in cortex and superior colliculus), this explanation fails to consider the effect of nervous system lesions on these other centers. An experiment by Sprague (1966) provides an interesting demonstration of a way in which systems involved in multiple control can affect each other. Cats with persistent hemianopia up to 1 year following unilateral visual cortex lesions showed a marked improvement immediately following subsequent removal of the contralateral superior colliculus. These results were taken to indicate that following the unilateral visual cortex lesion the contralateral superior colliculus had an inhibitory effect on the intact ipsilateral superior colliculus, and removal allowed for a return of function and consequent recovery.

Additional evidence for the fact that there is a new organization may be derived from studies demonstrating the existence of latent synapses following removal of one input. In one study with cats, Dostrovsky *et al.* (1976) mapped the receptive fields of the cells projecting from the body surface to the gracile and cuneate nuclei of the medulla. These workers carefully mapped those cells that responded most effectively to stimulation of the hindpaw. They then cooled the spinal cord at the lumbar level and, by doing so, completely blocked any response to stimulation of the paw during the cooling period. However, within a few minutes, and recording from the same neurons, these researchers were able to locate totally new receptive fields on the abdomen. When the cord was returned to normal by removal of the ice, the original receptive field reemerged, and the one on the abdomen disappeared. Because of the rapidity and temporary nature of these changes, it seems most likely that "silent" pathways were activated when the primary pathways that normally control activity in certain nuclei of the brain are temporarily (or permanently) suppressed.

More recently Merzenich *et al.* (1983), using monkeys, cut the median nerve, which projects from the medial portion of the hand to area 3b of the somatosensory cortex, and recorded the activity of individual neurons in the cortex in response to tactile stimulation. Immediately after nerve section, most neurons were unresponsive, but some neurons were activated by stimulation of the dorsal surface of the first two fingers—stimulation that under normal conditions did not activate these cells. Over time the initially unmasked inputs grew in extent until all of the dorsal surface of the first two digits was represented in detail. Analysis of the sequence of changes that occurred over time revealed a very interesting phenomenon. Throughout the expansion of the unmasked inputs, the internal topography of this new representation was maintained, and cortical sites were seen to shift in location hundreds of micrometers across the cortex. The authors drew the following conclusions:

> Relative stability of maps is a consequence of balanced competition. Changing the balance results in an orderly sequence of change that produces a new balance and a new organization. It should be possible to change this balance not only by removing an input, but also by increasing or decreasing the levels of activity produced by given skin surfaces, or by a given set of receptors. Thus, somatosensory cortical fields may be subject to alteration related to increased sensory use. Learning a new sensory-motor skill could actually modify a cortical map, and the modifications could be responsible for the increased skill (p. 436).

This study has enormous significance because it demonstrates that the unmasked inputs are not merely a stable representation of secondary underlying connections. A model in which unmasked inputs sprout to contact the deafferented area would account for the expansion of the representation but would not explain the reorganization that was observed over time.

With the initial demonstration by Raisman (1969) that there is sprouting of intact neurons following central nervous system damage, there has been much emphasis on this process as a mechanism for recovery of function. However, most evidence for sprouting as a mechanism for recovery of function merely relies on the fact that these events follow a similar time course (Loesche and Steward, 1977). Although experiments have been successful in determining the conditions under which there is anatomic or biochemical evidence of sprouting in the brain, there has been little evidence directly linking these phenomena to recovery of function. For example, studies have demonstrated that following a unilateral entorhinal cortex lesion there is sprouting of fibers from the dorsal psalterium, a fiber bundle that carries decussating fibers from the intact entorhinal cortex to the hippocampus (Ramirez and Stein, 1984). If this sprouting were responsible for the recovered function, the deficit should be reinstated following a secondary lesion of the sprouted fibers. However, when rats with unilateral lesions of the entorhinal cortex were allowed to recover on a spatial alternation task, severing the fibers of the dorsal psalterium did not result in any degradation of performance. Since animals with bilateral lesions do not recover (Ramirez and Stein, 1984; Scheff and Cotman, 1977), it is likely that the intact entorhinal cortex plays a role in recovery through some process that does not require input from the sprouted fibers. Alternatively, the sprouting could be said to initiate the recovery process, but the anomalous growth might not be essential to maintain the recovered behavior once it has occurred.

Through this brief review of some of the different theories for recovery, it becomes apparent that no theory by itself at this time provides a completely adequate explanation. Static theories such as redundancy and multiple control propose that recovery occurs as a result of a continued functioning of the remaining intact brain tissue. However, these theories do not address the fact that following injury there is a significant alteration in the function of this tissue. Other theories such as reactive synaptogenesis suggest that the dynamic changes that occur in response to injury are responsible for recovery. However, evidence for the role of these specific changes in recovery is at best circumstantial.

4. ATTEMPTS TO LOCATE RECOVERED FUNCTION

Hermann Munk (1839–1912) is most often associated with introducing the notion that one brain region can take over the function of a damaged area. As a result of his skilled surgery and careful postoperative testing, he was able to observe not only the loss of function that was associated with visual cortex lesions but the subsequent recovery (Benton, 1978). He described a condition that he termed "mind blindness" in which animals could see but were unable to recognize familiar people and objects. He noted that this was a temporary condition lasting a few weeks and proposed that new images were laid down in other undamaged areas of the occipital cortex. Thus, we can see that the concept of vicariation is at least as old as the observation of recovery of function. Attempts to validate or refute vicariation have relied on the demonstration that functions lost as a result of lesions are localized in another area following recovery. The experiments described in this section have used the following procedure. A lesion is made with a subsequent loss of function, and following recovery another area of the brain is damaged. If a reinstatement of the deficit occurs, it is presumed that the secondarily damaged area took over the function of the initially injured site and was responsible for the recovery.

In 1917, Leyton and Sherrington removed the motor area for the upper extremity of a chimpanzee and observed paralysis of the limb with a subsequent recovery of motor functions. In this animal, the lesion was extended, and there was no loss of the recovered motor function. In another animal, the same procedure was repeated with the additional removal of the corresponding contralateral motor area, and again, no loss of the recovered motor function was observed. Thus, recovery was observed, but attempts to locate the specific area that contributed to this function were unsuccessful.

Recently, other investigators have been more effective in identifying the possible role of the remaining intact tissue in recovery. A series of elegant experiments on recovery of visual discrimination following visual cortex lesions has identified a specific area that may very well be involved in the recovery. Baumann and Spear (1977) found that cats with visual cortex lesions (areas 17, 18, and 19) lost the ability to perform tasks requiring brightness, form, and pattern discriminations that had been learned preoperatively. These animals recovered to preoperative levels of performance, but when secondary lesions of the lateral suprasylvian (LS) gyrus were made, the recovered function was once again lost, and relearning was more prolonged than following damage to the visual cortex alone. It is important to note that lesions to the lateral suprasylvian gyrus in animals with intact visual cortices did not result in any alteration in visual performance (Spear et al., 1983). This study appears to identify a specific area in which the recovered function is now located, thus providing support for the theory of vicariation. In two other studies, however, the notion of vicariation is challenged by a lack of specific changes in the areas that were proposed as mediating the recovered function. When Spear and Baumann (1979a) attempted to determine if the recovery occurred through changes in the response properties of the neurons in the LS, they found that receptive field properties of neurons in the LS examined from a few hours after visual cortex removal to up to 5 weeks after the injury showed no change. Moreover, there was no difference in the response properties of neurons in cats that had recovered the capacity for brightness, form, and pattern discrimination when compared to those that had not (Spear and Baumann, 1979b).

The area responsible for recovery of function following sensorimotor cortex damage was investigated in a similar study by Glassman (1971). Unilateral electrolytic lesions were made in the sensorimotor cortex at multiple points, and there was a subsequent deficit in the use of the contralateral forelimb in a food retrieval task. This function was recovered in the same time period during which there was a return to normal measurements of somatosensory evoked potentials and movements generated by brain stimulation. These results and those discussed in the preceding paragraph indicate that mediation of recovered function may not occur as the result of physiological alterations in the area proposed to mediate the recovery. Such results are interpreted as supporting the notion that recovery occurs through the use spared neural tissue. A local reorganizational process such as sprouting would be necessary to change the response properties of cells, and these studies provide strong evidence against the involvement of such a process.

However, Spear and Baumann (1979a) emphasize that the response properties of the LS gyrus neurons do not return to their preoperative state even after recovery. Therefore, though the neurons in this area were spared, they are certainly not functioning as they do in the normal intact brain. In addition, the results do not give any indication of changes in the response of the visual system that may have occurred at other levels. It is known that visual cortex damage results in changes in response properties at many levels of the nervous system, including the retina, thalamus, dorsolateral geniculate nucleus, and superior colliculus. Therefore, it may be misleading to state that recovery occurs merely by resumption of activity in the remaining undamaged portion of the system. The effects of the initial lesion are widespread, and certainly the resumption of neuronal function involves activity of a nervous system with a new organization, be it neurochemical, structural, or behavioral.

A more recent study suggests that there may be age differences in the capacity for changing response properties. Spear and his colleagues (1984) removed areas 17, 18, and 19 in newborn cats and compared them on brightness and form discrimination to cats with similar lesions made at maturity. The animals with early brain damage were able to learn both the brightness and two-choice pattern discrimination tasks, whereas those given lesions as adults remained very impaired. Although the neonatal cats showed greater transneuronal degeneration of the retinal ganglion cells than adults, they also evidenced a marked increase in the number of anomalous projections to surviving neurons in the lateral geniculate nucleus. Single-unit recordings of neurons in the lateral geniculate nucleus were also very similar to those seen in normal cats. In turn, the neurons in the lateral geniculate nucleus showed a tenfold increase in the number of anomalous projections to the suprasylvian gyrus, an event that is never seen after lesions in adults. The neurons in the suprasylvian gyrus that received the anomalous inputs had normal receptive field characteristics when the lesions of visual cortex were made in neonates. In contrast, animals given lesions as adults never developed normal receptive fields in this area of the brain.

In the very young animals, alteration of response properties and behavioral recovery may have resulted from the preservation of neurons that would ordinarily have "died back" as development proceeded. However, following early lesions the projections survived and expanded into new (anomalous) target sites in the suprasylvian gyrus. The new innervation led to physiological activity practically identical to normal neurons in the visual cortex and to almost normal restoration of pattern and form discrimination.

Other experiments have used behavioral manipulations to support the use of spared neural tissue as a mechanism for recovery of function. The difference between the process of sparing and a process that would be involved in vicariation is evident in LeVere's (1975) conclusion that ". . . recovery of function following brain damage does not reflect the functional reorganization of neural tissue but rather the survival of the neural tissue that mediated the behavior patterns prior to the neurological lesion" (p. 345). In an experiment by LeVere and Morlock (1973), rats with posterior neocortex lesions that were trained on the reversal of an original black–white discrimination task took a longer time to reach criterion than those retrained on the original problem. These results were interpreted as providing support for the theory that some portion of the learned behavior remains following damage and is capable of influencing recovery.

Additional support for sparing is provided by an experiment showing that rats trained on visual discrimination prior to posterior neocortex lesions showed improved performance when given d-amphetamine postoperatively, even though preoperative administration of the drug did not result in improved performance (Jonason *et al.*, 1970). This study also indicated that some portion of the damaged system remained intact and that some element of the preoperative training persisted after the injury was responsible for the observed recovery.

However, other studies demonstrate that the process underlying recovery is more complicated. In one study (Davis and LeVere, 1982), rats were pre-operatively trained on a simultaneous two-choice brightness discrimination task, and following striate decortication some of the animals were required to demon-strate the functionality of the visual system in an operant chamber prior to postoperative testing. It was thought that this group would perform better since "forcing the decorticate rat to behave on the basis of visual cues should serve to reinstate any specific visual memories spared by the visual lesion" (p. 68). In fact, this group did not perform significantly better. The theory that recovery occurs through the continued function of spared tissue is not supported by this experiment. Establishing the functionality of this tissue is not a sufficient condi-tion for recovery. The results also shed light on the dilemma of the studies on adult cats by Spear and Baumann (1979a,b) cited in the beginning of this section, in which there was an area that was clearly critical to the recovered function although no specific changes in the physiological response of this area was evident. Perhaps there are specific areas critical to recovery, but the recovered behavior is not now "located" is these areas. It may be more appropriate to consider that it is the use of a system in conjunction with the rest of the nervous system that is required for the establishment of the disrupted behavior.

A study by Scheff and Wright (1977) indicates that behavioral experiences can have direct effects on the spared system. In this study, rats were trained on a black–white discrimination T maze for food reward prior to removal of the posterior neocortex in two stages. About half of the group received training between the surgeries. Though all animals learned the task postoperatively, only those animals with interoperative training were able to relearn the task with fewer trials than were required with initial training. In addition, the animals with interoperative training showed evoked responses in 70% of the cortex contiguous to, but outside of, the visual area as compared to 24–39% evoked response in the animals without interoperative training. In conclusion, although it is likely that recovery depends on the continued function of some spared component of the damaged system, there is also evidence that sparing of neural tissue by itself is not a sufficient condition for recovery. There is something else that is also critical: the postoperative experiences influence the interaction of the spared neural tissue with the rest of the nervous system.

Several clinical cases lend indirect support to the notion that recovery can occur through vicariation. Following removal of the temporal lobe of the dominant hemisphere as a treatment for epilepsy, patients demonstrated an impairment on auditory verbal paired-associate learning tasks. Long-term studies of these patients show that there was no change in the first 3 years postoperatively but after that time there was rapid improvement and that after the fifth year the mean performance reached preoperative levels (Blakemore and Falconer, 1967). Significant long-term changes in performance were also reported in a single case study of a commissurotomy patient. Initially, following commissurotomy for an intractable seizure disorder, the patient demonstrated an inability to verbally identify stimuli presented to the right hemisphere. However, there was considerable and steady improvement in language functions over a $3\frac{1}{2}$-year period of testing (Gazzaniga *et al.*, 1979). Norman Geschwind (1973) describes one case of an aphasic who showed no improvement in speech function between 3 months and 1 year who had completely recovered 1 year later without specific treatment. In another case, a patient had an aphasia that was still evident 6 years after onset but had substantial improvement after 18 years, and postmortem examination revealed a complete destruction of the classical speech area.

These studies indicate that there may be an underlying process involving some form of reorganization that is accomplished over very long periods of time and can provide a basis for recovery. Perhaps new areas of the brain may be incorporated into the accomplishment of a task much in the same manner that occurs in initial acquisition of complex functions through development. Rosner (1970) suggested that the time needed for recovery "increases with the difficulty or complexity of behavior, the degree of involvement of relevant tissue, and perhaps with the life span of the species" (p. 584). Andrew Kertesz (Chapter 19, this volume) discusses vicariation and compensation in terms of right hemisphere "takeover" after damage to left hemisphere areas thought to mediate language, and his chapter should be consulted for a thoughtful analysis of this problem.

5. CHANGING CONCEPTS OF BRAIN FUNCTION AND ANOTHER LOOK AT VICARIATION

Much of the objection to the notion of vicariation comes from the remnants of localizationist thinking. If functions are localized in specific areas, the relocalization of a function in a new area of the brain would require either radical reorganization of anatomic structure following injury or major "unblocking" of extensive "latent" pathways in regions thought to mediate different aspects of behavior. This sort of reorganization is not observed in adults. However, there is no categorical definition of vicariation that demands such reorganization. The notion that vicariation may occur in a nonlocalized way is well expressed in the following quote by Karl Lashley (1929):

> There is much incorrect speculation in the older literature concerning the parts functioning vicariously, as the assumption that the precentral gyrus of one side can assume

the functions of the other; but there is no certain evidence that the reacquired functions
are carried out vicariously by any specific loci. Attempts to discover such loci have
been in almost all cases fruitless (Lashley, 1922) and point rather to a reorganization
of the entire neural mass than to an action of specific areas (p. 25).

In addition, there is no evidence to suggest that completely unrelated areas of the
brain take over the function of the damaged area. However, there is no reason
that vicariation must necessarily be interpreted in this fashion. Even Lashley
(1938), who is identified with the notion of equipotentiality, maintained that
"there is not an unlimited capacity for vicarious function but a limitation to the
system which is more or less directly concerned with the same function under
normal conditions" (p. 741).

The idea that another area of the brain can take over the function of a
damaged area may be better understood when brain function is viewed from a
systems approach. The eminent neurologist John Hughlings Jackson (Jackson,
1939) first presented such an approach for understanding brain function. He
observed that lesions of a circumscribed area of the brain never led to a complete
loss of function. For example, a patient may not be able to speak a word on
command voluntarily yet may do so spontaneously. He proposed a model for
brain function based on a hierarchical organization of the brain rather than
specific centers.

Recently, the notion of hierarchical organization has been enhanced by
Vernon Mountcastle's articulation of a distributed function theory as expressed
in the following quote (1977):

Developments of recent decades require new formulations that include but transcend
the hierarchical principle of brain organization. Prominent among them is the concept
that the brain is a complex of widely and reciprocally interconnected systems and that
the dynamic interplay of neural activity within and between these systems is the very
essence of brain function (p. 21).

According to this theory, a function is not located in any one of the parts of the
system but "is a property of the dynamic activity of the system: it resides in the
system as such" (p. 38). In addition, it is proposed that it is normal for information
to flow through such a distributed system by following a number of different paths,
and the dominance of one pathway or another may be a dynamic changing
property. A similar view has been proposed by Luria (1966), who noted that "no
formation of the central nervous system is responsible solely for a simple function;
under certain conditions, a given formation may be involved in other functional
systems and may participate in the performance of other tasks (p. 27)."

A study by John et al. (1986) provides empirical support for the notion of
such widely distributed systems. Split-brain cats were trained on a visual discrimination task using illuminated geometric figures. Following training, red and
green contact lenses were used in combination with red and green illuminated
stimuli to localize stimuli selectively to one hemisphere. The animals were then
given injections of ^{14}C-labeled 2-fluorodeoxyglucose and continued the visual

discrimination activity. After a 1-hr interval, the animals received an injection of ^{18}F-labeled 2-fluorodeoxyglucose and performed the task at a random level without any learned cue information. This experimental design allowed one hemisphere to be used as a control for nonspecific stimuli related to the task while the other hemisphere received input relevant to the learned task. In this manner, all increases in activity that were unrelated to the task could be identified and eliminated.

The results demonstrated that the hemisphere that had received learned cue information had a marked increase in uptake of the radioactive label. When the effects of nonspecific stimuli were eliminated, it was estimated that 5 to 100 million neurons were involved in the aspect of the task that involved learned cue information. The significance of these findings to a theory of brain function is evident in the following quote (John *et al.*, 1986):

> The assumption in current efforts to localize plastic changes with learning is that those changes occur in discrete pathways. It is further assumed that after learning, augmented neural firing in those discrete pathways represents each specific memory, no matter how redundant or anatomically distributed such firing might be. The vast number and anatomical extensiveness of the neurons here shown to participate in representation of a single and rather simple memory is hard to reconcile with these assumptions. So many neurons seem to be involved in the mediation of one specific learned discrimination that there simply are not enough neurons available to represent any reasonable store of memories with this degree of redundancy (p. 1174).

The evidence presented in this section points toward a new direction in concepts of brain function that hopefully transcends the localization-of-function controversy. Articulation of concepts of normal brain function is critical because recovery from brain damage must be viewed within the context of normal neural, behavioral, and psychological processes. The ability to accomplish a behavior in a new way in normal subjects or following brain injury is indicative of the goal-directed nature of organisms. Indeed, if following brain injury new areas of the brain that were not normally used are now mobilized in the accomplishment of goals, this could be used as an argument for recovery occurring through vicariation. As noted by Mountcastle (1977), "the remarkable capacity for improvement of function after partial brain lesions is viewed as evidence for the adaptive capacity of such distributed systems to achieve a goal, albeit slowly and with error, with the remaining neural apparatus" (p. 8). It is important to note that the suggestion here is that the entire neural apparatus is involved in recovery, not merely one specific area that was spared from damage.

The many theories proposed to account for recovery all emphasize a specific event and have not been shown experimentally to provide a sufficient explanation. It is this search for the "something else" that underlies recovery of function that makes the theory of vicariation so important to consider. Looking at recovery in terms of vicariation requires a significant change in perspective. It requires that we look toward behavioral and clinical models that require analysis of the response of the whole organism to injury rather than the response of a small group of cells. It also requires that changes occurring in other areas of the nervous system in response to injury be considered as having a role in the recovery of function. We must look at the new organization that emerges in the entire nervous system as a result of damage and relate that to recovery processes.

6. CONCLUSIONS

The use of the term vicariation to explain recovery following brain damage may be subject to the criticism of being vague and failing to identify a specific mechanism that can be experimentally tested. However, none of the specific mechanisms that have been proposed thus far have been demonstrated to be essential to recovery. Vicariation cannot be considered as one of many theories or concepts that compete for an explanation of recovery. Rather, it cuts across most of these approaches. Whether or not the term itself survives, the fundamental concept it serves is required if we are to tie these seemingly disparate approaches together. In short, most hypotheses concerning recovery of function reflect the disciplines that have spawned the original research, and what is required is an approach that encompasses the vast range of perspectives from neurochemistry to psychiatry.

Does recovery have a legitimate status as a separable field of study, or is it simply a convenient rubric for a fundamentally disparate but apparently related set of phenomena? And if recovery does in fact refer to a definable process, is there a limited set of concepts that can explain and promote it? Here we have attempted to answer these questions by defining recovery as a set of vicarious operations. In so doing, we see vicariation not as one among a number of competing explanations of recovery but rather as a unifying concept consistent with the known complexities of human motivation, emotion, learning, and cognition. We therefore place the underlying physiological and biochemical mechanisms that accompany or cause recovery within a conceptual context. We contend that by placing these mechanisms within the overall explanatory umbrella of vicariation, their true contribution to recovery can be better understood and promoted. More explicitly, we see vicariation as an extension of the fundamental distinguishing characteristic of the human brain; the ability to use symbolic operations to accomplish invariant goals despite significant changes in both internal and external conditions.

As an approach to recovery, vicariation takes for granted the goal-oriented nature of human behavior and accepts this as an organizing principle around which the manifold events of behavior and cognition unfold. It incorporates physiological theory but does not compete with it. It allows for intervention at the level of behavior and seeks to define recovery at this same level. Vicariation suggests that losses following brain damage are real and, in the most specific of

178 MARY D. SLAVIN et al.

ways, irreplaceable. Nevertheless, the capacity to develop alternative strategies is a normal part of neural operations, and recovery following brain damage is one more manifestation of that ability.

REFERENCES

Baumann, T. P., and Spear, P. D., 1977, Role of the lateral suprasylvian visual area in behavioral recovery from effects of visual cortex damage in cats, *Brain Res.* **138**:445–468.

Benton, A., 1978, The interplay of experimental and clinical approaches in brain lesion research, in: *Recovery from Brain Damage* (S. Finger, ed.), Plenum Press, New York, pp. 49–68.

Blakemore, C. B., and Falconer. M. A., 1967, Long-term effects of anterior temporal lobectomy on certain cognitive functions, *J. Neurol. Neurosurg. Psychiatry* **30**:364–367.

Davis, N., and LeVere, T. E., 1982, Recovery of function after brain damage: The question of individual behaviors or functionality, *Exp. Neurol.* **75**:68–78.

Dostrovsky, J. O., Millar, J., and Wall, P. D., 1976, The immediate shift of afferent drive of dorsal column nucleus cells following deafferentation: A comparison of acute and chronic deafferentation in gracile nucleus and spinal cord, *Exp. Neurol.* **52**:480–495.

Finger, S., and Almli, C. R., 1985, Brain damage and neuroplasticity: Mechanisms of recovery or development? *Brain Res. Rev.* **10**:177–186.

Galambos, R., Norton, T. T., and Frommer, G. P., 1967, Optic tract lesions sparing pattern vision in cats, *Exp. Neurol.* **18**:8–25.

Gazzaniga, M. S., Volpe, B. T., Smylie, C. S., Wilson, D. H., and Le Doux, J. E., 1979, Plasticity in speech organization following commissurotomy, *Brain* **102**:805–815.

Geschwind, N., 1974, Late changes in the nervous system: An overview, in: *Plasticity and Recovery of Function in the Nervous System* (D. G. Stein, J. J. Rosen, and N. Butters, eds.), Academic Press, New York. pp. 467–508.

Glassman, R. B., 1971, Recovery following sensorimotor cortical damage: Evoked potentials, brain stimulation and motor control, *Exp. Neurol.* **33**:16–29.

Jackson, J. H., 1939, *Selected Writings* (J. Taylor, ed.), Hodder and Stoughton, London.

John, E. R., Tang, Y., Brill, A. B., Young, R., and Ono, K., 1986, Double-labeled metabolic maps of memory, *Science* **233**:1167–1175.

Jonason, K. R., Lauber, S. M., Robbins, M. J., Meyer, P. M., and Meyer, D. R., 1970, Effects of amphetamine on relearning pattern and black–white discriminations following neocortical lesions in rats, *J. Comp. Physiol. Psychol.* **73**:47–55.

Lashley, K. S., 1929, *Brain Mechanisms and Intelligence: A Quantitative Study of Injuries to the Brain*, University of Chicago Press, Chicago.

Lashley, K. S., 1938, Factors limiting recovery after central nervous system lesions, *J. Nerv. Ment. Dis.* **88**:733–755.

LeVere, T. E., 1975, Neural stability, sparing, and behavioral recovery following brain damage, *Psychol. Rev.* **82**:344–358.

LeVere, T. E., and Morlock, G. W., 1973, Nature of visual recovery following posterior neodecortication in the hooded rat, *J. Comp. Physio. Psych.*, **83**:62–67.

Leyton, A. S. F., and Sherrington, C. S., 1917, Observations on the excitable cortex of the chimpanzee, orangutan, and gorilla, *Q. J. Exp. Physiol.* **11**:135–222.

Loesche, J., and Steward, O., 1977, Behavioral correlates of denervation and reinnervation of the hippocampal formation of the rat: Recovery of alternation performance following unilateral entorhinal cortex lesions, *Brain Res. Bull.* **2**:31–39.

Luria, A. R., 1966, *Human Brain and Psychological Processes*, Harper & Row, New York.

Merzenich, M. M., and Kaas, J. H., 1982, Reorganization of mammalian somatosensory cortex following peripheral nerve injury, *Trends Neurosci.* December: 434–436.

Mountcastle, V. B., 1977, An organizing principle for cerebral function: The unit module and the distributed system, in: *The Neurosciences Fourth Study Program* (F. O. Schmitt and F. G. Worden, eds.), MIT Press, Cambridge, pp. 21–42.

Raisman, G., 1969, Neuronal plasticity in the septal nuclei of the adult rat, *Brain Res.* **14**:25–48.

Ramirez, J. J., and Stein, D. G., 1984, Sparing and recovery of spatial alternation performance after entorhinal cortex lesions in rats, *Behav. Brain Res.* **13**:53–61.

Rosner, B. S., 1970, Brain functions, *Annu. Rev. Psychol.* **21**:555–594.

Scheff, S. W., and Cotman, C. W., 1977, Recovery of spontaneous alternation following lesions of entorhinal cortex in adult rats: Possible correlation to axon sprouting, *Behav. Biol.* **21**:286–293.

Scheff, S. W., and Wright, D. C., 1977, Behavioral and electrophysiological evidence for cortical reorganization of function following serial lesions of the visual cortex, *Physiol. Psychol.* **5**:103–107.

Spear, P. D., 1984, Consequences of early visual cortex damage in cats, in: *Early Brain Damage,* Volume 2 (S. Finger and C. R. Almli, eds.), Academic Press, New York. pp. 229–252.

Spear, P. D., and Baumann, T. P., 1979a, Effects of visual cortex- removal on receptive-field properties of neurons in lateral suprasylvian visual area of the cat, *J. Neurophysiol.* **42**(1):31–56.

Spear, P. D., and Baumann, T. P., 1979b, Neurophysiological mechanisms of recovery from visual cortex damage in cats: Properties of lateral suprasylvian visual area neurons following behavioral recovery, *Exp. Brain Res.* **35**:177–192.

Spear, P. D., Millar, S., and Ohman, L., 1983, Effects of lateral suprasylvian visual cortex lesions on visual localization, discrimination, and attention in cats, *Behav. Brain Res.* **10**:339–359.

Sprague, J. M., 1966, Interaction of cortex and superior colliculus in mediation of visually guided behavior in the cat, *Science* **153**:1544–1547.

von Monokow, C., 1914, Die lokalisation im grosshirn und der abbau der funktion durch kortikale herde, J. F. Bergmann, Wiesbaden. Translated and excerpted by G. Harris in: *Mood, States and Mind,* K. H. Pribram (ed.), Penguin, London, 1969, pp. 27–37.

Wall, P. D., 1976, Plasticity in the adult mammalian central nervous system, *Prog. Brain Res.* **45**:359–379.

12

Hughlings Jackson's Theory of Localization and Compensation

SAMUEL H. GREENBLATT

1. INTRODUCTION

Hughlings Jackson might well have objected to the title of this book, but not because he eschewed either theory or controversy. Rather, I think he would have taken issue with the implications of the word "recovery." It seems to imply the theoretical possibility of complete recovery; this he denied, at least at the physiological level in adults. "Compensation," on the other hand, implies that one part may substitute for another to a greater or lesser degree. This concept arose almost inevitably from the theoretical base that he constructed for all of modern neurology.

2. SOME ASPECTS OF JACKSON'S ROLE IN THE DEVELOPMENT OF MODERN NEUROLOGY

Clinical neurology as we know it began in the 1860s with the work of Jean-Martin Charcot (1825–1893) in Paris (Guillain, 1959) and John Hughlings Jackson (1835–1911) in London (Clarke, 1973). Charcot was the more predominant figure in his own day, but Jackson's influence has grown steadily with the passage of time, probably because he worked more explicitly at the theoretical foundations of his undertaking. Jackson wove the 19th century themes of associationist psychology and evolutionary biology into a conceptual framework that

SAMUEL H. GREENBLATT • Department of Neurological Surgery, Medical College of Ohio, Toledo, Ohio 43699.

is still being explored and utilized (Greenblatt, 1965, 1970, 1977). The very presence of this chapter in this book is evidence of this continuing historical trend.

Among the many aspects of Jackson's theoretical contributions to modern neurology, only four of his most basic ideas require explication as background for compensation: (1) "graded" localization (the adjective is mine, not Jackson's), (2) evolution and dissolution in the nervous system, (3) the hierarchical organization of the nervous system, "levels" in Jacksonian terms, and (4) the idea of "positive" and "negative" lesions (the adjectives are Jackson's). Jackson developed his earliest concept of localization in the 1860s and published it in 1870 (Jackson, 1870). However, he did not derive a full-blown theory of compensation from it until later in the 1870s. By that time, the other ideas just mentioned had already been worked out. Accordingly, I provide brief discussions of the latter three ideas before returning to localization and compensation.

During the 1860s Jackson came under the increasingly strong influence of the pre-Darwinian evolutionist Herbert Spencer (1820–1903). Spencer's concept of biological evolution was essentially similar to Darwin's. However, it lacked the power of Darwinian theory, because Spencer did not see the importance of natural selection (survival of the fittest) until after Darwin had proposed this mechanism. On the other hand, Spencer applied evolutionary principles much more widely than Darwin. For our purposes, this was especially important in the physiological realm. In the course of Jackson's intellectual development, Spencer's emphasis on the basically sensorimotor nature of the central nervous system reinforced the earlier influence of one of Jackson's teachers, Thomas Laycock (1812–1876). Laycock had proposed that the entire neuraxis functions on the "reflex principle," i.e., according to the sensorimotor paradigm (Laycock, 1845).

Building on the ideas of Laycock and Spencer, Jackson developed a theory of "evolution and dissolution" in the nervous system. Eventually he came to a tripartite conception of three hierarchical "levels." Although he was somewhat vague about the details, the lowest level included, on the motor side, the anterior spinal nuclei, the oculomotor nuclei, and other brainstem structures associated with vegetative functions (Jackson, 1958, pp. 41–42). At the middle level, the details are less specified, but it certainly includes what we would now think of as the motor system above the anterior horn cell as well as the basal ganglia. The highest level would be roughly comparable to what we now call the prefrontal areas (Jackson, 1958, p. 99). Physiologically, Jackson thought of the lowest level as "representing" certain parts or functions of the body. The middle level "re-re-presented" these parts neurologically, and the highest level "re-re-represented" them (Jackson, 1958, p. 99). Using Spencerian concepts, he stated that the lowest level is the least differentiated and least specialized, whereas the highest level is most differentiated and most specialized. From his hierarchical theories, Jackson then drew some implications that are still part and parcel of contemporary neurological thought.

"Dissolution" is a Spencerian term that Jackson used to denote loss (or regression) of function in his hierarchically organized system. It is immediately obvious that this idea implies a notion of release of function in lower levels, i.e., disinhibition. Jackson drew this conclusion quite explicitly (Jackson, 1958, p. 30). What he did not point out clearly, to the best of my knowledge, is another implication that follows from his hierarchical theory: since each level is unique in its degree of specialization, its functions cannot really substitute for the related functions in another level. In other words, in Jackson's hierarchically organized nervous system, exactly equivalent functions are not available to be released by disinhibition of a lower level. For that matter, neither can a higher level substitute for the absent functions of a lower one. However, within a given level, compensation may be available for the dissolution that results from a "positive" or a "negative" lesion.

The idea of "positive" and "negative" lesions is another of Jackson's fundamental contributions to contemporary neurological thought. Like his theory of localization, it arose from his lifelong interest in epilepsy. The seizure itself is indicative of the existence of a "positive" lesion—abnormal hyperfunction. On the other hand, the Todd's paralysis that often follows a seizure indicates an absence or hypofunction of nervous tissue. Jackson was also interested in how the nervous system compensates for negative lesions caused by destruction of tissue. He was one of the first neurologists who understood the phenomenon of embolization in a vascular distribution, specifically the middle cerebral artery (Greenblatt, 1970). But stroke and experimental ablations were not his real models. In addition to seizures and movement disorders, syphilitic disease and mass lesions were much more common in his time and in his own practice. In essence, the time frame in which he thought about positive and negative lesions was shorter (minutes to weeks) than that of our contemporary models of recovery, both clinical and experimental (days to years).

3. JACKSON'S THEORY OF LOCALIZATION AND ITS DERIVATIVE: COMPENSATION

In 1873, David Ferrier (1843–1928) confirmed Jackson's original ideas about motor localization by performing electrical stimulation experiments on the cortices of monkeys (Ferrier, 1873). Both Jackson and Ferrier continued their work throughout the 1870s and beyond, but the clinical and experimental traditions diverged to some extent (Greenblatt, 1984). The latter became more strictly localizationist through the laboratory techniques of stimulation, ablation, and cytoarchitectonics. At the conceptual level, Jackson's theory of localization actually contained a much more subtle dynamism than its experimentally derived counterparts. But the nuances of his theory were lost to most of his contemporaries, because the subtleties were swept aside in the excitement of the experimentalists' heyday. Jackson himself tried to steer a middle course by declaring that he was neither a "universalist" (holist) nor a "localiser" (Jackson, 1958, pp. 33–34), but to no avail. Despite the acknowledged brilliance of his theories, they were almost entirely clinically based and therefore not confirmable by the relatively crude physiological methodologies of his time.

Nonetheless, by the later 1870s and 1880s, Jackson had a major standing among his contemporaries in Britain and abroad. He gave several prestigious lectures and published many rather lengthy papers on epilepsy and on evolution and dissolution in the nervous system. One of his more succinct contributions during this time of his ascendency was published in 1882 and 1883, "On some implications of dissolution of the nervous system" (Jackson, 1958, pp. 29–44). One of the dissolution's "implications" was compensation, whose explication also required a substantial discussion of his localization theory. This paper of 1882–1883 is used as the textual vehicle for my succeeding discussion of Jacksonian localization and compensation. In summarizing the paper, I ignore its rather extensive references to the "positive" lesions of epilepsy. Jackson's ideas in that area were very important in the historical development of his localization theory, but they are not essential to an understanding of his thoughts on compensation for negative lesions.

Near the beginning of his paper, Jackson stated 15 basic propositions about localization and compensation. Among the most important of these are the following, quoted directly:

1. Nervous centres represent movements, not muscles. From negative lesions of motor centres there is not paralysis of muscles, but loss of movements. . . .

2. There is no localization in the sense that every part of a centre represents an external muscular region in the same way as all other parts of the centre do. There is, to illustrate, no cerebral centre every part of which represents the arm and leg equally.

3. There is no localisation in the sense that each part of a centre . . . represents exclusively any limited region. To illustrate, there is no cerebral centre for the arm only. . . .

4. Each part of a centre represents the whole of a muscular region, and each part of it represents the whole region differently. Thus, in a hypothetical centre representing the face, arm, and leg, . . . in one part the face is most represented, in another the arm, and in another the leg.

5. The higher the centre the more numerous, different, and more complex, and more special movements it represents, and the wider region it represents—evolution. The highest centres represent innumerable, most complex, and most special movements of the whole organism, and in accord with Number 4 each unit of them represents the whole organism differently. In consequence, the higher the centre the more numerous, different,

complex, and special movements of a wider region are lost from a negative lesion of equal volume—dissolution.

6. It is an obvious corollary from the preceding that the higher the centre, whilst the wider spread the paralysis, the less in degree is it in any one part from a negative lesion of equal volume.

7. There is compensation, never absolute, in cases of partial destructive lesions of centres.

8. From increasing extent of negative lesions paralysis increases in compound degree. There is more paralysis of the parts first affected, and there is greater range of paralysis.

10. We have, however, to consider not only the extent of the destruction, but the rapidity of the destruction, the "gravity" of negative lesions. The more rapid the destruction of part of a centre, the wider spread is the paralysis, some of which is temporary.

15. The higher nervous arrangements inhibit (or control) their lower, and thus, when the higher are suddenly rendered functionless, the lower rise in activity (Jackson, 1958, pp. 29–30).

Before proceeding with Jackson's further formulation of the foregoing ideas, it is necessary to define a key term: "centre." Histologically, a center is a (presumably) macroscopic aggregate of neural tissue—gray matter. Since the neuron theory was not formulated by Waldeyer until 1891 and confirmed physiologically thereafter by Sherrington (Swazey, 1969), all of Jackson's theories presupposed that neural tissue was composed of cell bodies and separate "fibres." Physiologically, of course, Jackson has just given his own "associationist" definition of a center, as distinct from the "faculty" definition of the more naive localizationists. In other words, a cerebral center is not a self-contained little aggregate of tissue that performs complete functions by itself; it is only part of a wider integrative mechanism.

My impression is that Jackson's formulation of his localization theory was not seriously affected by the absence of the neuron theory, but it was impeded by his lack of understanding of the motor unit and the central role of the final common pathway (ventral horn cell). Without this later Sherringtonian concept, Jackson was forced to think in terms of direct connections from the cortex to the muscles. In some ways, Jackson's ideas about the function of the cortical centers make more sense if the final common pathway is intercollated into them. For the present purposes, however, we deal with the details of Jackson's localization theory in its own original context.

To illustrate his concept of how a cerebral motor center functions, Jackson posited the simplified example of a peripheral "region" of three muscles represented by a single hypothetical "center" in the cerebrum. If the muscles are X, Y, and Z, then their cerebral representations are x, y, and z. He then supposed that the cerebral center is able to direct the muscles to make seven different movements. In each of these seven movements, each muscle would participate to a different degree. Moreover, in general, muscle X is more represented than Y, and Y is represented more than Z. Hence, one of the movements, in Jackson's own notation system, would be represented in the center by $x^{21}y^2z^1$, another by $x^{14}y^3z^1$, etc. The problem with Jackson's notation system is that it appears to be algebraic, but it is not: x, y, and z are not meant to be multiplied together, and the superscript numbers are merely arithmetic multipliers, not geometric. If we were to translate Jackson's notation into proper algebraic form, it would read: $(21x + 2y + 1z)$ and $(14x + 3y + 1z)$, etc. In other words, one of the movements of the muscle region X, Y, and Z requires 21 hypothetical units of energy of muscle X, 2 units of muscle Y, and 1 unit of muscle Z, where "units of energy" is left undefined ("units" is my word, and "energy" is Jackson's). Thus, x^{21}, y^2, and z^1 are notations for *degrees of cerebral representation*.* Jackson referred to each combination of x, y, and z as a "term," so our hypothetical center contains seven "terms" (using Jackson's notation system): $x^{21}y^2z^1$ plus $x^{14}y^3z^1$ plus five others. Jackson did not specify whether the x^{14} in the second term really consists of two-thirds of the same cell bodies as the x^{21} in the first, or whether the x functions in the two different "terms" are carried out by entirely different pieces of tissue. However, by implication from what follows, it appears that he intended the latter possibility; that is, x^{21} and x^{14} are representative of movements in different pieces of tissue within the same center.

If the preceding "artificial scheme" of Jacksonian localization is understood and accepted, the specifics of dissolution and compensation follow from it quite logically. What is lost in dissolution is the ability of the organism to "will" certain movements. Individual muscles are not specifically paralyzed because they are not specifically represented. However, this may not be obvious on clinical examination, because different muscles are affected to different degrees. If the term $x^{21}y^2z^1$ is destroyed, it will appear that muscle X is severely paralyzed, because so much of its cerebral representation is lost. In fact, what really has been lost is the movement represented by $x^{21}y^2z^1$. Obviously, if a few more "terms" containing large representations of X are lost, then muscle X will be severely paralyzed in practical terms.

The concept of compensation that Jackson derived from his localization theory was rather rigidly stated. Obviously, if the hypothetical center discussed above consists of seven terms and only $x^{21}y^2z^1$ is lost, there remain six other terms, each of which has some representation of X, Y, and Z in it.

> The loss of the movement $x^{21}y^2z^1$ is largely compensated in time by greater activity of the six remaining terms. Compensation does not, in this scheme, mean that nervous arrangements take on duties they never had before, but that, having more or less closely similar duties, they serve . . . next as well as those destroyed. Plainly, the compensation is in no case of destruction of part of a centre absolute, and in the

---

*The reader is also cautioned that Jackson used different notation systems in different papers at different times to express similar ideas. However, the details of the underlying concepts changed somewhat as his ideas developed.

illustration given it is very imperfect. On the other hand, in many cases we may say
that compensation is "practically" perfect. (Jackson, 1958, p. 36.)

The unique nature and inherent limitations of Jackson's compensation theory can best be understood by comparing it to our own contemporary theories. It
fits none of them exactly. Behavioral compensation *per se* does not describe
Jackson's concept because he said nothing about behavioral strategies. He was
positing strictly physiological compensation. Jacksonian compensation is not
classical vicariation, because the latter implies "the idea that functions could be
taken over by structures not previously concerned with them . . ." (Finger and
Stein, 1982, p. 290).

At the physiological level, there is a contemporary model to which Jackson's localization theory can be compared. Mountcastle (1979) has proposed the
idea of a "distributed system," made up of a set of cortical columns that is
coordinated for a particular function. The columns need not be immediately
contiguous to each other, though obviously they must be connected in a coherent
and specific manner. Superficially at least, it does seem to me that Mountcastle's
model is an excellent expression of how Jackson's localization theory might
work in contemporary physiological terms. It should follow that if Jackson's
view of normal physiology is workable, then his theory of compensation is likely
to be workable as well.

4. CRITIQUE: HISTORICAL AND CONTEMPORARY

Sketchy though it is, the foregoing outline of Jackson's localization and
compensation theory is sufficient to allow some points of critique. In fact, all of
the remarks that follow can be subsumed under two headings: (1) differences in
timeframes and (2) the influence of the motor model.

4.1. Time Frames

I have remarked that the time scale of Jackson's compensation theory is
somewhat foreshortened in comparison to our own. There are at least two reasons for this. One is Jackson's interest in positive as well as negative lesions,
i.e., in the deficits associated with seizures and movement disorders. These
events occur in minutes and hours rather than days and months. In fact, given the
shorter survival times of stroke and tumor patients in Jackson's day, the negative
lesions associated with those diseases could not have been studied for as long as
they are now. In any case, I am not aware of any current efforts to understand
recovery from epileptic "lesions" within the general context of contemporary
recovery theory. Were Jackson here to plead his own case, I think he would
recommend this to us. Such a perspective might well be quite fruitful.

The other factor in Jackson's shorter time frame is the striking absence of any ontogenetic considerations. At the theoretical level, this is quite surprising, because Spencer certainly used such considerations in his biological theorizing, though it was less conspicuous in his psychological theories. At the practical level, in reading Jackson's published case reports, I have had the impression that he saw relatively few infants and small children. Whatever the reason, the fact remains that Jackson did not use ontogenetic data in formulating his theory of localization and compensation. By contrast, of course, studies on immature animals are a major part of contemporary efforts to understand recovery phenomena. Our ontogenetic orientation also has the effect of making the time scale of contemporary studies longer than Jackson's.

4.2. The Motor Model

Although Jackson was working within the theoretical framework of the associationists' sensory–motor paradigm (Greenblatt, 1965, 1970), it is obvious that the actual statement of his localization theory was given entirely in terms of the efferent part of the nervous system. The question is, does this fact really make any difference for localization or compensation? This question can be addressed (though probably not answered) by asking what would happen if sensory functions were substituted in the x, y, and z example that was given above.

For a single sensory modality, there appears to be only modest difficulty with Jackson's paradigm of localization. Presumably x, y, and z would be the cerebral representations of the same modality from different parts of the body, each part having stronger or weaker degrees of perception. For example, light touch is much more highly represented on the fingertips than on the forearm. But the question does arise, exactly what is represented in the cerebrum? That is, what is the sensory homologue of a motor movement? Is it differing degrees of sensitivity to the same stimulus, depending on where over the body the stimulus is applied? Surely it is not the sensory perception of a particular stimulus (e.g., a coin or a pen), because that degree of identification is not innate. But does it need to be innate?

Without attempting to solve the foregoing problem, let me add to the difficulty. What happens when the stimuli are multimodal? Geschwind (1965) has claimed that it is humans' possession of large fields of multimodal cortex that sets us apart from other animals in terms of our mental capabilities. Clearly such cortex is a late development in evolutionary terms, so Jackson would place it in his highest level of the nervous system. But how would it function in the physiological sense according to Jackson's paradigm? Do we now propose that x, y, and z each represent different modalities? To a first approximation, this is easier to conceptualize than the preceding situation in which each letter represented the same sensory modality. In other words, Jackson's motor-based paradigm does seem to account for multimodal integration to some extent. How far it might go, of course, is well beyond the scope of this chapter.

More to the present point, it seems inherently obvious that Jackson's theory of compensation could be pushed as far as his theory of localization will allow. Whether the inherently obvious is really valid is always an interesting and open question. The prospect of further exploring Jackson's localization theory in terms of Mountcastle's (1979) model raises the intriguing possibility that both models may be fruitfully applied to contemporary theorizing about the nature of recovery in the nervous system. In short, the "graded" nature of Jacksonian localization can be understood in terms of Mountcastle's distributed system. For example, in Jacksonian localization, some part of the cortex may appear to be a center for hand movement because it contains a number of distributed systems for different hand movements. As one moves (by millimeters) away from this area, some of the distributed systems in adjacent areas will still be related to hand movements, albeit intermixed with distributed systems for other motor functions. If the original center is destroyed, some hand movements will remain to the patient (or animal) because of the existence of distributed systems for some hand movements in the adjacent tissue. Thus, in contemporary terminology, Jackson would claim that his theory of compensation is a particular form of Finger and Stein's (1982, p. 291) "hypothesis 2" of vicariation theory. In other words, compensation occurs primarily because adjacent tissue can take over some of the lost functions.

But wait! If the organism has to use different movements to try to carry out the same motor task, is that not really behavioral compensation? As the latter term is presently used, I think it is. Bringing up the issue of behavioral compensation naturally leads to the ontogenetic issues inherent in the "Kennard principle" (Finger and Stein, 1982, pp. 136–139). In this light, information about the ontogeny of distributed systems becomes critically important. Without belaboring these issues further, I submit that their exploration might well lead to improved integration of various contemporary theories of recovery, which presently seem to be separate issues.

ACKNOWLEDGMENT. I am much indebted to C. Robert Almli for his critique of the manuscript of this chapter.

REFERENCES

Clarke, E., 1973, Jackson, John Hughlings, in: *Dictionary of Scientific Biography*, Volume 7 (C. C. Gillepsie, ed.), Charles Scribner's Sons, New York, pp. 46–50.

Ferrier, D., 1873, Experimental researches in cerebral physiology and pathology, *West Riding Lunatic Asylum Med. Rep.* 3:30–96.

Finger, S., and Stein, D. G., 1982, *Brain Damage and Recovery. Research and Clinical Perspectives,* Academic Press, Orlando, FL.

Geschwind, N., 1965, Disconnexion syndromes in animals and man, *Brain* **88**:237–294.

Greenblatt, S. H., 1965, The major influences on the early life and work of John Hughlings Jackson, *Bull. Hist. Med.* **39**:346–376.

Greenblatt, S. H., 1970, Hughlings Jackson's first encounter with the work of Paul Broca: The physiological and philosophical background, *Bull. Hist. Med.* **44**:555–570.

Greenblatt, S. H., 1977, The development of Hughlings Jackson's approach to disease of the nervous system 1863–1866: Unilateral seizures, hemiplegia and aphasia, *Bull. Hist. Med.* **51**:412–430.

Greenblatt, S. H., 1984, The multiple roles of Broca's discovery in the development of the modern neurosciences, *Brain Cognit.* **3**:249–258.

Guillain, G., 1959, *J.-M. Charcot 1825–1893. His Life—His Work* (translated by P. Bailey), Hoeber, New York.

Jackson, J. H., 1870, A study of convulsions, *Trans. St. Andrews Med. Grad. Assoc.* **3**:162–204. Reprinted in Jackson, J. H., 1958, *Selected Writings*, Volume 1 (J. Taylor, G. Holmes, and F. M. R. Walshe, eds.), Basic Books, New York, pp. 8–36.

Jackson, J. H., 1958, *Selected Writings*, Volume 2 (J. Taylor, G. Holmes, and F. M. R. Walshe, eds.), Basic Books, New York.

Laycock, T., 1845, Reflex function of the brain, *Br. For. Med. Rev.* **19**:298–311.

Mountcastle, V. B., 1979, An organizing principle for cerebral function: The unit module and the distributed system, in: *The Neurosciences. Fourth Study Program* (F. O. Schmitt and F. G. Worden, eds.), MIT Press, Cambridge, pp. 21–42.

Swazey, J. P., 1969, *Reflexes and Motor Integration: Sherrington's Concept of Integrative Action*, Harvard University Press, Cambridge, pp. 35–39, 70–78.

13

The Parcellation Theory and Alterations in Brain Circuitry after Injury

SVEN O. E. EBBESSON

1. INTRODUCTION

The last 20 years have opened the eyes of all neurobiologists to the wonderful news that neural connections in the mammalian nervous system change through-out the life span of the organism and that structural reorganization within the CNS is possible, although limited, after injury.

This new insight has resulted from the development of accurate methods for tracing axonal trajectories in the CNS. Nauta (1950, 1957) led the way, and today scores of methods are available that allow one to very accurately define synaptic relationships. Thus, thousands of studies on hundreds of pathways and in hundreds of vertebrate species, at different ages, have resulted in a very broad, very complex picture of interspecific variability in connections and of changes in circuitry during ontogenetic development. The new methods have also allowed the accurate definition of changes in circuitry after injury to the nervous system. It is no wonder that the mountain of new data has yielded few comprehensive theories that tie the observations together, explain the variabilities observed, and predict the range of variability. The parcellation theory was the first attempt to do so. It has been thoroughly applauded and criticized (Ebbesson, 1984a) yet has stood the test of time as the original language (Ebbesson, 1980, 1981) has been altered (Ebbesson, 1984a,b) to emphasize that parcellation is but one of many events in the evolution, ontogeny, and repair of the nervous system and that the opposite to parcellation, namely, invasion, does appear to occur in evolution, but rarely.

SVEN O. E. EBBESSON • Institute of Marine Science, University of Alaska–Fairbanks, Fairbanks, Alaska 99775-1080.

The theory originated from comparative studies we carried out for some 20 years in which we systematically compared the pattern of projections of a given system in a broad spectrum of species, sometimes including all vertebrate classes. We studied five systems with modern neuroanatomic methods (Ebbesson, 1970; Ebbesson *et al.*, 1981), namely, the olfactory, visual, ascending spinal, tectal efferents, and thalamic efferents. Scattered data were also available on other systems. To our surprise, our studies revealed very similar distributions of a given system in all vertebrate classes. Contrary to common belief, we found, for example, in primitive species (1) spinothalamic projections, (2) visual and somatic thalamotelencephalic projections, (3) telencephalic projections to thalamus, brainstem, and spinal cord, reminiscent of cortical projections in mammals, (4) olfactory bulb projections similar to those described in mammals, and (5) structures comparable to neocortex on the basis of connections (labeled "neocortical equivalents" by us).

Such observations led us to the conclusion that the basic neural systems in mammals have been present from the beginning of vertebrate evolution, and what we see in modern vertebrates represents specializations on a common theme. As we extended the analysis we noted other closely related facts that solidified the concept of parcellation.

The parcellation theory is a general theory that explains and predicts certain aspects of interspecific variability of brain organization and a broad spectrum of neuronal interrelationships and their changes with time and injury (Ebbesson, 1972, 1980, 1981, 1984a,b). The theory bridges and connects several neuroscience disciplines and provides a theoretical framework for analysis, prediction, and explanation of brain circuitry.

The parcellation theory allows certain predictions about the range of variation of a given system at all levels of analysis including the cellular and aggregate levels. For example, the interspecific variability in organization of cortical columns, thalamic nuclei, cortical areas, and tectal layers can be explained. The available data suggest that diffuse, undifferentiated systems existed in the beginning of vertebrate evolution and that during the evolution of complex behaviors and analytic capacities related to these behaviors, a range of patterns of neural systems evolved that relate to these functions. One principle underlying the growth, differentiation, and multiplication of neural systems appears to be the process of parcellation as defined by the theory.

The theory states that one aspect of evolutionary and ontogenetic development is the selective loss of connections of newly formed daughter cells, aggregates, and subsystems. The new structures thus have fewer types of input, are more specialized, and are more finely tuned in a specific function (i.e., the available data suggest that axons very rarely invade unknown territories during evolutionary or ontogenetic development, but follow in their ancestor's path to their ancestral targets; if the connection is later lost, it reflects the specialization of the circuitry). The ontogenetic parcellation process can be manipulated, to a certain extent, by deprivation or surgically induced sprouting. Injury to neurons, whether surgical or accidental, results in a predictable growth of axons that is not random but is, in fact, a restoration of ancestral and developmental connections. We refer to this process here for the first time as "traumatic resynaptogenesis" (TRS).

2. INVASION

We found no evidence of neural systems invading others in evolution (with the possible exception of the pyramidal system). In fact, it became clear that the opposite was true. In many systems in adult lower vertebrates, projections of a given cell group are more extensive than in higher vertebrates, i.e., a given cell group is sometimes related to a greater variety of cells and aggregates in the primitive condition. An example of this is the telencephalic projection to the cerebellar cortex in salamanders (Kokoros and Northcutt, 1977) and sharks (Ebbesson, 1980). This connection does not exist in adult mammals but is found in neonate mammals and disappears during ontogeny (Distel and Hollander, 1980).

Before our studies and conclusions (Ebbesson, 1980), it was generally believed that "invasion" was a key mechanism in the evolutionary development of neocortex, for example. It was thought that nonolfactory sensory modalities invaded the thalamus and the olfactory telencephalon to form neocortex. The evidence against the invasion hypothesis comes mainly from the recent discoveries of (1) neocortical equivalents in the telencephalon of anamniotes (Ebbesson and Schroeder, 1971; Cohen *et al.*, 1973; Graeber *et al.*, 1973), (2) the absence of olfactory input to large areas of the elasmobranch and teleost telencephalon (Ebbesson and Heimer, 1968, 1970; Scalia and Ebbesson, 1971), and (3) the discoveries of nonolfactory inputs to dorsal thalamus in sharks, frogs, and reptiles (Ebbesson, 1966, 1967, 1969, 1978; Ebbesson and Hodde, 1981; Ebbesson *et al.*, 1972; Ebbesson and Goodman, 1981).

3. OVERLAP OF CONNECTIONS IS A FEATURE OF PRIMITIVE AND DEVELOPING BRAINS

We found a spectrum of connectional arrangements in various vertebrates in which either one system overlapped with another in a presumed primitive condition or there was no, or only limited, overlap of systems in the more highly developed species. An example of this is the variability observed in the visual systems. In the simplest arrangement (shark), one thalamic aggregate that projects to the telencephalon receives retinal and tectal inputs, whereas in the highly differentiated adult birds or mammals, two separate cell groups exist for the two parcellated inputs. Less-differentiated species often have two cell groups with partial overlap of retinal and tectal inputs. This overlap is more extensive during ontogenesis.

The parcellation theory grew out of an analysis of such patterns of arrangement, and an explanation of the interspecific variability was found. Figure 1 illustrates the proposed sequence of evolutionary and ontogenetic parcellation. In this hypothetical model, neurons in a single cluster (Fig. 1a) have two overlapping inputs, A and B. This primitive arrangement changes as selective connections are lost during development (Fig. 1b,c) until two distinct parcels with

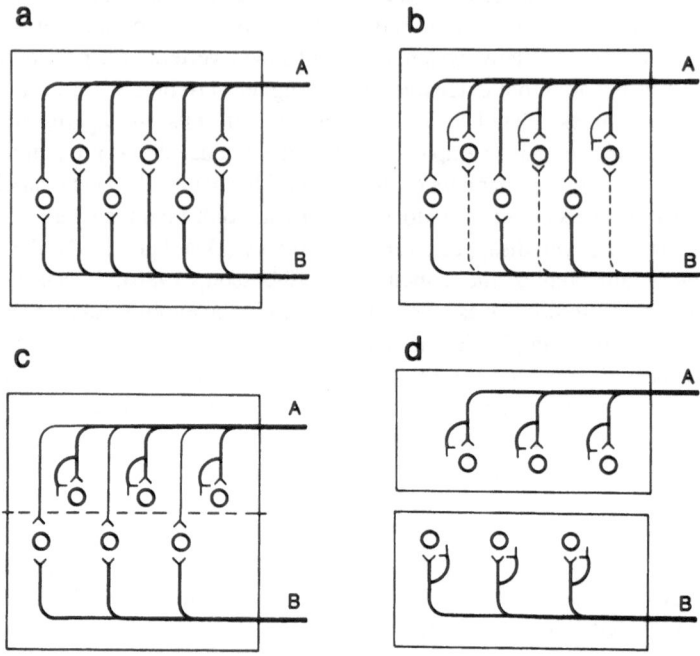

Figure 1. Schematic diagrams show how evolutionary and ontogenetic parcellation is thought to occur. The parcellation process involves the loss of one or more inputs to a cell, or aggregate of cells, as new subcircuits and more specialized cell clusters are formed. The neurons in the hypothetical aggregate a become less and less influenced by a given input (A or B) in b and c, until two new distinct parcels are formed (d). Each new parcel contains neurons with more restricted inputs. The parcellation theory explains many observations related to the evolution and ontogenetic develoment of new aggregates. This includes the development of the monocular laminae in the dorsal lateral geniculate nucleus, ocular dominance columns, and specialized cortical laminae. "Traumatic re-synaptogenesis" is the reverse of parcellation and is defined as the recapture of vacated "ancient" synaptic sites and/or of "transient" synaptic sites that were briefly occupied during ontogenetic development. This process can be imagined as the reverse of what is shown in c and d; i.e., the reduction of input B in c as compared to d, resulting from injury, could result in sprouting of A fibers to the lower aggregate in c.

different inputs develop (Fig. 1d). Parcellation then is depicted as the evolutionary and ontogenetic loss of select inputs.

4. ONTOGENETIC PARCELLATION

There was a time not long ago that we thought that newborns had the same
brain connections and structures as adults. In fact, young animals were often
used for convenience in anatomic studies because the cells "stained better" and
it was easier to get axonal transport of various substances used for tract tracing.
Since then, we have learned that connections and morphology change drastically
during development and that the connections change in a manner very similar to
that proposed to occur in evolutionary development (Ebbesson, 1980). It is, of
course, no surprise that young specimens of various species share more characteristics than adults and that young specimens of highly specialized species may
present aspects of primitive organization.

The connectional trends during ontogenetic development are predictable by
the parcellation theory: connections are selectively lost during the differentiation.
This is referred to as ontogenetic parcellation. The transient corticocerebellar
projections have already been mentioned, but many other examples are now
known. For example, Stanfield and O'Leary (1985) report transient corticospinal
projections from the occipital cortex during postnatal development in the rat. The
development of ocular dominance columns (Hubel et al., 1977; Rakic, 1976,
1977) also appears to involve the parcellation process (Ebbesson, 1980). Commissural connections of the neocortex are significantly altered by the selective
loss of such connections to the more specialized cortical regions (Innocenti,
1979). This mimics precisely the presumed evolutionary sequence (Ebner and
Myers, 1965; Ebbesson, 1980).

5. CYTODIVERSIFICATION

Another dimension of the parcellation theory lies in its capacity to explain,
to a certain extent, the evolutionary and ontogenetic changes in morphology of
neurons. Evolutionary development and specialization of a given neuron are
accompanied by a reduction of specific inputs that are often characteristic of the
given cell aggregate. The observed variability of cell types is then, to a certain
degree, a reflection of evolutionary cytodiversification (Ebbesson, 1980, 1981,
1984a; Ebbesson et al., 1975). This explanation is in agreement with the conclusions of Ramon-Moliner (1962) and Ramon-Moliner and Nauta (1966), who
concluded that specialization of dendritic trees reflects "a high degree of homo-
geneity of input" (e.g., the superior olive), whereas generalized or isodendritic

types of neurons prevail in brain regions characterized by afferent connections of heterogeneous origin (e.g., the reticular formation).

Ontogenetic parcellation also involves changes in neuronal morphology. At least some neurons go through developmental stages involving extensive dendritic formation followed by retraction (Jhaveri and Morest, 1982), and other neurons go through a developmental phase with a larger number of dendritic spines (presumably followed by a distinct reduction of the spines). This presumably reflects ontogenetic parcellation.

6. EXPERIMENTALLY INDUCED SPROUTING AND ACCIDENTAL BRAIN INJURY

The last 20 years have produced many insights into the details of brain injury and the limits of functional recovery and repair of circuits. An understanding of the restorative mechanisms and of the factors contributing to repair is crucial to all who deal with brain damage. It has become even more important in the last few years since it was discovered that even minor head injury results in cell death and degeneration of axons (Jane *et al.*, 1985).

Once it was learned that sprouting takes place even in the adult CNS, a great deal of effort has been made to determine the extent of such growth and the factors that play a role. Since the "regeneration" is indeed minimal in most circumstances, it has become important to discover which factors enhance the growth and repair. An example of the latter is the elegant work by Borgens *et al.* (1981), who showed that the application of electric fields can enhance spinal cord regeneration in lamprey.

The limits of regeneration and collateral sprouting have been determined in a number of systems in various species. The early work of Liu and Chambers (1958) in adult cat spinal cord, the pioneering work by Schneider (1973), and Goodman and Horel (1966), and the work of Cotman and Lynch (1976) on the rat hippocampus stand out. As one reviews the literature on the subject, one is struck by the nonrandomness of axonal sprouting. Although the early work suggested random growth, it is now clear that only selective synaptogenesis takes place. But the choice of synaptic sites was not explained until the "parcellation theory" suggested that the choice may have to do with the evolutionary and ontogenetic histories of the systems studied. The theory suggests that in case of injury, axons grow to target cells with which they have had synaptic relationships in their evolutionary or ontogenetic past. This aspect of the theory has not been labeled before, but it is here called "traumatic resynaptogenesis" (TRS) to emphasize the apparent reason for the choice of synaptic sites of axons following injury. The sprouting of retinal fibers to the lateral posterior nucleus following tectal lesions in the neonate hamster (Schneider, 1973) is interpreted as sprouting to a nucleus that is postulated to have had, once in evolution, a considerable retinal input (Ebbesson, 1972, 1980). It is noteworthy that transient retinal projections to the lateral posterior nucleus have been observed in neonate hamsters. This observation provides yet another example of ontogenetic parcellation. Other examples of "traumatic resynaptogenesis" have been discussed elsewhere (Ebbesson, 1980, 1984a). Since the data are still meager, we need to emphasize that other factors may also play a role in sprouting.

It is not clear from the available data, but one would presume that traumatic synaptogenesis is more likely to occur in instances in which parcellation occurred most recently in the evolutionary history. Thus, one would guess that regeneration, or recoupling of preinjury connections, would occur most easily, followed by a gradient of resynaptogenesis determined by the ancestral relationship of previous synapses.

The result of such selective TRS explains, perhaps, why so little if any sprouting occurs between modalities. Not only is the parcellation of a modality-specific circuit thought to be ancient, but other factors may also play a role. Inputs from another modality with a different message, out of phase temporally and located in a functionally meaningless position on the neuron, would predictably interfere with the normal functions of the given cell. In fact, such a contact might predictably lead to rejection of the synapse by some yet unknown mechanism.

Why do neurons make connections selectively? Although the parcellation theory and the concept of TRS explain what we observe, we are far from understanding the underlying mechanisms. It appears likely that the selectivity increases as a function of the degree of development. It is also likely that the same chemotrophic mechanisms operate at the time of injury as during normal development.

In terms of survival value, there is no reason to assume that the brain has evolved regenerative capacities just to function in case of injury (except for specializations such as the loss of the tail in lizards). Whatever regeneration or reactive synaptogenesis takes place as a result of injury, it must occur "accidentally" and must be based on whatever dormant structural and chemical substrates are left over from developmental days.

7. CONCLUSION

This chapter deals with one general theory of synaptology that evolved from a comparison of all hodological data and has been looked at from various perspectives. The "parcellation theory" explains to a certain extent (1) interspecific variability in brain organization as reflected in synaptic relationships, (2) evolutionary development of more "sophisticated circuits," and (3) ontogenetic development of connectional arrangements, and (4) predicts the "new connections" resulting from CNS injury. The latter is thought to be the reverse of

198 SVEN O. E. EBBESSON

parcellation and is here called "traumatic resynaptogenesis" (TRS). This is defined as the recapture of vacated "ancient" synaptic sites and/or of "transient" synaptic sites that were briefly occupied during ontogenetic development. Thus, knowledge of the evolutionary and ontogenetic histories of neural systems should provide insights into the predictions of limits and patterns of regeneration.

REFERENCES

Borgens, R. B., Roederer, E., and Cohen, M. J., 1981, Enhanced spinal cord regeneration in lamprey by applied electrical fields, *Science* **213**:611–617.

Cohen, D. H., Duff, T. A., and Ebbesson, S. O. E., 1973, Electrophysiological identification of a visual area in shark telencephalon, *Science* **182**:492–494.

Cotman, C. W., and Lynch, G. S., 1976, Reactive synaptogenesis in the adult nervous system, in: *Neuronal Recognition* (S. H. Barondes, ed.), Plenum Press, New York, pp. 69–108.

Distel, H., and Hollander, H., 1980, Autoradiographic tracing of developing subcortical projections of the occipital region in fetal rabbits, *J. Comp. Neurol.* **192**:505–518.

Ebbesson, S. O. E., 1966, Ascending fiber projections from the spinal cord in the tegu lizard (*Tupinambis nigropunctatus*), *Anat. Rec.* **154**:341–342.

Ebbesson, S. O. E., 1967, Ascending axon degeneration following hemisection of the spinal cord in the tegu lizard (*Tupinambis nigropunctatus*), *Brain Res.* **5**:178–206.

Ebbesson, S. O. E., 1969, Brain stem afferents from the spinal cord in a sample of reptilian and amphibian species, *Ann. N.Y. Acad. Sci.* **167**:8–101.

Ebbesson, S. O. E., 1970, Selective silver impregnation of degenerating axoplasm in poikilothermic vertebrates, in: *Contemporary Research Methods in Neuroanatomy* (W. J. H. Nauta and S. O. E. Ebbesson, eds.), Springer-Verlag, New York, pp. 132–161.

Ebbesson, S. O. E., 1972, A proposal for a common nomenclature for some optic nuclei in vertebrates and the evidence for a common origin to two such cell groups, *Brain Behav. Evol.* **6**:75–91.

Ebbesson, S. O. E., 1978, Somatosensory pathways in lizards: The identification of the medial lemniscus and related structures, in: *Behavior and Neurology of Lizards* (P. MacLean and N. Greenberg, eds.), DHEW Publication No. (ADM) 77-491, U.S. Government Printing Office, Washington, pp. 91–104.

Ebbesson, S. O. E., 1980, The parcellation theory and its relation to interspecific variability in brain organization, evolutionary and ontogenetic development and neuronal plasticity, *Cell Tissue Res.* **213**:179–212.

Ebbesson, S. O. E., 1981, Interspecific variability in brain organization and its possible relation to evolutionary mechanisms, in: *Brain Mechanisms of Behavior in Lower Vertebrates* (P. Laming, ed.), Cambridge University Press, Cambridge, pp. 59–76.

Ebbesson, S. O. E., 1984a, Evolution and ontogeny of neural circuits, *Behav. Brain Sci.* **7**:321–366.

Ebbesson, S. O. E., 1984b, An update of the parcellation theory, *Behav. Brain Sci.* **7**:350–366.

Ebbesson, S. O. E., and Goodman, D. C., 1981, Organization of ascending spinal projections in *Caiman crocodilus*, *Cell Tissue Res.* **215**:383–395.

Ebbesson, S. O. E., and Heimer, L., 1968, Olfactory bulb projections in two species of sharks (*Galeocerdo cuvieri* and *Ginglymostoma cirratum*), *Anat. Res.* **160**:469.

Ebbesson, S. O. E., and Heimer, L., 1970, Projections of the olfactory tract fibers in the nurse shark (*Ginglymostoma cirratum*), *Brain Res.* **17**:47–55.

Ebbesson, S. O. E., and Hodde, K., 1981, Ascending spinal systems in the nurse shark (*Ginglymostomas cirratum*), *Cell Tissue Res.* **216**:313–331.

Ebbesson, S. O. E., and Schroeder, D. M., 1971, Connections of the nurse shark's telencephalon, *Science* 173:254–256.

Ebbesson, S. O. E., Jane, J. A., and Schroeder, D. M., 1972, A general overview of major interspecific variations in thalamic organization, *Brain Behav. Evol.* 6:92–130.

Ebbesson, S. O. E., Schroeder, D. M., and Butler, A. B., 1975, The Golgi method and the revival of comparative neurology, in: *Golgi Centennial Symposium* (M. Santini, ed.), Raven Press, New York, pp. 153–160.

Ebbesson, S. O. E., Hansel, M., and Scheich, H., 1981, An "on the slide" modification of the de Olmos HPR method, *Neurosci. Lett.* 22:1–4.

Ebner, F. F., and Myers, R. E., 1965, Distribution of the corpus callosum and anterior commissure in cat and racoon, *J. Comp. Neurol.* 124:353–366.

Goodman, D. C., and Horel, J., 1966, Sprouting of optic tract projections in the brain stem of the rat, *J. Comp. Neurol.* 127:71–88.

Graeber, R. C., Ebbesson, S. O. E., and Jane, J. A., 1973, Visual discrimination in sharks without optic tectum, *Science* 180:413–415.

Hubel, D. H., Wiesel, T. N., and LeVay, S., 1977, Plasticity of ocular dominance columns in monkey striate cortex, *Phil. Trans. R. Soc. Lond. [Biol.]* 278:377–409.

Innocenti, G. M., 1979, Adult and neonatal characteristics of the callosal zone at the boundary between area 17 and 18 in the cat, in: *Structure and Function of Cerebral Commissures* (J. S. Russel, N. W. van Hof, and G. Berlucchi, eds.), Macmillan, New York, pp. 244–258.

Jane, J. A., Rimel, R. W., Alves, W. M., Dacey, R. G., Jr., Winn, H. R., and Colohan, A. R., 1985, Minor and moderate head injury: Model systems, in: *Trauma of the Central Nervous System* (R. G. Dacey, Jr., H. R. Winn, R. W. Rimel, and J. A. Janes, eds.), Raven Press, New York, pp. 27–33.

Jhaveri, S., and Morest, D. K., 1982, Sequential alterations of neuronal architecture in nucleus magnocellularis of the developing chicken: A Golgi study, *Neuroscience* 7:837–853.

Kokoros, J. J., and Northcutt, R. G., 1977, Telencephalic efferents of the tiger salamander, *Ambystoma tigrinum tigrinum* (Green), *J. Comp. Neurol.* 173:613–628.

Liu, C. M., and Chambers, W. W., 1958, Intraspinal sprouting of dorsal root axons, *Arch. Neurol. Psychiatry* 79:46–61.

Nauta, W. J. H., 1950, Uber die sogenannte terminale Degeneration im Zentralnervensystem und ihre Darstellung durch Silberimprägnation, *Arch. Neurol. Psychiatry* 66:353–376.

Nauta, W. J. H., 1957, Silver impregnation of degenerating axons, in: *New Research Techniques of Neuroanatomy* (W. F. Windle, ed.), Charles C. Thomas, Springfield, IL, pp. 17–26.

Rakic, P., 1976, Prenatal genesis of connections subserving ocular dominance in the rhesus monkey, *Nature* 261:467–471.

Rakic, P., 1977, Prenatal development of the visual system in rhesus monkey, *Phil. Trans. R. Soc. Lond. [Biol.]* 278:245–260.

Ramon-Moliner, E., 1962, An attempt at classifying nerve cells on the basis of their dendritic patterns, *J. Comp. Neurol.* 119:211–227.

Ramon-Moliner, E., and Nauta, W. J. H., 1966, The isodendritic core of the brain stem, *J. Comp. Neurol.* 212:311–336.

Scalia, F., and Ebbesson, S. O. E., 1971, The central projections of the olfactory bulb in a teleost (*Gymnothorax funebris*), *Brain Behav. Evol.* 4:376–399.

Schneider, G. E., 1973, Early lesions of superior colliculus: Factors affecting the formation of abnormal retinal projections, *Brain Behav. Evol.* 8:73–109.

Stanfield, B. B., and O'Leary, D. D., 1985, The transient corticospinal projection from the occipital cortex during the postnatal development of the rat, *J. Comp. Neurol.* 238:236–248.

14

Trophic Hypothesis of Neuronal Cell Death and Survival

FRED H. GAGE and SILVIO VARON

1. INTRODUCTION

Recovery of some components of function generally occurs following damage to
the brain. Depending on the extent and location of the damage and the age of the
organism at the time of the damage, several different mechanisms have been
proposed for the observed recovery. These mechanisms include regeneration,
collateral sprouting, receptor supersensitivity, and enzymatic hyperactivity
(Aguayo, 1985; Marshall, 1984; Cotman and Nieto-Sampedro, 1984). Conceptually, the recovery will occur if there is sufficient residual system to subserve
the function—compensation—or if a similar anatomically intact system substitutes its functional activity for that of the damaged system—substitution. In
many instances, however, there is no or little functional recovery because the
system underlying the lost function is sufficiently destroyed to have no re-
cuperative mechanisms on which to fall back and no other anatomic system is
available to take over the lost function. Under these circumstances large numbers
of neurons die. A major cause of the loss of function following brain damage is
attributed to the death or loss of neurons. Protection of damaged neurons from
death, or revitalization of damaged neurons, can be most rationally attempted or
planned with a more complete understanding of the basic principles involved in
normal survival of neurons during development and aging.

    Throughout normal development the brain makes more neurons than are
needed. Some neurons undergo cell death at times when their growing axons

FRED H. GAGE • Department of Neurosciences, University of California at San Diego, La Jolla,
California 92093.   SILVIO VARON • Department of Biology, University of California at San
Diego, La Jolla, California 92093.

connect with target territories, reducing the final number of neurons to that found in the adult (Cunningham, 1982; Cowan *et al.*, 1984). The extent of neuronal cell death can be affected by experimentally manipulating the target area of the developing neurons to add or subtract target tissue (Prestige, 1970; Landmesser and Pilar, 1978; Hamburger and Oppenheim, 1982). These observations have led several investigators to propose that developmental cell death and survival are regulated by proteins presumably supplied by the target territory, which have been called neuronotrophic factors (NTF) (Varon and Adler, 1980, 1981; Levi-Montalcini, 1982; Berg, 1984; Thoenen and Edgar, 1985). These NTFs are presumably taken up into axon terminals and retrogradely transported to the cell bodies, where they play key roles in cell maintenance.

The purpose of this chapter is to explore the hypothesis that the presence or absence of NTFs is also responsible for cell death and survival in the adult damaged central nervous system (Appel, 1981; Hefti, 1983; Varon, 1985; Hefti and Weiner, 1986). Specifically we (1) state the hypothesis in theoretical terms, (2) present the data that support the hypothesis, and (3) discuss the implications of this hypothesis for protection against cell death and, subsequently, induction or improvement of recovery of function.

2. KEY TERMS AND CONCEPTS

Trophic refers to the ability of one tissue or cell to support and/or nourish another. A *neurotrophic agent* is a chemical or molecule that is made in a neuron and supports the survival of or nourishes any cell or tissue (i.e., muscle). The term *neuronotrophic agent* refers to a chemical or molecule that is made in a neuron and supports the survival of or nourishes, exclusively, a neuron. A protein endowed with such a competence is designated a *neuronotrophic factor* (NTF).

Trophic differs markedly from *tropic*, which refers to the influence of one cell or tissue on the direction of movement or outgrowth of another. Thus, a *neurotropic factor* is a chemical or molecule that can influence the direction and/ or growth of a neuronal axon.

Only one NTF has been sequenced and characterized, and that is the nerve growth factor (NGF) (Levi-Montalcini, 1966, 1982; Varon, 1975; Greene and Shooter, 1980; Thoenen and Barde, 1980; Ullrich *et al.*, 1983). Two other NTFs have been isolated to purification and have a defined trophic activity, and those are brain-derived growth factor (BDGF) (Barde *et al.*, 1982) and ciliary neuronotrophic factor (CNTF) (Barbin *et al.*, 1984; Manthorpe and Varon, 1985; Manthorpe *et al.*, 1986b). Neuronotrophic activities have also been proposed for two well-defined proteins isolated for their growth-promoting properties on nonneural cells, basic fibroblast growth factor (bFGF) (Walicke *et al.*, 1986) and epidermal growth factor (EGF) (Herschman *et al.*, 1983).

Neuronal targets of NTF are neurons that respond to the presence of the NTF with changes in their behaviors, e.g., survival, growth, transmitter functions. Dependence on NTFs for survival may only manifest itself over a particular period of development or following trauma to the neuron. Such a dependence is best demonstrated *in vivo* through experiments that remove the NTF source and reveal subsequent cell death or that add additional NTF sources and cause a reduction in neuronal death. The same neurons, *in vitro*, will fail to survive unless supplied with exogenous NTF, a feature that has been critical for the isolation and characterization of the NTF protein itself. In other circumstances, however, the NTF target characteristics of a neuron may be revealed not by its survival but by changes in neurite extension or the amounts of transmitter-producing enzymes. For example, NGF has its traditional targets in the peripheral nervous system (PNS), the sensory neurons of the dorsal root ganglia (DRG), and the catecholamine-containing principal neurons of sympathetic ganglia (Black, 1978; Patterson, 1978). The contact point between the NTF and the target neuron is thought to be a specific receptor molecule on the neuronal surface. These NGF high-affinity binding sites have been identified on specific neuronal targets (Greene and Shooter, 1980; Tanuichi and Johnson, 1985).

The source of a specific NTF is generally the area innervated by the axons of the NTF neuronal targets, though other cells and tissues may have high concentrations of the NTF. Thus, the submaxillary gland is one source of NGF in that it receives innervation from ganglionic sympathetic neurons. That some organs or tissues have higher concentrations of the NTF does not necessarily imply a more important role for that NTF in the function or viability of the corresponding set of target neurons. Thus, the male mouse submaxillary gland has ten times higher concentrations of NGF than that in the female mouse (Varon, 1975; Greene and Shooter, 1980; Levi-Montalcini, 1982), but the significance of this difference is not readily apparent at the level of the innervating neurons.

The mechanisms by which NTFs exert their effects on target neurons are still unknown (Varon, 1975, 1985; Greene and Shooter, 1980; Varon and Skaper, 1980, 1983). Among several possibilities postulated from experimental observations are direct or indirect actions on membrane and cytoskeletal properties, ion fluxes, nutrient transport, energy metabolism, and/or nucleic acid and protein synthesis.

Figure 1 summarizes some possible features of NTF production, delivery, and actions with its neuronal target. For example, there are cells that produce and can secrete the NTF. One such source could be the postsynaptic partner of the neurons (muscle for a spinal motoneuron, other neurons for an intrinsic CNS neuron). Another source can be a glial cell (astroglia, Schwann). The NTF, synthesized on the endoplasmic reticulum and packaged in the Golgi of the source cell, is presumably released by exocytosis into the interstitial fluid at the synaptic site. The model presumes regulatory mechanisms that control produc-

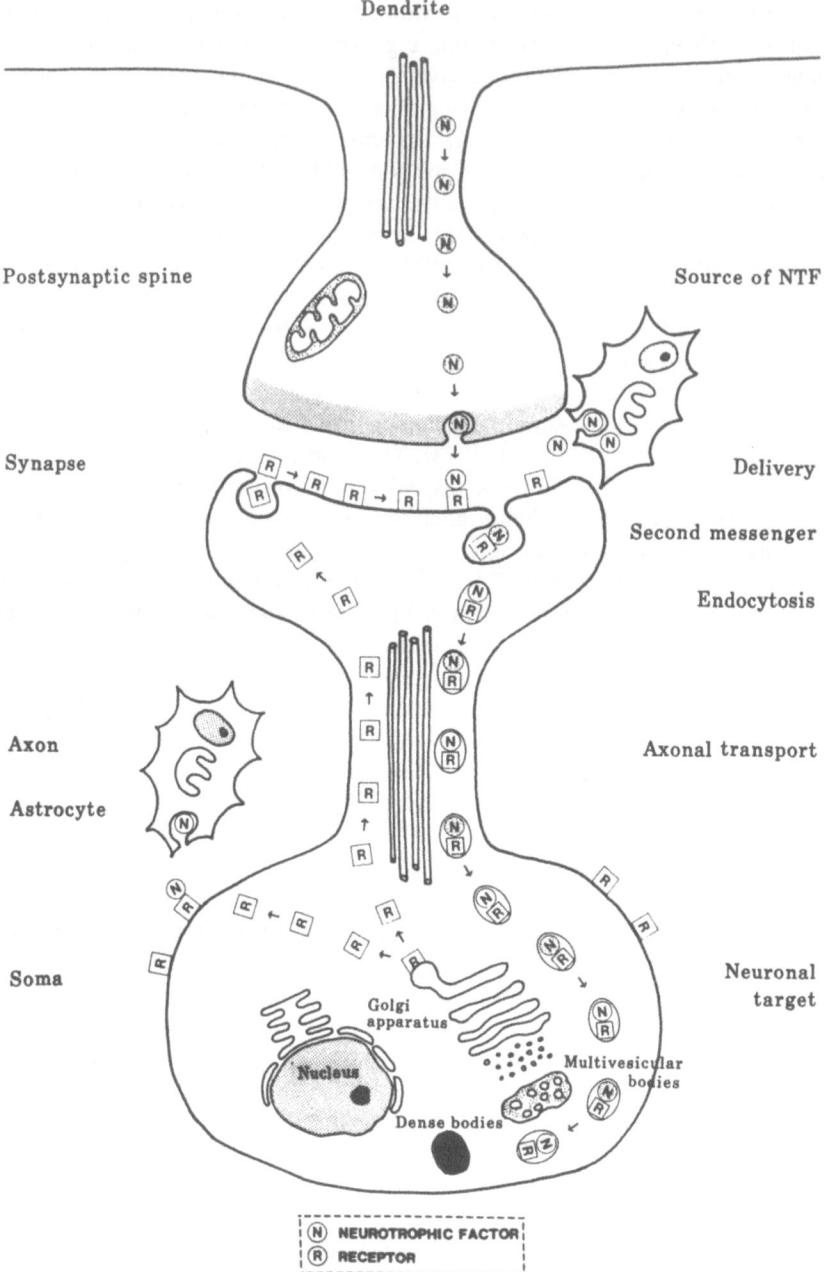

Figure 1. Schematic diagram of prototypic pre- and postsynaptic neurotrophic-factor-reactive neurons. On the left side of the diagram are structure-related terms, and on the right are function-related terms.

tion and secretion of the NTF from the source cells. From the interstitial fluid, the released NTF complexes with specific receptors on the surface membrane of the NTF neuronal target. The receptors are presumed to be produced in the target cell body, anterogradely transported to the presynaptic or other sites, and externalized there to become available to the NTF. At this point the NTF could act on the receptor, affect the neuron indirectly via a second message, and then be released from or degraded at the target surface. The NTF/receptor complexes could also be internalized in an endocytotic vessicle to be degraded by the lysosomal system. Alternatively, endocytosis of the NTF/binding site complex could serve as an intracellular transport system to deliver the NTF to other receptors in one or more cell organelles for direct action there.

3. THE NEURONOTROPHIC FACTOR HYPOTHESIS

The neuronotrophic hypothesis (Appel, 1981; Hefti, 1983; Varon, 1985) postulates that (1) adult CNS neurons *in situ* are supported and regulated by their respective NTFs, (2) proper maintenance of these neurons depends on adequate supply and utilization of the NTFs, (3) an interference with the NTF support, or "trophic deficit," will result in defective performance or even degeneration and death of the target neurons, and (4) such trophic deficits may be at the basis of degenerative CNS diseases (e.g., Parkinson's, motor neuron, or Alzheimer's diseases) or normal aging.

The general model of Fig. 1 makes it clear that there are many critical points in the chain of NTF-related events, the perturbation of which would lead to the introduction of a trophic deficit and a decrease in trophic benefits to the target neuron. Some of these are summarized in Table 1.

Extraneuronally (i.e., not in the target neuron), a reduction in the synthesis and/or secretion of NTF by the source cells would obviously deprive the target neurons of a normal supply of the trophic factor. Equally devastating, however, would be an interference with the transfer of NTF from source to target receptors. One dramàtic example in the developing PNS is the immunosympathectomy (that is, the destruction of the peripheral sympathetic system) that results

Table 1. Trophic Deficits

Extraneuronal: NTF	Neuronal: receptor	Intraneuronal: NTF–receptor complex
Synthesis	Binding	Incorporation
Production	Affinity	Transport
Secretion	Synthesis	Second messenger
Delivery		Organelle

from the introduction into the animal of antibodies against NGF (Black, 1978; Patterson, 1978; Levi-Montalcini, 1982; Johnson, 1983).

Neuronally (i.e., at the level of the target neuron), the synthesis of the NTF receptors and their anterograde transport along the axon are also vital. For example, problems in axonal transport through disturbances of microtubules, microfilaments, or neurofilaments—cytoskeletal structures underlying axonal transport—would decrease the number of available NTF receptors. Incorporation of the receptor into the neuronal membrane and its subsequent selective exposure to the NTF will also greatly influence trophic effectiveness.

Intraneuronally (i.e., within the target neuron), putative second messengers would take over and distribute the influence of the NTF to various neuronal machineries. Surface degradation, intake, and intracellular destruction of the NTF would have to be considered possible causes of trophic deficits. The production of second messages (i.e., the "transduction" process) and their distribution inside the neuron may be regulated by other intrinsic or extrinsic influences, unrelated yet concurrent with the NTF ones. Lastly, the cellular machineries influenced by the NTF may also operate to different extents depending on other regulatory conditions (e.g., vascular or local supplies of oxygen, nutrients, hormones). It is at this point that our lack of fundamental information on these regulatory forces and mechanisms prevents further formulations of the neuronotrophic hypothesis.

The majority of the information that has contributed to the ability to make this formulation has been derived from studies of developing peripheral nervous system neurons *in vivo* and *in vitro*. Recent evidence concerning the presence and function of nerve growth factor in the central nervous system of adult mammals has made the above hypothesis relevant to the adult damaged nervous system and potentially relevant to the recovery of function following damage.

4. NERVE GROWTH FACTOR: ITS PRESENCE AND COMPETENCE IN THE CNS

Nerve growth factor occurs in the male mouse submaxillary gland (the most abundant source for its purification) as a 7-S complex of three different subunits (Varon *et al.*, 1968) and Zn^{2+} (Greene and Shooter, 1980; Yankner and Shooter, 1982). The NGF activity resides within the β subunit, a 25.3-kDa basic dimer protein. The β-NGF (also known as 2.5-S NGF) from the mouse and human has been sequenced and cloned (Scott *et al.*, 1983; Ullrich *et al.*, 1983). As previously stated, NGF has been extensively studied in the peripheral nervous system and has been found to be a potent NTF for sensory and sympathetic neurons. Because of this influence on peripheral catecholamine neurons, several attempts have been undertaken to determine whether NGF affects central catecholamine neurons as well (Bjorklund and Stenevi, 1972). To date the conclusion is that NGF is not effective on central catecholamine neurons *in vivo* (Drey-fus *et al.*, 1980) or *in vitro* (Konkol *et al.*, 1978). Recent experiments, however, provide strong evidence that NGF may act as a neuronotrophic factor for central cholinergic neurons in very much the same way that it affects peripheral sympathetic neurons.

Nerve growth factor and the mRNA coding for it are present in the rat brain, and their concentrations correlate well with the innervation territory of rat brain cholinergic neurons (Korsching *et al.*, 1985). For example, the levels of NGF and its mRNA are highest in the hippocampal formation and neocortex, the cholinoceptive target areas for the septum and nucleus basalis cholinergic neurons, respectively, whereas cerebellum and optic tectum (areas of little or no cholinergic innervation) are low in NGF. Furthermore, when radioactively labeled NGF was injected into either the hippocampus or the cortex, it was taken up into presynaptic terminals and retrogradely transported to the cholinergic cells of origin in the septal/diagonal band and nucleus basalis regions, respectively, but not to other brain areas projecting to the same injection sites such as the locus ceruleus or substantia nigra (Schwab *et al.*, 1979; Seiler and Schwab, 1984). Together these observations support the idea that NGF can be synthesized in the cholinergic target tissues and selectively transported retrogradely to the cholinergic neuronal bodies.

Actual synthesis of NGF in the hippocampus has recently been shown by several experimental approaches. Collins and Crutcher (1985) measured NGF-like activity in medium that had been conditioned by slices of rat hippocampus taken from animals that received either a septal lesion or a control lesion not interrupting the septohippocampal pathway. One week following the lesion, there was a significant increase in NGF-like activity in the medium containing the slice from the denervated hippocampus as opposed to the slice from the intact, nonlesioned hippocampus. Korsching and colleagues (1985) measured NGF and mRNA for NGF in hippocampal tissue extracts following fimbria–fornix transection and found that NGF protein levels, but not mRNA for NGF, were significantly increased by 2 weeks following the lesion and returned to normal levels within 4 weeks. Yoshida *et al.* (1986) also found that septal deafferentation resulted in an increased growth-promoting effect of hippocampal extracts when tested on central nervous system cell cultures. In order to assess the functionality of this increase in NGF-like activity in the brain, Gage *et al.* (1984) implanted peripheral sympathetic neurons from neonatal superior cervical ganglia (which are dependent on NGF for their survival for the first 5 postnatal days) in the adult rat hippocampus with or without a fimbria–fornix transection. Only in the presence of a fimbria–fornix lesion did the neonatal superior cervical ganglion neurons survive, and under these conditions the neurons grew to adult size and innervated the partially denervated hippocampus. In a subsequent study Gage and Bjorklund (1986), using the same paradigm, grafted embryonic septal/diagonal band neurons to the hippocampus in the presence or absence of fimbria–fornix lesions and found that the central cholinergic neurons as well as the rest of the grafted neurons survived and matured better in the presence of a fimbria–fornix transection.

These studies not only support the idea that NGF is synthesized in the source area of the hippocampus and neocortex and retrogradely supplied to the cholinergic cells of origin, they also indicate that interruption of this transport pathway results in an accumulation of NGF in the source region (i.e., hippocampus). Such an accumulation of NGF has been proposed to be responsible for the anomalous ingrowth of host peripheral sympathetic fibers into the hippocampal formation that has also been observed following fimbria–fornix deafferentation (Crutcher and Davis, 1981). This link was recently supported by Springer and Loy (1985) when they demonstrated that injecting anti-NGF into the denervated hippocampus prevented sympathetic sprouting in the vicinity of the injection site.

It must be emphasized that the cellular sources of the NGF in the hippocampus or neocortex are at present unknown. Some evidence supports a role for granule cells in the dentate gyrus and their mossy fibers as a source of NGF in the hippocampal formation. Zinc, which participates in the 7-S complex of the NGF molecule from mouse submaxillary gland, occurs at high levels in granule cells and mossy fibers. Zinc content also rises in the hippocampus following septal deafferentation, peaking at 2 weeks and returning to normal at 4 weeks (Stewart *et al.*, 1984), a time course similar to the already discussed rise of NGF following fimbria–fornix transection (Collins and Crutcher, 1985; Korshing *et al.*, 1985). Additional, though more controversial, support comes from data showing that colchicine injections, which destroy granule cells, also prevent the septohippocampal sprouting induced by septal damage (Peterson and Loy, 1983; Crutcher and Davis, 1982; Kesslak *et al.*, 1987). Another possible source is the astroglial cells, which, like Schwann cells of the PNS, have the capacity of supporting NGF-dependent neurons in culture, and their activity is blocked by anti-NGF antibodies (Varon *et al.*, 1974; Lindsay, 1979; Varon and Somjen, 1979; Rudge *et al.*, 1985; Manthorpe *et al.*, 1986). Thus, astrocytes could be an alternate or additional CNS source of NGF in the intact as well as denervated hippocampus.

5. NERVE GROWTH FACTOR'S FUNCTIONAL ROLES IN THE CNS

The now well-documented presence of NGF in the CNS and its ability to provide trophic support for host or implanted peripheral sympathetic neurons do not by themselves demonstrate that NGF plays an actual role for CNS cholinergic or other neurons. Evidence in this direction has, however, also accumulated in recent times. Johnson and colleagues (Tanuichi and Johnson, 1985; Tanuichi *et al.*, 1986) have demonstrated and measured the occurrence of NGF receptors in the rat brain. Autoradiographic techniques have shown these receptors to be localized to, among other regions, the septum (Schwab *et al.*, 1979) and nucleus basalis (Seiler and Schwab, 1984). It is not yet known, however, whether the NGF receptors are selectively restricted to cholinergic neurons or are also present on other neuronal types (and possibly glial cells) in the same brain regions.

Support for a trophic role of NGF on central cholinergic neurons also comes from the demonstration of functional responses by these cells to NGF. Recent *in vitro* experiments suggest that NGF supports the survival and outgrowth of fetal central cholinergic neurons (Martinez *et al.*, 1985), though this had previously been stated not to be the case (Hefti *et al.*, 1985). More directly, *in vivo* experiments confirm a direct role for NGF on central cholinergic neurons. Specifically, NGF injected directly into the brains of newborn animals results in the increase in choline acetyltransferase (ChAT) activity in the cortex, hippocampus, septum, and striatum (Mobley *et al.*, 1985). In addition, injections of anti-NGF into 15-day embryonic rats results in a decrease in levels of ChAT in the adult animals, suggesting an important role for NGF in the development of the CNS cholinergic system (Otten *et al.*, 1985).

In order to determine more directly the effects of NGF on adult neurons, several laboratories have begun to test the effects of chronic administration of NGF on cell survival following damage to the CNS. Gage, Bjorklund, Varon, and their colleagues (Gage *et al.*, Williams *et al.*) set out specifically to determine whether the adult septal cholinergic neurons were responsive to NGF, and more directly whether their survival could be influenced by the exogenous administration of NGF. As a first step, a model was set up that demonstrated a loss of cholinergic (acetylcholinesterase-positive) neurons as well as noncholinergic neurons in the septal–diagonal band region following fimbria–fornix transection (Gage *et al.*, 1986). This loss amounted to 70% of the cholinergic neurons in the medial septum and 50% of the cholinergic neurons in the vertical limb of the diagonal band. In the next set of experiments, NGF (11 μg/ml of 7-S NGF: 10,000 trophic units/ml) was infused continuously into the lateral ventricle adjacent to the septal area (Williams *et al.*, 1986). The animals were sacrificed at 2 weeks following the fimbria–fornix lesion (a time when cholinergic cell death would reach maximal levels), and the cell loss was evaluated. The NGF had prevented the death of 70% of the endangered cholinergic neurons in the medial septum and 100% of those in the diagonal band. In addition, NGF also reduced the death of noncholinergic neurons by 50% and 60%, respectively. Hefti (1986) has independently observed cholinergic cell protection by NGF using repeated injections of NGF intraventricularly into animals with partial fimbria–fornix lesions. This procedure consequently results in a more modest initial cell death in the septum, though a comparable cell savings was observed.

These results clearly demonstrate that NGF can have a protective effect on damaged cholinergic neurons that would otherwise die. At present it is difficult to make a definitive statement concerning the specificity of the survival-promoting effect of NGF, i.e., how NGF also supports the survival of noncholinergic neurons. However, cell counts indicate that the cholinergic neurons in the septal area only account for about 10–15% of the total neuronal population in the area. Thus, estimates based on these measurements raise the intriguing possibility that NGF may be acting directly or indirectly as a general NTF on septohippocampal projecting neurons.

In addition to the effects of NGF on cell survival, effects of NGF on behavioral recovery have also been assessed. In several different lesion paradigms, it has been demonstrated that a single or multiple injections will increase the rate at which predicted recovery of function will take place (Hart *et al.*, 1978; Kimble *et al.*, 1979; Stein and Will, 1983; Will and Hefti, 1985). In all these situations the damaged animals would have subsequently recovered, so at present it can only be said that NGF had a facilitatory effect on processes already at work. In addition, no evidence was presented as to any one specific system or set of systems on which NGF could be acting to effect the recovery cited in these studies.

6. SUMMARY AND CONCLUSIONS

There is growing evidence that there are macromolecules in the central nervous system that are directly responsible for viability of CNS neurons. These special proteins, generically called neuronotrophic factors (NTF), are present in high concentrations during development and are more than likely responsible for the initial growth and development of neurons and their processes. In the adult the presence of these factors may play a maintenance role in the health of normal cells. During a trauma that interrupts the normal function of these factors, the dependent cells will be in jeopardy. Age-related neurodegenerative diseases may be in part related to dysfunction in specific NTFs.

Neuronotrophic factors may be produced in neurons or glia or both and are thought to interact with specific neuronal receptors. The interaction with a receptor at the presynaptic axonal ending can lead to endocytotic incorporation of the factor/receptor complex and active retrograde transport to the cell soma. Whether the primary action of the NTF is at the receptor level, leading to a cascade of intracellular events via a secondary messenger, or via direct action of the NTF on intracellular molecular targets is not yet fully ascertained.

At present only a few NTFs have been identified, characterized, and purified; however, more are already at various stages of characterization. It appears likely that individual NTFs will be active on selective classes of neurons. Other NTFs may address broad classes of cells and perhaps elicit from them specific cell behaviors, e.g., survival, axonal elongation, transmitter regulation. The purification, characterization, and eventual sequencing of these molecules will undoubtedly lead to methods for obtaining sufficient quantities of the purified factors and the opportunity to test directly the influence of these molecules on cell function and the behavior of whole organisms *in vivo*. We are at the dawn of a field of study in the neurosciences that will bring dramatic new insights to basic neurobiology that may potentially provide direct clinical applications.

ACKNOWLEDGMENTS. We thank Sheryl Christenson for the excellent typing and illustration of Fig. 1. The research leading to the review was supported in our laboratories by the Office of Naval Research, the Margaret and Herbert J. Hoover, Jr. Foundation, and NIA (AGO6088).

REFERENCES

Aguayo, A. J., 1985, Axonal regeneration from injured neurons in the adult mammalian central nervous system, in: *Synaptic Plasticity* (C. W. Cotman, ed.), Guilford Press, New York, pp. 457–484.

Appel, S. H., 1981, A unifying hypothesis for the cause of amyotrophic lateral sclerosis, parkinsonism, and Alzheimer disease, *Ann. Neurol.* **10**:499–505.

Barbin, G., Manthorpe, M., and Varon, S., 1984, Purification of the chick eye ciliary neuronotrophic factor (CNTF), *J. Neurochem.* **43**:1468–1478.

Barde, Y. A., Edgar, D., and Thoenen, H., 1982, Purification of a new neurotrophic factor from mammalian brain, *EMBO J.* **1**:549–553.

Berg, D. K., 1984, New neuronal growth factors, *Annu. Rev. Neurosci.* **7**:149–170.

Bjorklund, A., and Stenevi, U., 1972, Nerve growth factor: Stimulation of regenerative growth of central noradrenergic neurons, *Science* **175**:1251–1253.

Black, I. B., 1978, Regulation of autonomic development, *Annu. Rev. Neurosci.* **1**:183–214.

Collins, F., and Crutcher, K. A., 1985, Neurotrophic activity in the adult rat hippocampal formation: Regional distribution and increase after septal lesion, *J. Neurosci.* **5**:2809–2814.

Cotman, C. W., and Nieto-Sampedro, M., 1984, Cell biology of synaptic plasticity, *Science* **225**:1287–1294.

Cowan, W. M., Fawcett, J. W., O'Leary, D. D. M., and Stanfield, B. B., 1984, Regressive events in neurogenesis, *Science* **225**:1258–1265.

Crutcher, K. A., and Davis, J. N., 1981, Sympathetic noradrenergic sprouting in response to central cholinergic denervation, *Trends Neurosci.* **4**:70–72.

Crutcher, K. A., and Davis, J. N., 1982, Target regulation of sympathetic sprouting in the rat hippocampal formation, *Exp. Neurol.* **75**:347–359.

Cunningham, T. J., 1982, Naturally occurring neuron death and its regulation by developing neural pathways, *Int. Rev. Cytol.* **74**:163–186.

Dreyfus, C., Peterson, E. R., and Crain, S. M., 1980, Failure of nerve growth factor to affect fetal mouse brain stem catecholaminergic neurons in culture, *Brain Res.* **194**:540–547.

Gage, F. H., and Bjorklund, A., 1986, Enhanced graft survival in the hippocampus following selective denervation, *Neuroscience* **17**:89–98.

Gage, F. H., Bjorklund, A., and Stenevi, U., 1984, Denervation releases a neuronal survival factor in adult rat hippocampus, *Nature* **308**:637–639.

Gage, F. H., Wictorin, K., Fischer, W., Williams, L. R., Varon, S., and Bjorklund, A., 1986, Cell loss and sprouting of cholinergic neurons in medial septum and diagonal band following fimbria-fornix transection: Quantitative temporal analysis, *Neuroscience* **19**(1):241–255.

Greene, L. A., and Shooter, E. M., 1980, The nerve growth factor: Biochemistry, synthesis, and mechanism of action, *Annu. Rev. Neurosci.* **3**:353–402.

Hamburger, V., and Oppenheim, R. N., 1982, Naturally occurring neuronal death in vertebrates, *Neurosci. Comment* **1**:39–55.

Hart, T., Chaimas, N. B., Moore, R. Y., and Stein, D. G., 1978, Effects of nerve growth factor on behavioral recovery following caudate nucleus lesions in rats, *Brain Res. Bull.* **3**:245–251.

Hefti, F., 1983, Alzheimer's disease caused by a lack of nerve growth factor? *Ann. Neurol.* **13**:109–110.

Hefti, F., 1986, Nerve growth factor (NGF) promotes survival of septal cholinergic neurons after injury, *J. Neurosci.* **6**:2155–2162.

Hefti, F., and Weiner, W. J., 1986, Nerve growth factor and alzheimer's disease, *Ann. Neurol.* **20**:275–281.

Hefti, F., Hartikka, J. J., Eckenstein, R., Gnahn, H., Heumann, R., and Schwab, M., 1985, Nerve growth factor increases choline acetyltransferase but not survival or fiber outgrowth of cultured fetal septal cholinergic neurons, *Neuroscience* **14**:55–68.

Herschman, H. R., Goodman, R., Chandler, C., Simpson, D., Cawley, D., Cole, R., and De Vellis, J., 1983, Is epidermal growth factor a modulator of nervous system function? *Birth Defects* **19**:79–94.

Johnson, E. M., 1983, An autoimmune approach to the study of nerve growth factor and other factors, in: *Growth and Maturation Factors*, Volume 5 (G. Guroff, ed.), John Wiley & Sons, New York, pp. 55–72.

Kesslak, J. P., Frederickson, C. J., and Gage, F. H., 1987, Quantification of hippocampal noradrenaline and zinc changes after selective cell destruction, *Exp. Brain Res.* **67**:77–84.

Kimble, D. P., BreMiller, R., and Perez-Polo, J. R., 1979, Nerve growth factor applications fail to alter behavior of hippocampal lesioned rats, *Physiol. Behav.* **23**:653–657.

Konkol, R. J., Mailam, R. B., Bendeich, E. G., Garrison, A. M., Mueller, R. A., and Breese, G. R., 1978, Evaluation of the effects of nerve growth factor and anti-nerve growth factor on the development of central catecholaminergic neurons, *Brain Res.* **144**:277–285.

Korsching, S., Auberger, G., Heumann, R., Scott, J., and Thoenen, H., 1985, Levels of nerve growth factor and its mRNA in the central nervous system of the rat correlate with cholinergic innervation, *EMBO J.* **4**:1389–1393.

Landmesser, L., and Pilar, A., 1978, Interactions between neurons and their targets during *in vivo* synaptogenesis, *Fed. Proc.* **37**:2016–2021.

Levi-Montalcini, R., 1966, The nerve growth factor: Its mode of action on sensory and sympathetic neurons, *Harvey Lect.* **60**:217–259.

Levi-Montalcini, R., 1982, Developmental neurobiology and the natural history of nerve growth factor, *Annu. Rev. Neurosci.* **5**:341–362.

Lindsay, R. M., 1979, Adult rat brain astrocytes support survival of both NGF-dependent and NGF-insensitive neurons, *Nature* **282**:80–82.

Manthorpe, M., and Varon, S., 1985, Regulation of neuronal survival and neuritic growth in the avian ciliary ganglion, in: *Growth and Maturation Factors*, Volume 3 (G. Guroff, ed.), John Wiley & Sons, New York, pp. 77–117.

Manthorpe, M., Skaper, S. D., Williams, L. R., and Varon, S., 1986, Purification of adult rat sciatic nerve ciliary neuronotrophic factor, *Brain Res.* **367**:282–286.

Marshall, J. F., 1984, Brain function: Neural adaptation and recovery from injury, *Annu. Rev. Psychol.* **35**:277–308.

Martinez, H. J., Dreyfus, C. F., Jonakait, G. M., and Black, I. B., 1986, Nerve growth factor promotes cholinergic development in brain striatal cultures, *Proc. Natl. Acad. Sci. U.S.A.* **82**:7777–7781.

Mobley, W. C., Rutkowski, J. L., Tennekoon, G. I., Buchanan, K., and Johnston, M. V., 1985, Choline acetyltransferase activity in striatum of neonatal rats increased by nerve growth factor, *Science* **229**:284–287.

Otten, U., Weskamp, G., Schlumpf, M., Lichtensteiger, W., and Mobley, W. C., 1985, Effects of antibodies against nerve growth factor on developing cholinergic forebrain neurons in the rat, *Soc. Neurosci. Abstr.* **11**:661

Patterson, P. H., 1978, Environmental determination of autonomic neurotransmitter functions, *Annu. Rev. Neurosci.* **1**:1–17.

Peterson, G. M., and Loy, R., 1983, Sprouting of sympathetic fibers in the hippocampus in the absence of major target cell candidates, *Brain Res.* **264**:21–29.

Prestige, M. C., 1970, Differentiation, degeneration and the role of the periphery: Quantitative considerations, in: *The Neurosciences: Second Study Program* (F. O. Schmitt, ed.), Rockefeller University Press, New York, pp. 73–82.

Rudge, J. S., Manthorpe, M., and Varon, S., 1985, The output of neuronotrophic and neurite promoting agents from rat brain astroglial cells: A microculture method for screening potential regulatory molecules, *Brain Res.* **19**:161–172.

Schwab, M. E., Otten, U., Agid, Y., and Thoenen, H., 1979, Nerve growth factor (NGF) in the rat CNS absence of specific retrograde axonal transport and tyrosine hydroxylase induction in locus coeruleus and substantia nigra, *Brain Res.* **168**:473–483.

Scott, J., Selby, M., Urdea, M., Quiroga, M., Bell, G. I., and Rutter, W. J., 1983, Isolation and nucleotide sequence of cDNA encoding the precursor of mouse nerve growth factor, *Nature* **302**:538–540.

Seiler, M., and Schwab, M. E., 1984, Specific retrograde transport of nerve growth factor (NGF) from cortex to nucleus basalis in the rat, *Brain Res.* **300**:33–36.

Springer, J. E., and Loy, R., 1985, Intrahippocampal injections of antiserum to nerve growth factor inhibit sympathohippocampal sprouting, *Brain Res. Bull.* **15**:629–634.

Stein, D. G., and Will, B. E., 1983, Nerve growth factor produces a temporary facilitation of recovery from entorhinal cortex lesions, *Brain Res.* **261**:127–131.

Stewart, G. R., Frederickson, C. J., Howell, G. A., and Gage, F. H., 1984, Cholinergic denervation-induced increase of chelatable zinc in mossy-fiber region of the hippocampal formation, *Brain Res.* **290**:43–51.

Tanuichi, M., and Johnson, E. M., 1985, Characterization of the binding properties and retrograde axonal transport of a monoclonal antibody directed against the rat nerve growth factor receptor, *J. Cell Biol.* **101**:1100–1106.

Tanuichi, M., Schweizer, J. B., and Johnson, E. M., 1986, Nerve growth factor receptor molecules in rat brain, *Proc. Natl. Acad. Sci. U.S.A.* **83**:1950–1954.

Thoenen, H., and Barde, Y.-A., 1980, Physiology of nerve growth factor, *Physiol. Rev.* **60**:1284–1335.

Thoenen, H., and Edgar, D., 1985, Neurotrophic factors, *Science* **229**:238–242.

Ullrich, A., Gray, A., Bermen, C., and Dull, T. J., 1983, Human beta-nerve growth factor gene sequence highly homologous to that of mouse, *Nature* **303**:821–825.

Varon, S., 1975, Nerve growth factor and its mode of action, *Exp. Neurol.* **48**:75–92.

Varon, S., 1985, Factors promoting the growth of the nervous system, *Discuss. Neurosci.* **2**(3):1–62.

Varon, S., and Adler, R., 1980, Nerve growth factors and control of nerve growth, *Curr. Top. Dev. Biol.* **16**:207–252.

Varon, S., and Adler, R., 1981, Trophic and specifying factors directed to neuronal cells, *Adv. Cell. Neurobiol.* **2**:115–163.

Varon, S., and Skaper, S. D., 1980, Short-latency effects of nerve growth factor: An ionic view, in: *Tissue Culture in Neurobiology* (E. Giacobini, A. Vernadadkis, and A. Shahar, eds.), Raven Press, New York, pp. 333–347.

Varon, S., and Skaper, S. D., 1983, The Na^+, K^+-pump may mediate the control of nerve cells by nerve growth factor, *Trends Biochem. Sci.* **8**:22–25.

Varon, S., and Somjen, G., 1979, Neuron–glia interactions, *Neurosci. Res. Prog. Bull.* **17**:1–239.

Varon, S., Nomura, J., and Shooter, E. M., 1968, Reversible dissociation of the mouse nerve growth factor into different subunits, *Biochemistry* **7**:1296–1303.

Varon, S., Raiborn, C., and Norr, S., 1974, Association of antibody to nerve growth factor with ganglionic non-neurons (glia) and consequent interference with their neuron-supportive action, *Exp. Cell Res.* **88**:247–256.

Wallicke, P., Cowan, W. M., Veno, N., Baird, A., and Guillemin, R., 1986, Fibroblast growth factor promotes survival of dissociated hippocampal neurons and enhances neurite extension, *Proc. Natl. Acad. Sci. U.S.A.* **83**:3012–3016.

Will, B., and Hefti, F., 1985, Behavioral and neurochemical effects of chronic intraventricular injections of nerve growth factor in adult rats with fimbria lesions, *Behav. Brain Res.* **17**:17–24.

Williams, L. R., Varon, S., Peterson, G., Wictorin, K., Fischer, W., Bjorklund, A., and Gage, F. H., 1986, Continuous infusion of nerve growth factor prevents basal forebrain neuronal death after fimbria–fornix transection, *Proc. Natl. Acad. Sci. U.S.A.* **83**:9231–9235.

Yankner, B. A., and Shooter, E. M., 1982, The biology and mechanism of action of nerve growth factor, *Annu. Rev. Biochem.* **51**:845–868.

Yoshida, K., Kohsaka, S., Idei, T., Otani, M., Toya, S., and Tsukada, Y., 1986, Septal deafferentation enhances the neurotrophic effects of rat hippocampus on cultured neural cells from the central nervous system, *Neurosci. Lett.* **66**:181–186.

15

Sensory Cortical Reorganization following Peripheral Nerve Injury

ROBERT W. DYKES and RAJU METHERATE

1. SOMATOTOPIC ORDER IN THE PRIMARY SOMATOSENSORY CORTEX

The somatotopic order of the body representation found in somatosensory cortex has been described in great detail in the monkey (Merzenich *et al.*, 1978; Kaas *et al.*, 1983), cat (Dykes *et al.*, 1980; Sretavan and Dykes, 1983; Felleman *et al.*, 1983), and several other species (Chapin and Lin, 1984; Wall and Cusick, 1984; Sur *et al.*, 1978, 1980). In each case, a linear progression of recording sites in the cortex produces a precisely organized sequence of receptive fields across the body. The pattern of receptive fields creates a map of the body with a surprising degree of resolution. For example, by recording from small clusters of neurons, Merzenich *et al.* (1978) observed neural activity elicited from receptive fields only a few millimeters in diameter. As the electrode was moved in horizontal steps as small as 50 μm, the receptive field loci shifted progressively until, when a cortical distance of 500 μm had been traversed, the receptive fields no longer overlapped the first ones encountered. In the rat, comparable groups of neurons may serve only one vibrissa (Welker, 1971). In many mammalian species, several maps of the body have been reported (cf. Dykes and Ruest, 1986). The conclusion seems inescapable that the cortex contains several high-resolution, precisely organized representations of the body surface.

The anatomic evidence for a comparable precision of anatomic connections is less compelling. A high-resolution, orderly progression of afferent terminals or intrinsic connections in somatosensory cortex that might form the basis of

ROBERT W. DYKES and RAJU METHERATE • Departments of Physiology, Surgery, and Neurology and Neurosurgery, McGill University, Montreal, Quebec H3A 1A1, Canada.

these functionally defined maps has not been demonstrated. Anatomic and physiological studies of somatosensory neurons show that individual cortical neurons often receive inputs (Dykes *et al.*, 1984) or distribute intracortical efferent connections (Landry and Deschênes, 1981; Landry *et al.*, 1986, 1987a,b) that are more extensive than those needed to give the observed resolution to the somatosensory map. That is, the individual functional units of the cortex have connections extending well beyond the dimensions of the resolution of a particular part of the map. Why this does not degrade the observed detail of the map is not clear (Fig. 1).

For example, the cortical region serving the dorsal surface of the first digit of the cat's forepaw may encompass an area of cortex measuring 0.5 by 0.75 mm (Sretavan and Dykes, 1983; Dykes and Gabor, 1981), but the thalamic arbors bringing somatosensory information to that region may extend anteroposteriorly for 1.0 mm and mediolaterally for 1.5 to 2.0 mm (Landry and Deschênes, 1981). Similarly, the basket cells of Marin-Padilla (1969) may extend several millimeters anteroposteriorly. Thus, although many cells in somatosensory cortex have more restricted arborizations (Jones, 1975), at least some important axons and cells distribute their influences to areas far greater than the distances required to produce the high-resolution somatotopy found in the cortex.

These observations imply that the somatotopic map found in cortex is not created by an order imposed on the cortex through its thalamic input or intrinsic anatomic structure but rather is an order somehow created, or at least sharpened in an important manner, by the functions of cortical neurons.

Recently, the degree to which cortical neuronal activity can control the details of the somatotopic map was emphasized in a dramatic way by the experiments of Merzenich and his colleagues (Merzenich *et al.*, 1983a,b, 1984; Jenkins *et al.*, 1984) in which they showed that the somatotopic map can be altered by manipulation of the afferent input. Their experiments demonstrated that (1) many details of somatotopically ordered maps are generated dynamically by the cortex and (2) somatotopic maps will change in adult animals if the dynamics of the sensory input are altered through deafferentation or through alterations in the use of a sensory surface. For example, experiments in owl monkeys showed that within 3 months after median nerve transection, a new somatotopic order replaced the old order in the region previously serving the median nerve. This new map, representing the back of the hand, the palm, and the digits served by the ulnar nerve, was as detailed and as precisely organized as the entirely different map that had been there prior to nerve transection.

To account for these changes, Edelman and Finkel (1984) invoked a concept termed degeneracy. They argued that the less precisely organized anatomic substrate of the cortical maps, because of the broad divergence of the thalamocortical fibers and the wide-ranging arbors of some cortical neurons, is capable of participating in more than one map. Conversely, the same map could potentially be created from a number of different synaptic arrangements. To

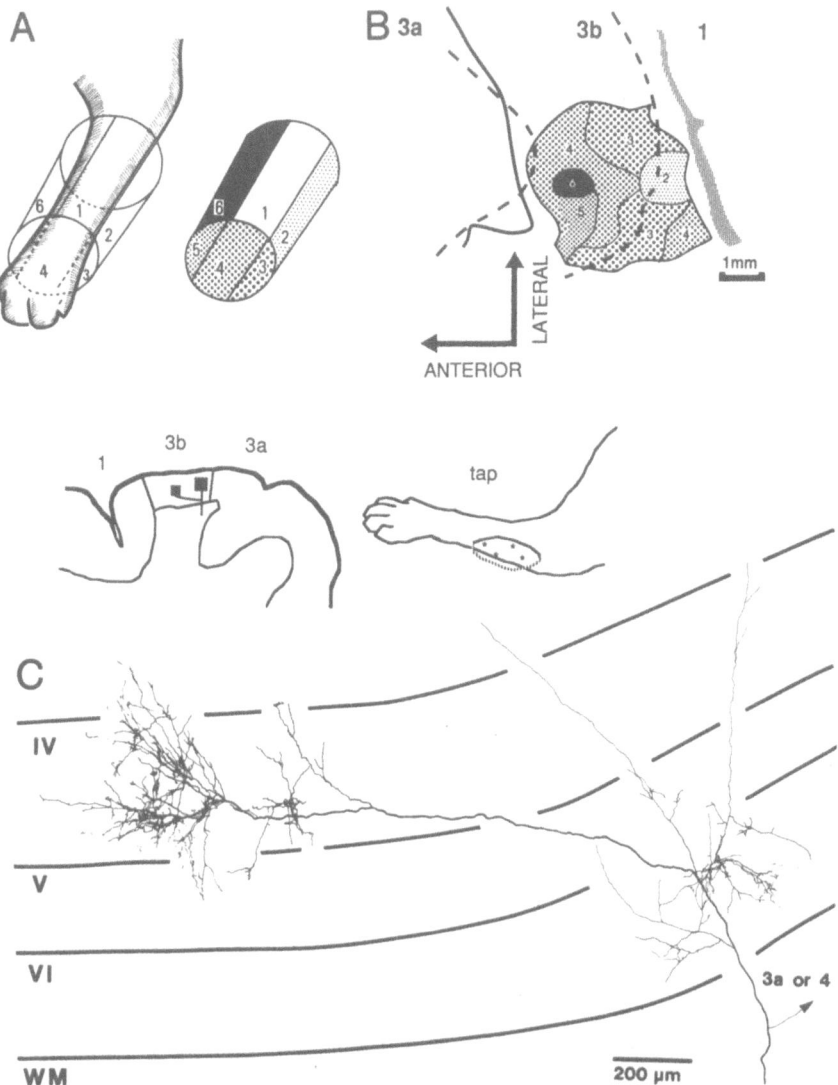

Figure 1. (A) A model of the forearm with the sectors of the arm shaded differently so that the region serving each sector may be indicated on the somatotopic map in the cortex. (B) A high-resolution map of part of the somatosensory cortex of the cat serving the forelimb region (modified from Sretavan and Dykes, 1983). (C) A camera lucida reconstruction of a single thalamocortical afferent fiber filled with horseradish peroxidase in the forelimb representation of the cat somatosensory cortex (modified from Landry *et al.*, 1987a,b). The first inset shows the arbor in relation to the cytoarchitectonic borders. The second inset shows the receptive field. By comparing the scales in B and C, it is possible to see that the thalamocortical fiber delivers information from a restricted forearm receptive field to a large cortical region. The areal distribution of the fiber is much greater than the area representing a cutaneous site corresponding to the receptive field area. This extensive dispersion of the afferent signal is not observed in the cortical maps or in the receptive field size of single cells, presumably because intracortical processing of the afferent signal allows only the central portion of the afferent drive to be reflected in the functional receptive field.

examine this very interesting idea, it is necessary to determine if the organization and behavior of individual neurons within the somatosensory cortex have the requisite properties to interact in the manner specified by Edelman and Finkel (1984). The following sections review evidence consistent with such a model and suggest a cellular mechanism that might participate in map formation and map reorganization.

2. CONTROL OF SOMATOTOPIC ORDER

If the somatosensory cortex can form more than one map, it is important to determine what controls the details of the actual map and the conditions under which the map can be changed to a different one. There seem to be two main possibilities. One, preferred by Edelman and Finkel (1984), is that the map is under a dynamic control mechanism that allows the details to be changed continuously on a temporal scale of hours, minutes, or even seconds. A similar model was suggested by McKenna et al. (1982) as a way to resolve the differences between their data, showing one somatosensory map, and other published data suggesting that multiple maps exist. They proposed that receptive field size and the locations of receptive field centers were quite labile and were influenced by motivational states and anesthetic levels to a degree that one map could fragment into several maps under anesthesia. It is known that some receptive fields of somatosensory cortical neurons do change with attention and motivational states (Hyvarinen et al., 1968; Sakata and Iwamura, 1978). In other cases receptive field thresholds change during volitional movements (Chapin and Woodward, 1981, 1982a,b), and active exploration of a surface changes perceptual thresholds (Coquery, 1978). However, these observations are not sufficient to conclude that the somatotopic order of a cortical map undergoes a major reorganization with changes in motivational state, attention, or the onset of a somatic stimulus.

If the maps changed with each motivational state, or with changes in anesthetics it would be difficult to explain how it is possible for receptive fields to be mapped repeatedly on the same body site in behaving animals (Mountcastle et al., 1975; Iwamura et al., 1985a,b) and how it is possible to obtain an orderly somatosensory map from an anesthetized animal when it requires hours or days to obtain that map (Merzenich et al., 1978; Dykes et al., 1980; Sretavan and Dykes, 1983; Kaas et al., 1983; Felleman et al., 1983). Nor would it be possible to explain how the same map can be obtained months later after a crushed median nerve has regenerated (Wall et al., 1983).

It appears that over extended time periods most neurons in somatosensory cortex retain most of their functional attributes, including a particular receptive field, and that the map retains its somatotopic order. Thus, major changes in somatotopy and novel receptive fields arise only under special circumstances. This does not preclude changes in levels of responsiveness of neurons with changes in behavioral state; however, changes leading to a new somatotopic order seem to be relatively slow and occur only under certain conditions. The best-documented cases of cortical reorganization (Kalaska and Pomerantz, 1979; Rasmusson, 1982; Rasmusson and Turnbull, 1983; Merzenich et al., 1983a, 1984; Wall and Cusick, 1984) suggest that even though some changes occur within minutes after deafferentation, the new somatotopic order requires many weeks to become stabilized as a new map. Further, the new map is attained only by progressing through a sequence of intervening steps (Merzenich et al., 1983b).

Some of the flexibility of responsiveness that has been observed in cortical neurons as a function of behavioral state (see references cited above) can be explained readily if there is an interplay of arousal mechanisms and the strong GABA-mediated inhibition that regulates receptive field size and excitability in somatosensory cortex (Hicks and Dykes, 1983; Dykes et al., 1984). How these rapid changes might come about is discussed elsewhere (Dykes, 1987). Longer-term, relatively permanent changes of excitability are discussed here.

3. ACETYLCHOLINE AS A PERMISSIVE AGENT FOR NEURONAL PLASTICITY

The rest of this chapter argues that some circumstances leading to a change in the somatosensory map produce an alteration in acetylcholine release or acetylcholine sensitivity of neurons in the cortical region undergoing reorganization. Acetylcholine is hypothesized to be a neuromodulator in somatosensory cortex, where it acts as a permissive agent allowing cortical neurons (1) to undergo changes in excitability that alter the magnitudes of their responses to afferent stimuli and (2) to develop receptive fields that were not previously present. This permissive agent is thought to act principally through muscarinic binding sites and to influence the excitability of cortical neurons through an already identified membrane process involving a reduction of potassium conductance (Krnjevic et al., 1971) that leads to an enhanced neuronal excitability in both anesthetized (Krnjevic et al., 1971) and conscious, behaving animals (Woody et al., 1978).

In the absence of evidence that sprouting of neuronal processes occurs in cortex undergoing reorganization, in this model it is assumed that the reorganization of the map occurs through preexisting synaptic contacts that were previously functional but did not cause the cell to discharge. The support for this assumption comes from three types of evidence. (1) Existing, widely branching thalamic arbors provide the anatomic basis for excitation from relatively distant cortical sites and consequently for excitation from distant body parts (Landry and Des-chênes, 1981; Landry et al., 1987b). (2) Iontophoretic application of bicuculline to single cortical cells blocks intracortical GABA-mediated inhibition, thereby uncovering strong and effective excitatory inputs to many neurons from relative large distances away on the body surface (Hicks and Dykes, 1983; Dykes et al., 1984). (3) The reorganization of somatosensory maps has a distance limit that corresponds roughly to the size limits of the thalamocortical arbors in somatosensory cortex, suggesting that this limit for reorganization depends on the size of preexisting thalamocortical arbors (Merzenich *et al.*, 1984).

The arguments that acetylcholine plays a key role in map reorganization arise from several different kinds of circumstantial evidence. (1) Acetylcholine release is increased during consciousness and increased alertness (Celesia and Jasper, 1966; Collier and Mitchell, 1967), two states that are correlated with changes in excitability. Its release has been shown to vary with EEG states (Jasper and Tessier, 1971) and learning behaviors (Rasmusson, 1975; Rasmusson and Szerb, 1975, 1976). (2) Interruption of pathways that bring the cholinergic and noradrenergic inputs to visual cortex prevent that sensory region from undergoing reorganization of ocular dominance bands (Bear and Singer, 1986). (3) Neuronal responses to iontophoretic applications of acetylcholine in deafferented cortex are abnormal in their time course and magnitude (Lamour and Dykes, 1988), suggesting an altered neuronal sensitivity to acetylcholine during the time that the cortex is undergoing reorganization. (4) Electrical stimulation of areas of the basal forebrain that release acetylcholine in somatosensory cortex, when paired with somatic stimuli, produce long-lasting enhancement of responses to subsequently applied somatic stimuli. This enhancement lasts for many hours after the forebrain stimulation (Rasmusson and Dykes, 1988).

4. NEURONAL RESPONSES FOLLOWING DEAFFERENTATION

To examine this hypothesis more carefully, consider the changes that occur in single cortical neurons in a deafferented zone following a major deafferentation. In the raccoon, changes begin immediately. Within minutes after the nerves to the fifth digit are interrupted, neurons in the fifth digit representation begin to respond to stimuli applied to the adjacent digit. At first the neurons, excited by stimuli to adjacent digits, respond to the removal of the stimulus as if the response had been triggered by a rebound from lateral inhibition (Fig. 2) (Rasmusson and Turnbull, 1983). After several weeks, the proportion of direct excitatory inputs begins to increase, and by 8 weeks most are direct excitatory responses. Multiunit recordings in owl monkeys show similar results; new receptive fields are uncovered so that previously ineffective inputs from adjacent skin regions begin to drive the cells (Merzenich *et al.*, 1983b). The uncovering of

Figure 2. (A) An illustration of the response of neurons in deafferented cortex. The response at the offset of the tactile stimulus suggests that afferent inhibition has been reduced and that a neuron not receiving significant excitatory drive will respond to the release from lateral inhibition. (B) The proportion of these responses (inhibitory) decreases with time, and 8–20 weeks after the injury, all responses occur at the onset of the stimulus (excitatory). (Reproduced with permission from Rasmusson and Turnbull, 1983.)

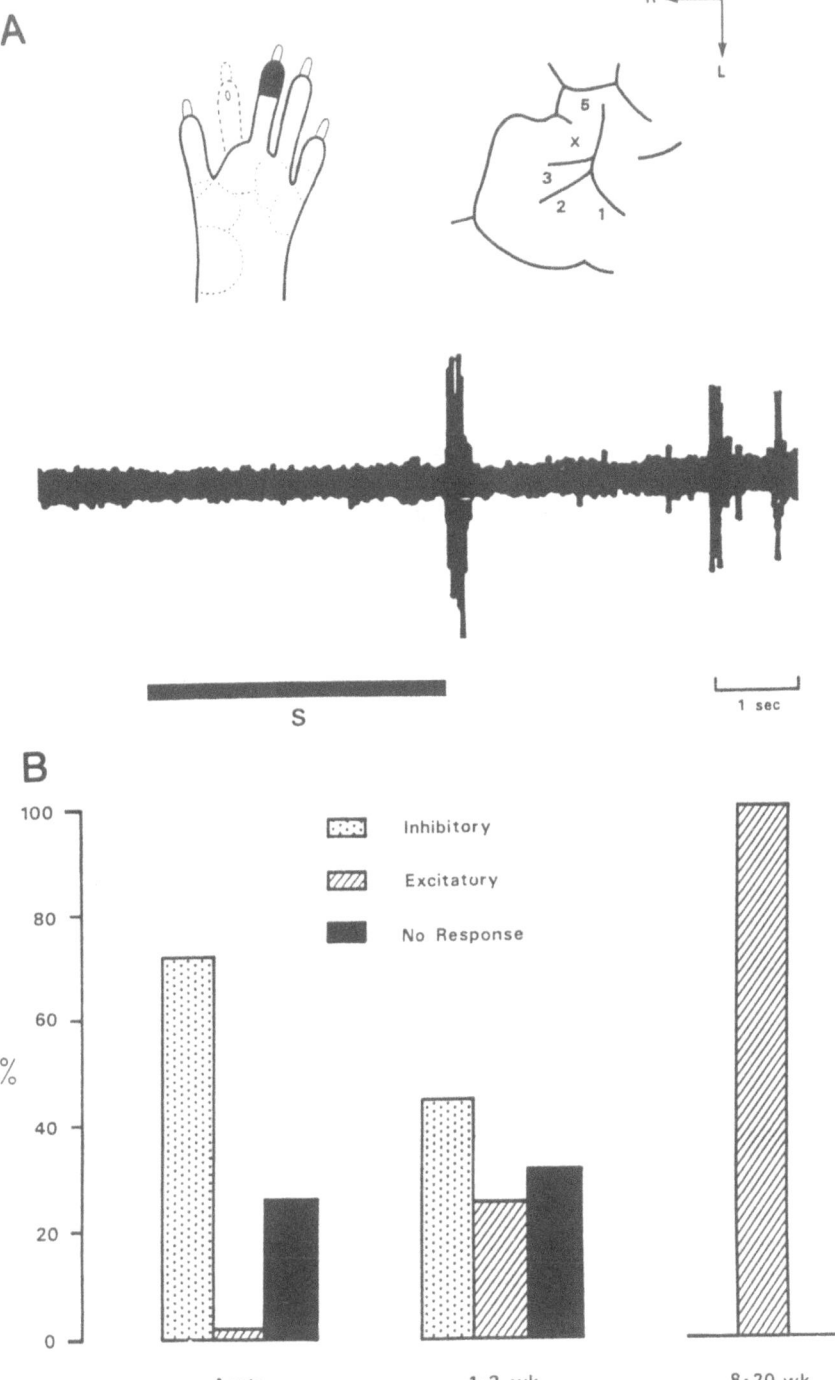

A

R ← ┐
      ↓
      L

5
X
3
2 1

S

1 sec

B

100 ┤

```
              ▫️ Inhibitory
              ▨ Excitatory
              ■ No Response
```

80 ┤

%

60 ┤

40 ┤

20 ┤

0 ┤

Acute   1-2 wk   8-20 wk

previously unexpressed inputs and the rebound responses, as well as comparable observations following manipulations at lower levels of the somatosensory pathway (Metzler and Marks, 1979; Wall, 1977), may be explained by hypothesizing that there is a removal of some form of tonic inhibitory control secondary to the deafferentation.

The loss of tonic inhibitory control following deafferentation has been documented in the hindlimb region of rat somatosensory cortex. Transection of the rat sciatic nerve removes about 85% of the normal afferent input to the hindlimb representation (Wall and Cusick, 1984). No studies of single cells in this area of the rat cortex are available for the time period immediately after deafferentation, but it is clear that within a few days many of the deafferented neurons have acquired new inputs, and within a few weeks most of the deafferented zone becomes part of an enlarged representation of the saphenous nerve input.

When individual neurons in this area containing the new map of the saphenous nerve input were studied with iontophoretic techniques, it became clear that deafferentation had removed inhibitory drive from the cortex undergoing reorganization. At least four observations suggested that there was less intracortical inhibition than was present in normal cortex. (1) The mean spontaneous activity of cortical neurons was at least twice normal throughout all cortical layers. (2) The number of spontaneously active cells was increased by 13%. (3) The proportion of cells depolarized by a fixed dose of glutamate was three times that observed in normal hindlimb cortex. (4) The number of neurons driven by somatic stimuli returned to the proportion seen in control animals despite the loss of 85% of the afferent input (Dykes and Lamour, 1988 a,b).

In normal rat somatosensory cortex, the response of a neuron to iontophoretically administered acetylcholine is a graded increase in discharge frequency that is generally slow in onset and tends to outlast the duration of the application (Krnjevic and Phillis, 1963). The response in layer V tends to have both a transient and a sustained component, whereas the response to cells in layer VIb has no transient (Lamour et al., 1982a). Following deafferentation this normal pattern was altered. In the granular cortex under reorganization, (1) the mean frequency of discharge produced by acetylcholine was lower than that in normal cortex, (2) more neurons were depolarized by acetylcholine, and (3) many more neurons were inhibited by acetylcholine (Lamour and Dykes, 1988).

Most striking, however, was the time course of the responses to acetylcholine observed in some neurons from deafferented cortex. These cells responded with a transient paroxysmal discharge that was suddenly arrested by a silent period that was, in turn, followed by another intense burst of impulses. These cycles, having a duration of about 12 sec, could be repeated for many minutes (Lamour et al., 1988). These data suggest that deafferentation altered the sensitivity of somatosensory cortical neurons to acetylcholine. Perhaps the reduced afferent drive produced by deafferentation changed the number or distribution of acetylcholine receptors on cortical neurons, thereby allowing acetylcholine to display a wider range of effects including some very dramatic ones. It is the hypothesis that these two effects observed in deafferented somatosensory cortex, the release from inhibition and the enhanced responsiveness to acetylcholine, interact in specific ways to produce long-term alterations in the excitability of cortical neurons seen as a new and different somatotopic order in the deafferented cortex. To explain how this occurs, it is necessary to discuss the effects of acetylcholine on normal somatosensory cortical neurons.

5. THE EFFECTS OF ACETYLCHOLINE ON NEURONS IN NORMAL SOMATOSENSORY CORTEX

As in other regions of cat cortex (Krnjevic and Phillis, 1961), 20% to 30% of the neurons in cat primary somatosensory cortex are excited by iontophoretic applications of acetylcholine (Tremblay et al., 1985; Spehlmann and Downes, 1974). However, this response was not a good predictor of whether or not acetylcholine could influence the response of a neuron to somatic stimulation; some neurons with no overt excitation by iontophoretically administered acetylcholine gave greater responses to somatic stimuli in the presence of acetylcholine than in control situations, and other cells excited by acetylcholine showed no change in responses to somatic stimuli during acetylcholine treatment (Metherate et al., 1986).

Even though the excitatory effect of acetylcholine could not be used as a predictor of its effect on responses to somatic stimuli, it was possible to show, in a series of 47 neurons from cat forelimb cortex, that 79% had an enhanced responsiveness to somatic stimulation during acetylcholine administration (Metherate et al., 1987). The enhanced responsiveness took several forms (Fig. 3). Most cells produced a larger response to somatic stimuli when acetylcholine was present. Others displayed a reduced threshold, and still others, previously without a receptive field, developed a cutaneous receptive field not obviously different from those seen in other neurons before the application of drugs. In contrast to the effects of iontophoretically administered bicuculline, the receptive fields were seldom enlarged by acetylcholine, and uncovered receptive fields were relatively normal in size. Comparable observations were made in rat cortex, where they were most frequently expressed in the middle layers (Lamour et al., 1988).

Similar data in visual cortex show that the responses to light can be altered by application of acetylcholine (Sillito and Kemp, 1983). The changes produced in receptive field properties by acetylcholine enhance the signal-to-noise ratio in visual cortical neurons.

In general, the effects described above are transient and disappear soon after the acetylcholine treatment is stopped. Yet, if acetylcholine is to play a role in cortical reorganization, it must be involved in long-term effects on neuronal excitability. Krnjevic et al. (1971) noted that the effects produced by acetylcholine did outlast the period of application, but for relatively short periods of

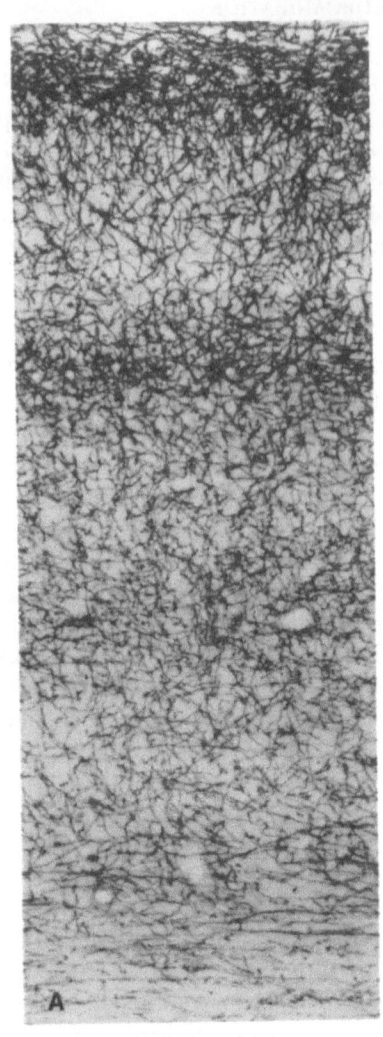

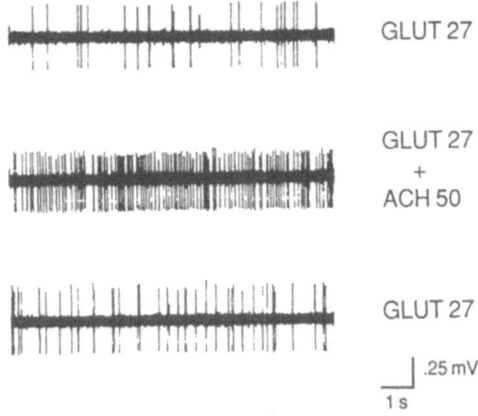

GLUT 27

GLUT 27
+
ACH 50

GLUT 27

.25 mV

1 s

only a few minutes, Woody *et al.* (1978) were the first to note that acetylcholine produced changes in membrane resistance that lasted for as long as the neuron was studied. The longest effect they recorded was 1.5 hr.

During a search for long-lasting effects of acetylcholine in cat somatosensory cortex, Metherate *et al.* (1987) demonstrated that when acetylcholine was presented alone, only 20% of 34 cells showed any enhanced excitability to subsequently applied doses of glutamate and that most of these enhancements disappeared within 5 minutes. In contrast, when acetylcholine was administered simultaneously with a depolarizing pulse of glutamate, 61% of the 41 cells tested demonstrated enhanced responses, and most of those were enhanced for periods of time greater than 5 min. In several cases the enhancement lasted beyond 1 hr.

When the glutamate test of excitability was replaced by a test of responsiveness to somatic stimulation, acetylcholine administered to somatosensory cortical neurons produced comparable enhancements of responsiveness to somatic stimuli if the acetylcholine was presented while the cell was being depolarized by somatic inputs. In the cat cortex, 37 of 47 cells treated in this way showed an enhanced responsiveness to somatic stimuli for periods up to 1.5 hr (Fig. 4).

These observations were confirmed in rat somatosensory cortex (Lamour *et al.*, 1987). When urethane anesthesia was used rather than sodium pentobarbital as had been used in the cat, about half of the neurons excited by somatic stimuli in the presence of acetylcholine showed prolonged enhancement of responses to somatic stimulation. In both of these studies previously silent neurons developed receptive fields if they could be excited in the presence of acetylcholine, and a significant proportion of the uncovered receptive fields remained effective for long periods of time. Perhaps most interesting was the observation, in both cat and rat somatosensory cortex, that if the enhanced responses that developed during acetylcholine treatment underwent a subsequent period of growth a few minutes later, then the enhanced responses continued for the duration of the time that the cell was studied (Lamour *et al.*, 1988).

6. CELLULAR MECHANISMS

The cellular mechanisms mediating the effects of acetylcholine are well studied. Several binding sites for acetylcholine can be found on neuronal cell

←

Figure 3. (A) Organization of cholinergic inputs to cat visual cortex as shown by histochemical staining of fibers containing acetylcholinesterase. Notice the differential density of the fibers throughout the cortex and their different orientations in some layers (reproduced with permission from Bear *et al.*, 1986). (B) The response of a neuron to glutamate is enhanced by simultaneous application of acetylcholine. Often this effect is transient and disappears within minutes after the acetylcholine is removed.

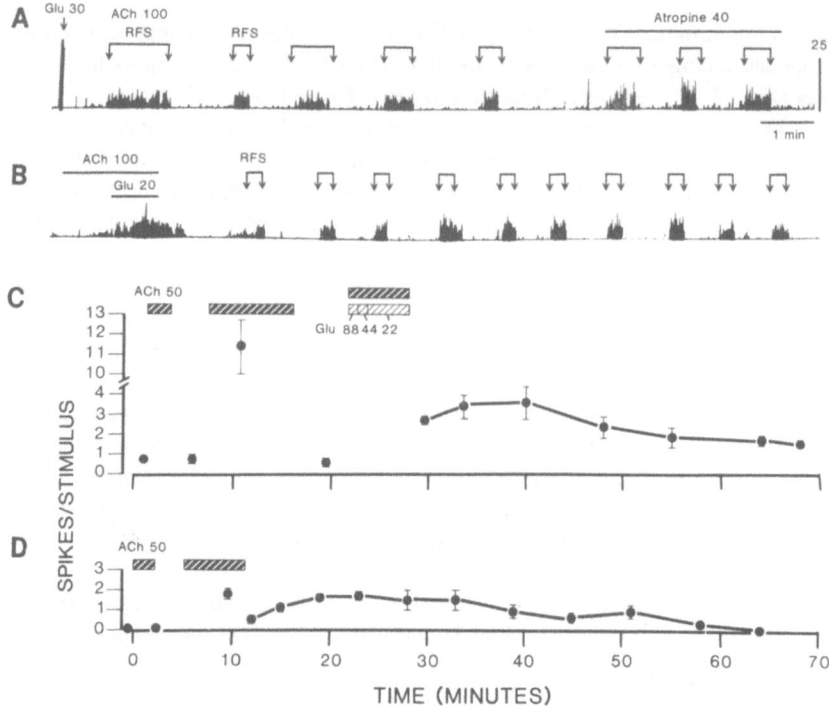

Figure 4. Examples of the effects of acetylcholine in somatosensory cortex. In a silent cell in rat somatosensory cortex, glutamate was administered iontophoretically from a multibarrel pipette to excite the cell. A receptive field was located during glutamate administration that subsequently disappeared. It reappeared when 100 nA of acetylcholine was released from another barrel of the pipette. The receptive field was stimulated (RFS) during acetylcholine treatment, and for the following 12 min the receptive field remained occult. This long-lasting effect of acetylcholine was not reversed by atropine applied iontophoretically 8 min after the treatment with acetylcholine. (B) A second example of an enhancement with acetylcholine that had a long-term effect. In this case the cell required both acetylcholine and glutamate to be enhanced but subsequently remained responsive for more than 15 min. (C) A spontaneously active cell from cat cortex with a previously existing receptive field. The responsiveness was enhanced during acetylcholine treatment, but the effect lasted only a few minutes. (D) A previously unresponsive cell in cat somatosensory cortex that displayed an increased responsiveness when the field was stimulated during iontophoretic application of acetylcholine. The enhanced responsiveness lasted for 50 min.

membranes. Each has different properties, not all are found on each cell, and the consequences of binding differ with the binding site involved (Burgen, 1984; MacIntosh, 1984). There are at least two different nicotinic binding sites found on neurons. One of these has been recently isolated, its amino acid sequence determined, and its orientation in the cell membrane determined (Changeux *et al.*, 1984). In the mammalian brain, however, the nicotinic binding sites, which allow acetylcholine to act as a traditional rapid-acting neurotransmitter, are far less common than muscarinic acetylcholine binding sites (Morley *et al.*, 1983).

There are two and perhaps three (MacIntosh, 1984) different muscarinic binding sites. In the brain the M_1 subtype may be located on postsynaptic surfaces and the M_2 subtype on presynaptic terminals (Mash *et al.*, 1985). They appear to have properties similar to those of muscarinic receptors found in peripheral tissues (Krnjevic, 1975). Acetylcholine binding to muscarinic receptors produces much slower responses than binding to nicotinic receptors. Typically, iontophoretic application of acetylcholine to a cell with muscarinic receptors produces a slow onset of firing over an interval of 5 to 50 sec. This is produced by a depolarizing shift in membrane potential that differs from most neuronal depolarizations because (1) it results from a selective increase in membrane resistance for potassium ions and (2) it does not inactivate in time (Krnjevic *et al.*, 1971).

Because of the reduced permeability to potassium ions, the cell membrane potential can move further away from the potassium equilibrium potential (about -100 mV), and it takes longer for the cell to repolarize and therefore to return to the resting state. As a result, cells with low potassium permeability are more readily excited than they would be otherwise. This mechanism has been confirmed to be active in awake, behaving cats studied with chronically implanted recording devices (Woody and Engel, 1972).

More important for the present discussion is the discovery that acetylcholine binding to muscarinic receptors can lead to a reduction of potassium permeability that outlasts the effective binding time of acetylcholine by many orders of magnitude. Woody *et al.* (1978) showed that if acetylcholine was administered to a cell while it was discharging, the increased membrane resistance, which normally dissipates in 4–5 min, could be prolonged for as long as the neuron could be held (up to 1.5 hr). The same effect could be produced by intracellular injections of cyclic GMP or calcium- and calmodulin-dependent protein kinase (Woody *et al.*, 1984). In contrast, prolonged effects did not follow either the application of acetylcholine by itself or depolarization by itself; the process critical for the prolonged change was the administration of acetylcholine to the cell while it was depolarized and discharging action potentials. Once this process has occurred, the membrane-permeability change is independent of acetylcholine, as demonstrated by the inability to reverse the effect with atropine. The long-term effect seems to be mediated by an intracellular second messenger.

The inhibitory effects of acetylcholine on cortical neurons are poorly understood and are complicated by the existence of both pre- and postsynaptic locations of acetylcholine receptors, perhaps having different effects. Recently, McCormick and Prince (1985) have suggested, from *in vitro* studies of guinea pig cingulate cortex, that the inhibition produced by acetylcholine is mediated by a rapid excitation of GABAergic interneurons. They argued that the acetylcholine receptors on the inhibitory interneurons are muscarinic receptors but differ in that they cause the cell to fire more quickly than do the muscarinic receptors on pyramidal cells, thereby causing a transient inhibition of the latter cells. This may be the factor that allows acetylcholine to change the balance between excitation and inhibition in cortical neurons.

7. THE HYPOTHESIS

The data reviewed above support the idea that changes in the levels of acetylcholine or changes in its efficacy could create a condition in which neurons in the somatosensory cortex would undergo changes in excitability. Specifically, in the presence of acetylcholine or some intracellular second messenger, there should occur long-term changes in potassium permeability in those neurons having excitatory drive sufficient to discharge them. The prolonged decrease in permeability would make the affected cell more responsive to subsequent sensory inputs. In both cat and rat cortex, acetylcholine treatments combined with depolarization have been shown to uncover new receptive fields and strengthen responses from existing ones. Thus, those neurons released from inhibition can now respond to new inputs, and, because of the enhanced sensitivity to acetylcholine in those areas, the responses that do occur will become strengthened further by the acetylcholine-mediated reduction of potassium permeability.

Note that not all possible inputs will be enhanced but only those that are actually activated by the use of the cutaneous surfaces remaining after a deafferentation. Thus, this mechanism has the virtue of enhancing only those responses that are elicited by the animal. There is the additional possibility, not explored here, that the release of acetylcholine from the basal forebrain that occurs during certain behaviors can further select, by the timing of its release, the particular afferent inputs that will be enhanced.

This hypothesis has the interesting attributes that (1) it does not require that a sprouting mechanism be invoked, (2) it enhances only certain inputs that are sufficient to drive the cell, and (3) neuronal change occurs only while the acetylcholine is present, thereby allowing a new somatotopic order to be locked into place by a subsequent reduction of the ambient acetylcholine levels or by a reduction of binding sites on the affected cell.

The need for the afferent input to drive the cell if it is to produce a long-term change in excitability is reminiscent of the requirements for learning set down by Hebb (1949), formalized by Edelman and Finkel (1984) among others, and recognized as a critical requirement in the reorganization described by Merzenich *et al.* (1983a,b; 1984). In this context, the release from inhibition that occurs after deafferentation may be a critical step in the process because this reduction of inhibition allows previously unexpressed inputs to drive a cell and thereby to become candidates for the new inputs.

The requirement for a cell to be excited before its responsiveness is enhanced may also account for the sequence of changes that pass over deafferented cortex while it is undergoing reorganization. Because of shared intracortical connections, some neurons may not be excited until their neighbors have been enhanced by acetylcholine. Then, as they begin to respond together with the neighbors, they facilitate the drive on other, adjacent cells. Subsequently enhanced cells may alter the drive on some previously enhanced cells, leading to a change in their excitatory drive and thereby further molding the somatotopic pattern. This mechanism may underlie the sequential reorganization of owl monkey cortex described by Merzenich et al. (1983b).

One major lacuna of this hypothesis is that there is no obvious way that inappropriate responses can be suppressed. Close examination of the sequence of changes that occur following deafferentation (Merzenich et al., 1983b) suggests that some inputs must also be suppressed during evolution of the new map. Perhaps this could come about through acetylcholine-mediated effects on inhibitory interneurons, but equally plausible is the possibility that one or more other neuromodulatory agents such as norepinephrine allow synaptic efficacy to be weakened in a manner analogous to the way acetylcholine strengthens excitatory drive. Norepinephrine is a compound with some of the requisite effects and has been implicated in neuronal plasticity in the visual cortex (Bear and Singer, 1986).

8. SUMMARY

The neuronal plasticity that requires minutes or hours to occur and that may lead to changes lasting indefinitely has been shown to have certain specific attributes. One that appears paramount but could be discussed only briefly here is that neural plasticity must be gated by a signal not involved in actually determining what the altered neuronal state will be. Singer (1983) has postulated a gating by brainstem mechanisms for the reorganization that occurs in visual cortex of young cats. In the case of reorganization following deafferentation, gating may arise from an intrinsic source triggered by a lack of afferent drive. In this case, the opening of the gate may result in a dramatic increase in muscarinic receptors for acetylcholine, which seem to appear subsequent to the loss of sensory input (Dykes and Lamour, 1988a). That the distribution of the acetylcholine binding sites can change in somatosensory cortex can also be inferred from the study by Lamour et al. (1982b) showing the spread of acetylcholine excitability in rat somatosensory cortex following lesions of the basal forebrain. In a study of 85 cells in rats 2 weeks after and 114 cells 3 weeks after bilateral electrolytic lesions of the basal forebrain, they saw an increase in the proportion of excited cells (60% and 51% versus 38% in the normal rat) and the appearance of excited cells in layers not normally containing cells affected by acetylcholine.

A second attribute of the neuronal plasticity that follows deafferentation is that it leads to a new but ordered state. In the auditory and visual cortex, altered sensory inputs combined with appropriate behavioral manipulations lead to changes in excitability that enhance responses to the auditory (Weinberger and Diamond, 1987) and visual stimuli (Singer and Rauschecker, 1982) presented during the gated period. In the somatosensory cortex, neuronal activity following deafferentation leads to a new somatotopic map establishing ordered relationships among the remaining innervated body surfaces. This requires that the new receptive fields of individual neurons fit into the new somatotopic order defined by the neuronal context in which they are located. Because a neuron must be driven by a somatosensory input before that input can be enhanced, the enhanced input will, by definition, have meaning in the context of the remaining sensory surface. If excitatory intracortical interactions play a significant role, then the receptive field of a neuron also will have meaningful relationships to the receptive fields of neighboring neurons. The result will be an orderly somatotopic map. This is a mechanism that might be called a group selection process by Edelman and Finkel (1984).

A third paramount characteristic of neuronal plasticity is that it has clearly defined limits; a cell is not omnipotent. In the visual cortex, after changes in ocular dominance columns, some cells remain sluggish or are still driven by the occluded eye; in the auditory system there are some structures that do not display neuronal plasticity at all (Weinberger and Diamond, 1987); and in the somatosensory cortex there are spatial boundaries beyond which reorganization will not occur (Merzenich *et al.*, 1984; Dykes and Ruest, 1986). These boundaries appear to be related to the distances covered by thalamic arbors, the presence of cytoarchitectonic boundaries, and the mechanoreceptor class involved. There are also limits to the number of neurons that can undergo plastic changes and limits to the enhancements that a particular cell can undergo (Lamour and Dykes, 1988). These limits imply that there will be only a certain degree of recovery following deafferentation or injury. These limits have important consequences for what should be expected from clinical rehabilitation programs (Dykes, 1984).

The working hypothesis that acetylcholine is at least one of the permissive agents involved in neuronal plasticity provides a theoretical framework for some of the recent electrophysiological, behavioral, and anatomic observations on plasticity in the somatosensory cortex and leads to testable predictions about this model.

ACKNOWLEDGMENTS. The authors are indebted to Mrs. Genevieve Holding for typographical and editorial work on the manuscript, to the Government of Quebec for salary support for Robert W. Dykes, and to the U.S. National Institute of Mental Health for salary support for Raju Metherate.

REFERENCES

Bear, M. F., and Singer, W., 1986, Modulation of visual cortical plasticity by acetylcholine and noradrenaline, *Nature* **320:**172–176.

Burgen, A. S. V., 1984, Muscarinic receptors—an overview, in: *Subtypes of Muscarinic Receptors* (B. I. Hirschowitz, R. Hammer, A. Giachetti, J. J. Keirns, and R. R. Levine, eds.), Elsevier, Amsterdam, pp. 1–3.

Celesia, G. G., and Jasper, H. H., 1966, Acetylcholine released from cerebral cortex in relation to state of activation, *Neurology* **16:**1053–1064.

Changeux, J. P., Devillers-Thiery, A., and Chemouilli, P., 1984, Acetylcholine receptor: An allosteric protein, *Science* **225:**1335.

Chapin, J. K., and Lin, C.-S., 1984, Mapping the body representation in the SI cortex of anesthetized and awake rats, *J. Comp. Neurol.* **229:**199–213.

Chapin, J. K., and Woodward, D. J., 1981, Modulation of sensory responsiveness of single somatosensory cortical cells during movement and arousal behaviors, *Exp. Neurol.* **72:**164–178.

Chapin, J. K., and Woodward, D. J., 1982a, Somatic sensory transmission to the cortex during movement: Gating of single cell responses to touch, *Exp. Neurol.* **78:**654–669.

Chapin, J. K., and Woodward, D. J., 1982b, Somatic sensory transmission to the cortex during movement: Phasic modulation over the locomotor step cycle, *Exp. Neurol.* **78:**670–684.

Collier, B., and Mitchell, J. F., 1967, The central release of acetylcholine during consciousness and after brain lesions, *J. Physiol. (Lond.)* **188:**83–98.

Coquery, J.-M., 1978, Role of active movement in control of afferent input from skin in cat and man, in: *Active Touch* (G. Gordon, ed.), Pergamon Press, New York, pp. 161–169.

Dykes, R. W., 1984, Central consequences of peripheral nerve injuries, *Ann. Plast. Surg.* **13:**412–422.

Dykes, R. W., 1987, Control of the neuronal receptive field in somatosensory cortex, *The Role of Neuroplasticity in the Response to Drugs* **78:**198–204.

Dykes, R. W., and Gabor, A. J., 1981, Magnification functions and receptive field sequences for submodality-specific bands in SI cortex of the cat, *J. Comp. Neurol.* **202:**597–620.

Dykes, R. W., and Lamour, Y., 1988a, An electrophysiological laminar analysis of single somatosensory neurons in deafferented rat hindlimb granular cortex subsequent to transection of the sciatic nerve, *Brain Res.*, (in press).

Dykes, R. W., and Lamour, Y., 1988b, An electrophysiological study of single somatosensory neurons in rat granular cortex serving the limbs: a laminar analysis, *J. Neurophysiol.* (in press).

Dykes, R. W., and Ruest, A., 1986, What makes a map in somatosensory cortex? in: *Cerebral Cortex*, Volume 5, *Sensory-Motor Areas and Aspects of Cortical Connectivity* (E. G. Jones and A. Peters, eds.), Plenum Press, New York, pp. 1–29.

Dykes, R. W., Rasmusson, D. D., and Hoeltzell, P., 1980, Organization of primary somatosensory cortex in the cat, *J. Neurophysiol.* **43:**1527–1546.

Dykes, R. W., Landry, P., Metherate, R. S., and Hicks, T. P., 1984, Functional role of GABA in cat primary somatosensory cortex: Shaping receptive fields of cortical neurons, *J. Neurophysiol.* **52:**1066–1093.

Edelman, G. M., and Finkel, L. H., 1984, Neuronal group selection in the cerebral cortex, in: *Dynamic Aspects of Neocortical Functions*, (G. M. Edelman, W. F. Gall, and W. M. Cowan, eds.), John Wiley and Sons, New York, pp. 653–695.

Felleman, D. J., Wall, J. T., Cusick, C. G., and Kaas, J. H., 1983, The representation of the body surface in S-I of cats, *J. Neurosci.* **3:**1648–1669.

Hebb, D. O., 1949, *The Organization of Behavior*, John Wiley & Sons, New York.

Hicks, T. P., and Dykes, R. W., 1983, Receptive field size for certain neurons in primary somatosensory cortex is determined by GABA-mediated intracortical inhibition, *Brain Res.* **274:**160–164.

Hyvarinen, J., Sakata, H., Talbot, W. H., and Mountcastle, V. R., 1968, Neuronal coding by cortical cells of the frequency of oscillating peripheral stimuli, *Science* **162:**1130–1132.

Iwamura, Y., Tanka, M., Sakamoto, M., and Hikosaka, O., 1985a, Diversity in receptive field properties of vertical neuronal arrays in the crown of the postcentral gyrus of the conscious monkey, *Exp. Brain Res.* **58:**400–411.

Iwamura, Y., Tanka, M., Sakamoto, M., and Hikosaka, O., 1985b, Vertical neuronal arrays in the postcentral gyrus signaling active touch: A receptive field study in the conscious monkey, *Exp. Brain Res.* **58:**412–420.

Jasper, J. J., and Tessier, J., 1971, Acetylcholine liberation from cerebral cortex during paradoxical (REM) sleep, *Science* **172:**601–602.

Jenkins, W. M., Merzenich, M. M., and Ochs, M. T., 1984, Behaviorally controlled differential use of restricted hand surfaces induce changes in the cortical representation of the hand in area 3b of adult owl monkey, *Soc. Neurosci. Abstr.* **10:**665.

Jones, E. G., 1975, Varieties and distribution of nonpyramidal cells in the somatic sensory cortex of the squirrel monkey, *J. Comp. Neurol.* **160:**205–268.

Kaas, J. H., Merzenich, M. M., and Killackey, H. P., 1983, The reorganization of somatosensory cortex following peripheral nerve damage in adult and developing mammals, *Annu. Rev. Neurosci.* **6:**325–356.

Kalaska, J., and Pomeranz, B., 1979, Chronic paw denervation causes an age-dependent appearance of novel responses from forearm in "paw cortex" of kittens and adult cat, *J. Neurophysiol.* **42:**618–633.

Krnjevic, K., 1975, Acetylcholine receptors in vertebrate CNS, in: *Handbook of Psychopharmacology,* Volume 6 (L. L. Iverson, S. D. Iverson, and S. H. Snyder, eds.), Plenum Press, New York, pp. 92–126.

Krnjevic, K., and Phillis, J. W., 1961, Sensitivity of cortical neurons to acetylcholine, *Experientia* **17:**469.

Krnjevic, K., and Phillis, J. W., 1963, Acetylcholine sensitive cells in the cerebral cortex, *J. Physiol. (Lond.)* **166:**296–327.

Krnjevic, K., Pumain, R., and Renaud, L., 1971, The mechanism of excitation by acetylcholine in the cerebral cortex, *J. Physiol. (Lond.)* **215:**247–258.

Lamour, Y., and Dykes, R. W., 1988, Somatosensory neurons in deafferented rat hindlimb granular cortex subsequent to transection of the sciatic nerve: Effects of glutamate and acetylcholine *Brain Res.* (in press).

Lamour, Y., Dutar, P., and Jobert, A., 1982a, Excitatory effect of acetylcholine on different types of neurons in the first somatosensory neocortex of the rat: Laminar distribution and pharmacological characteristics, *Neuroscience* **7:**1483–1494.

Lamour, Y., Dutar, P., and Jobert, A., 1982b, Spread of acetylcholine sensitivity in the neocortex following lesion of the nucleus basalis, *Brain Res.* **252:**377–381.

Lamour, Y., Dutar, P., Jobert, A., and Dykes, R. W., 1988, An iontophoretic study of single somatosensory neurons in rat granular cortex serving the limbs: A laminar analysis of glutamate and acetylcholine effects on receptive field properties *J. Neurophysiol.* (in press).

Landry, P., and Deschênes, M., 1981, Intracortical arborizations and receptive fields of identified ventrobasal thalamocortical afferents to the primary somatic sensory cortex in the cat, *J. Comp. Neurol.* **199:**345–371.

Landry, P., Diadori, P., and Dykes, R. W., 1987a, Postsynaptic responses evoked in primary somatosensory cortical neurons following stimulation of the ventroposterior lateral nucleus and the corpus callosum in the cat *Exp. Brain Res.* (in press).

Landry, P., Diadori, P., Leclerc, S., and Dykes, R. W., 1987b, Morphological and electrophysiological characteristics of somatosensory thalamocortical axons studied with intraaxonal staining and recordings in the cat, *Exp. Brain Res.* **65**:317–330.

MacIntosh, F. C., 1984, Subtypes of muscarinic receptors: A summary with comments, in: *Subtypes of Muscarinic Receptors* (B. D. Hurschowitz, R. Hammer, A. Jiachetti, G. G. Kevins, and R. R. Levine, eds.), Elsevier, Amsterdam, pp. 100–103.

Marin-Padilla, M., 1969, Origin of the pericellular baskets of the pyramidal cells of the human motor cortex: A Golgi study, *Brain Res.* **14**:633–646.

Mash, D. C., Flynn, D. D., and Potter, L. T., 1985, Loss of M_2 muscarinic receptors in the cerebral cortex in Alzheimer's disease and experimental cholinergic denervation, *Science* **228**:1115–1117.

McCormick, D. A., and Prince, D. A., 1985, Two types of muscarinic response to acetylcholine in mammalian cortical neurons, *Proc. Natl. Acad. Sci. U.S.A.* **82**:6344–6348.

McKenna, T. M., Whitsel, B. L., and Dryer, D. A., 1982, Anterior parietal cortical topographic organization in macaque monkey: A reevaluation, *J. Neurophysiol.* **48**:289–317.

Merzenich, M. M., Kaas, J. H., Sur, M., and Lin, C.-S., 1978, Double representation of the body surface within cytoarchitectonic areas 3b and 1 in "SI" in the owl monkey (*Aotus trivirgatus*), *J. Comp. Neurol.* **181**:41–74.

Merzenich, M. M., Kaas, J. H., Wall, J., Nelson, R. J., Sur, M., and Felleman, D., 1983a, Topographic reorganization of somatosensory cortical areas 3b and 1 in adult monkey following restricted deafferentation, *Neuroscience* **8**:33–55.

Merzenich, M. M., Kaas, J. H., Wall, J. T., Sur, M., Nelson, R. J., and Felleman, D. J., 1983b, Progression of change following median nerve section in the cortical representation of the hand in areas 3b and 1 in adult owl and squirrel monkeys, *Neuroscience* **10**:639–665.

Merzenich, M. M., Nelson, R. J., Stryker, M. P., Cynader, M. S., Schoppmann, A., and Zook, J. M., 1984, Somatosensory cortical map changes following digit amputation in adult monkeys, *J. Comp. Neurol.* **224**:591–605.

Metherate, R., Tremblay, N., and Dykes, R. W., 1986, Effects of acetylcholine on neuronal responses to somatic stimuli in cat somatosensory cortex, *Soc. Neurosci. Abstr.* **12**:797.

Metherate, R., Tremblay, N., and Dykes, R. W., 1987, Acetylcholine permits prolonged potentation of neural responsiveness in cat somatosensory cortex, *Neuroscience* **22**:75–81.

Metzler, J., and Marks, P. S., 1979, Functional changes in cat somatic sensory–motor cortex during short term reversible epidural blocks, *Brain Res.* **328**:97–104.

Morley, B. J., Farley, G. R., and Javel, E., 1983, Nicotinic acetylcholine receptors in the mammalian brain, *Trends Pharmacol.* **4**:225–227.

Mountcastle, V. B., Lynch, J. C., Georgopoulos, A., Sakata, H., and Acuna, C., 1975, Posterior parietal association cortex of the monkey: Command functions of operatious within extrapersonal space, *J. Neurophysiol.* **38**:871–908.

Rasmusson, D. D., 1975, The effect of stimulus cueing, stimulus modality and response inhibition on acetylcholine release from visual and sensorimotor corticies of awake, conditioned rabbits, Ph.D. Thesis, Dalhousie University, Halifax, Nova Scotia.

Rasmusson, D. D., 1982, Reorganization of raccoon somatosensory cortex following removal of the fifth digit, *J. Comp. Neurol.* **205**:313–326.

Rasmusson, D. D., and Dykes, R. W., 1988, Long-term enhancement of evoked potentials in cat somatosensory cortex produced by co-activation of the basal forebrain and cutaneous receptors *Exp. Brain Res.* (in press).

Rasmusson, D. D., and Szerb, J. C., 1975, Cortical acetylcholine release during operant behaviour in rabbits, *Life Sci.* **16**:683–690.

Rasmusson, D. D., and Szerb, J. C., 1976, Acetylcholine release from visual and sensorimotor cortices of conditioned rabbits. The effects of sensory cueing and patterns of responding, *Brain Res.* **104**:243–259.

Rasmusson, D., and Turnbull, B. G., 1983, Immediate effects of digit amputation of SI cortex in the raccoon: Unmasking of inhibitory fields, *Brain Res.* **288**:368–370.

Sakata, H., and Iwamura, Y., 1978, Cortical processing of tactile information in the first somatosensory and parietal association areas in the monkey, in: *Active Touch* (G. Gordon, ed.), Pergamon Press, Oxford, pp. 55–72.

Sillito, A. M., and Kemp, J. A., 1983, Cholinergic modulation of the functional organization of the cat visual cortex, *Brain Res.* **289**:143–155.

Singer, W., 1983, Neuronal mechanisms of experience-dependent self-organization of the mammalian visual cortex, *Acta. Morphol. Hung.* **31**:235–260.

Singer, W., and Rauschecker, J. P., 1982, Central Core control of developmental plasticity in the kitten visual cortex. II. Electrical activation of mesencephalic and diencephalic projections, *Exp. Brain Res.* **41**:223–233.

Spehlmann, R., and Downes, K., 1974, The effects of acetylcholine and of synaptic stimulation on the sensorimotor cortex of cats. I. Neuronal responses to stimulation of the reticular formation, *Brain Res.* **74**:229–242.

Sretavan, D., and Dykes, R. W., 1983, The organization of two cutaneous submodalities in the forearm region of area 3b of cat somatosensory cortex, *J. Comp. Neurol.* **213**:381–398.

Sur, M., Nelson, R. J., and Kaas, J. H., 1978, The representation of the body surface in somatosensory area I of the grey squirrel, *J. Comp. Neurol.* **179**:425–449.

Sur, M., Nelson, R. J., and Kaas, J. H., 1980, The representation of the body surface in somatic koniocortex in the prosimian, *Galago, J. Comp. Neurol.* **189**:381–402.

Tremblay, N., Metherate, R., and Dykes, R. W., 1985, Tactile stimulation and cholinergic modulation in the cat somatosensory cortex, *Can. J. Physiol. Pharmacol.* **63**:A.

Wall, J. T., and Cusick, C. G., 1984, Cutaneous responsiveness in primary somatosensory (S-I) hindpaw cortex before and after deafferentation in adult rats, *J. Neurosci.* **4**:1499–1515.

Wall. J. T., Felleman, D. J., and Kaas, J., 1983, Recovery of normal topography in the somatosensory cortex of monkeys after nerve crush and regeneration, *Science* **221**:771–773.

Wall, P. D., 1977, The presence of ineffective synapses and the circumstances which unmask them, *Phil. Trans. R. Soc. Lond. [B]* **278**:361–372.

Weinberger, N. M., and Diamond, D. M., 1987, Physiological plasticity in auditory cortex: Rapid induction by learning, *Prog. Neurobiol.* **29**:1–55.

Welker, C., 1971, Microelectrode delineation of fine grain somatotopic organization of SmI cerebral neocortex in albino rat, *Brain Res.* **26**:259–275.

Woody, C. D., and Engel, J., 1972, Changes in unit activity and thresholds to electrical microstimulation at coronal–pericruciate cortex of cat with classical conditioning of different facial movements, *J. Neurophysiol.* **35**:230–241.

Woody, C. D., Swartz, B. E., and Gruen, E., 1978, Effects of acetylcholine and cyclic GMP on input resistance of cortical neurons in awake cats, *Brain Res.* **150**:373–395.

Woody, C. D., Alkon, D. L., and Hay, B., 1984, Depolarization-induced effects of Ca^{+2}–calmodulin-dependent protein kinase injection, *in vivo*, in single neurons of cat motor cortex, *Brain Res.* **321**:192–197.

16

Is Dendritic Proliferation of Surviving Neurons a Compensatory Response to Loss of Neighbors in the Aging Brain?

PAUL D. COLEMAN and DOROTHY G. FLOOD

1. INTRODUCTION

The plastic capacities of the developing brain are well known, and it is the young brain that usually is emphasized when considering neuronal plasticity. The view taken here is that the plastic capacity of the developing brain does not suddenly cease as some developmental landmark is reached but that some degree of residual plasticity is maintained to the end of the developmental continuum (death). Functionally, this residual plasticity may be manifested in a variety of ways, including (1) recovery from strokes and lesions and (2) compensatory responses to the degenerative phenomena of the aging brain. We define neuronal "plasticity" as the adaptive response(s) of neurons to perturbations in their local environment. The perturbations may be in the chemical composition of the neuron's immediate surround, its afferent supply, its targets, or in its neighboring neurons and glia. The plastic response(s) to such perturbations may include alteration in dendritic and/or axonal morphology, in synapses, receptors, metabolism, even in genetic expression (e.g., Black *et al.*, 1984; Davis *et al.*, 1986).

In this chapter, and in our work, we concentrate on the dendritic component of neuronal plasticity. We emphasize dendrites for two major reasons. First, the dendrites of many cell types constitute as much as 95% of the receptor surface that a neuron offers for contact with other neurons (Schade and Baxter, 1960). Dendrites, therefore, constitute important determiners of the ability of neuronal

PAUL D. COLEMAN • Department of Neurobiology and Anatomy, University of Rochester Medical Center, Rochester, New York 14642. DOROTHY G. FLOOD • Department of Neurology, University of Rochester Medical Center, Rochester, New York 14642.

ensembles to receive and process information. Additionally, dendrites are among the more rapidly changeable elements of the light microscopic morphology of neurons. For example, in developing brain dendritic extent may increase by as much as hundreds of microns per day (e.g., Cowan *et al.*, 1980; Purves and Hadley, 1985). Altered dendritic extent, qualitatively similar to (but quantitatively decreased from) that seen at earlier ages, occurs in adult animals as a response to experimentally induced injury (e.g., Sumner and Watson, 1971; Standler and Bernstein, 1982; Caceres and Steward, 1983). Although the nature of the dendritic response to experimentally produced injury has not yet been established in aged brain, dendritic proliferation is seen in regions of brain that show age-related neuronal loss (e.g., Hinds and McNelly, 1977; Buell and Coleman, 1979, 1981; Flood *et al.*, 1985). It has been suggested that this dendritic proliferation may represent a plastic compensatory response to loss of neighbor neurons (Hinds and McNelly, 1977; Buell and Coleman, 1979).

2. THE AGING BRAIN

Although the term "plasticity" is usually applied to the neuronal response to some experimentally induced manipulation of the brain, the use of this term with regard to studies of the unperturbed, normally aging brain is taken here as the response of surviving neurons to age-related, naturally occurring, changes in the brain, such as loss of neighboring neurons, loss of afferent supply, alterations in trophic factors, etc., in other words, the adaptations of the brain to age-induced (rather than experimenter-induced) lesions of the brain. Static measures of instantaneous dendritic extent in postmortem tissue then serve as useful markers of plastic dendritic response to age-related neuron loss, etc., only when dendritic extent is examined in individual subjects from a range of ages that may be presumed to represent a range of age-associated lesion extent. Accordingly, age-related dendritic plasticity may be modeled as indicated in Fig. 1A, which indicates average dendritic extent per neuron as a function of age. If, on the other hand, we assume that the average neuron can not keep extending its dendritic tree indefinitely and that the neuron must, at some time, succumb to increasing age and be no longer able to compensate for the death of its neighbors, then model 2 (Fig. 1B) indicates a plastic increase in dendritic extent through old age, followed by dendritic regression in the "oldest old" ages.

The first quantitative data indicating an age-related increase in dendritic extent in the normally aging human brain were obtained from the parahippocampal gyrus (entorhinal cortex) (Buell and Coleman, 1979, 1981) These results conformed to model 1, which led to the suggestion that the net increase in dendritic extent in the average cell was a compensatory response to the death of neighboring neurons. The net increase in dendritic extent was determined to be 0.21 μm per terminal dendrite per year. Since the average neuron had 26 termi-

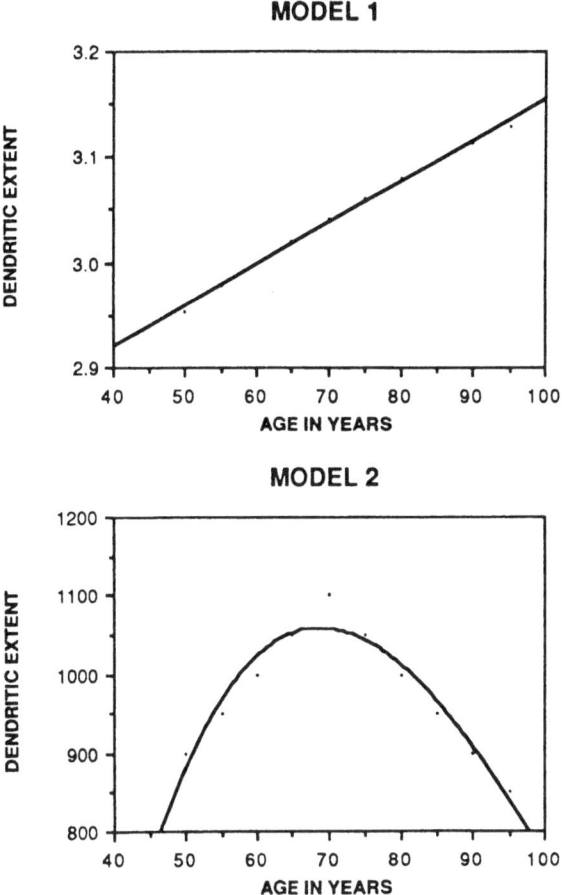

Figure 1. Two models of age-related dendritic extent. Increasing dendritic extent with increasing age is proposed to be a reflection of a plastic compensatory response of surviving neurons to death of their neighboring neurons. Models 1 and 2 represent normal aging in regions that lose neurons with increasing age.

nal dendrites, the average cell increased its dendritic extent by 5.46 μm per year. Although neuron numbers and density have not yet been studied in this region of the aging human brain, it was presumed that the age-related neuronal loss that has been found in a wide variety of neocortical regions of aging human brain could be generalized to the parahippocampal gyrus. [It should be emphasized, however, that not all brain regions of all species show age-related neuronal loss; see Curcio *et al.* (1982) for a recent review.]

Subsequent study of the dentate gyrus granule cells, which receive a significant portion of their input from the entorhinal cortex subregion of the parahippocampal gyrus, gave results that conform more closely to model 2. Whether this represents a regional difference in rate of change in the balance between proliferative and regressive influences remains to be seen. Model 2 is consistent with data obtained in other laboratories in aging rodent olfactory bulb (Hinds and McNelly, 1977) and monkey frontal cortex (Cupp and Uemura, 1980) and subiculum (Uemura, 1985).

If the age-related increase in dendritic extent is a compensatory response to death of neighboring neurons, it becomes of interest to determine the degree to which dendritic proliferation may compensate for the loss of dendritic material consequent to death of some neurons. Although neuronal numbers and dendritic extent have not been commonly studied in the same brain regions, there are data from the dentate gyrus that allow the necessary calculations. Mouritzen Dam (1979) has determined that there is a 15% decrease in neuron density in the human dentate gyrus between the ages of 40 and 80 years. This is a loss of 0.38% per year. If we assume a hypothetical set of 1000 neurons (since the total number of granule cells has not been estimated) at age 52 (the youngest age in our sample), then four neurons would be lost between the ages of 52 and 53. Each of these four neurons will have had a total dendritic extent of 784 μm (from Flood *et al.*, 1987), resulting in a loss of dendritic extent through neuron death of 3136 μm in our hypothetical population. The data of Flood *et al.* (1987) also indicate an increase in dendritic extent of 12.6 μm per neuron per year. Thus, the dendritic extent of the remaining 996 neurons will increase by 12,550 μm between the ages of 52 and 53. These calculations contain many implicit assumptions that limit their accuracy. Nevertheless, they indicate that an age-related increase in dendritic extent may be sufficient to compensate for loss of dendritic material through neuron death, up to a limit. However, we emphasize that death of neighbor neurons need not inevitably lead to proliferation of dendrites of the surviving neurons if there are regressive influences that are more potent than the proliferative influences, as appears to be the case with rat cerebellar Purkinje cells, where there is both death of Purkinje cells and partial denervation through loss of afferent parallel fiber input (Rogers *et al.*, 1980, 1981, 1984).

If the suggestion is correct that net age-related increase in dendritic extent in surviving neurons is a compensatory response to the death of neighboring neurons, then there should be no such net increase in dendritic extent in brain regions that do not lose neurons with increasing age. Because of the vast expanse and imprecise boundaries of most regions of neocortex, accurate determination of total neuron numbers and, therefore, age-related neuron loss is difficult and imprecise. However, the individual barrels of the mouse somatosensory cortex (Woolsey and Van der Loos, 1970) represent an ideal model cortex for the determination of neuron numbers as a function of age. Each barrel in the snout somatosensory cortex corresponds to the cortical representation of one of the long mystacial hairs of the mouse snout. The barrels generally appear in constant configuration from mouse to mouse, are of limited size with definite boundaries, and have a known functional correlate. Study of barrel C3 in C57B1/6 mice showed that there was no change in cortical height, barrel area or volume, or in neuron numbers at six ages from 4 to 33 months (older ages were not studied) (Curcio and Coleman, 1982). Subsequent study of dendritic extent of single layer IV stellate cells in the posteromedial barrel subfield cortex of C57B1/6 mice showed no change in dendritic extent at six ages from 4 to 45 months of age (Coleman et al., 1986). These data are consistent with the hypothesis that age-related increased dendritic extent may be a compensatory response to loss of neighboring neurons. In the absence of age-related death of neighboring neurons, there is no notable net increase in dendritic extent. This is not meant to deny that some individual dendritic trees may be proliferating while others are regressing. It only specifies that the net effect sums to zero.

The mechanism(s) by which the apparent compensatory dendritic proliferation may occur remain speculative. One set of speculations revolves around the potential role of glia as sources of trophic factors. A significant body of literature has established that the Schwann cells peripherally and the glia centrally are the sources of trophic factors. The Schwann cells are the peripheral source of nerve growth factor, and the glia, particularly astrocytes, have been shown to be a source of factors that promote both neuron survival (e.g., Banker, 1980) and neurite outgrowth (e.g., Müller and Seifert, 1982; Banker, 1980). Thus, one may propose a model based on a reactive state of astrocytes produced by a signal from dead or dying neurons. The reactive astrocytes then secrete trophic factor(s) that stimulate the growth of neuronal processes. This model is illustrated in Fig. 2.

Also illustrated in the model of Fig. 2 is the potential effect of the nor-adrenergic (NE) system as a trophic system. The early arrival of NE terminals in a number of regions of developing brain (at times before the formation of more specific afferent connections) led to suggestions that NE may have a wider role than that of transmitter alone and that it might regulate certain aspects of the early development and plasticity of the monoamine-receptive cells of the nervous system (e.g., Lauder and Bloom, 1974; Coyle and Molliver, 1977). Indeed, some studies of the role of NE during early development indicated that depletion of the NE system could apparently delay the development (usually measured as dendritic extent) of cerebral cortex (e.g., Maeda et al., 1974; Felten et al., 1982; Parnavelas and Blue, 1982) and cerebellum (e.g., Lovell, 1982; Robain et al., 1985). In addition, physiological study suggested that NE depletion could drastically retard plasticity (measured as ocular dominance shift in a monocular deprivation paradigm) in the postnatal kitten visual cortex and that NE supplementation could reverse this effect (e.g., Kasamatsu and Pettigrew, 1976; Pettigrew and Kasamatsu, 1978; Kasamatsu et al., 1981). However, there were also studies that failed to find any influence of the NE system in regulating morphological development or plasticity of cerebral cortex (e.g., Ebersole et al., 1981; Lidov and Molliver, 1982; Wendlandt et al., 1977), hippocampal dentate gyrus (Amaral et al., 1980), or cerebellum (e.g., Sievers et al., 1981), although

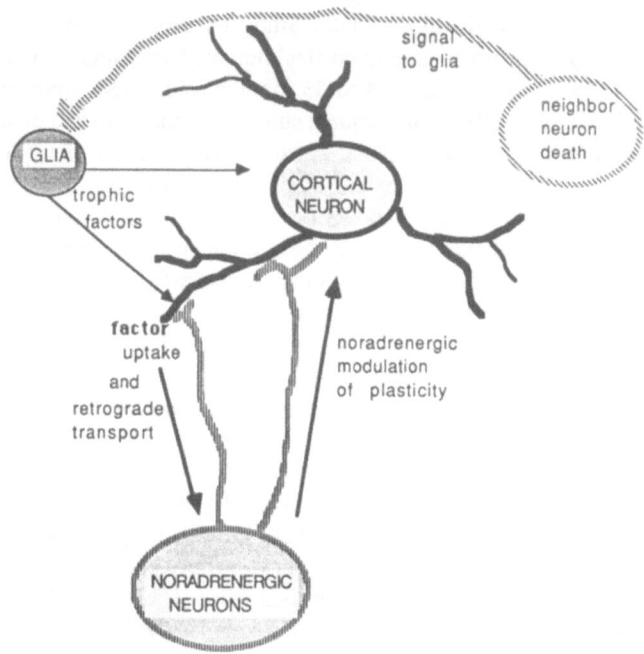

Figure 2. A model in which astrocytes respond to the death of neighboring neurons by secreting trophic factors that act either directly on target neurons or indirectly through elaboration of trophic factors that are taken up at NE terminals and retrogradely transported to act to modulate NE transmitter production.

these were not as extensive quantitative studies as investigations with positive data (e.g., Felten *et al.*, 1982). Negative results have also been reported concerning the NE effect on the physiologically determined ocular dominance shift in the visual cortex after monocular deprivation (e.g., Daw *et al.*, 1985). These conflicting results may be resolved by a consideration of possible effects of the NE lesion procedures on other transmitter systems that may interact with the NE system in the regulation of neuronal plasticity. It may be suggested that decreased neuronal plasticity in advanced old age may be partially related to impairment of the NE system. Certainly, a number of studies have established that advanced age is associated with decrements in the noradrenergic system (e.g., McGeer and McGeer, 1976; Goldman-Rakic and Brown, 1981).

Study of the effect of neighbor neuron death in the retina suggests that reduced competition for afferent supply may also be a regulating process in compensatory dendritic proliferation of surviving neurons. Perry and Linden (1982; Linden and Perry, 1982) have been able to produce loss of ganglion cells by restricted retinal lesion, leaving intact the afferents from other retinal layers.

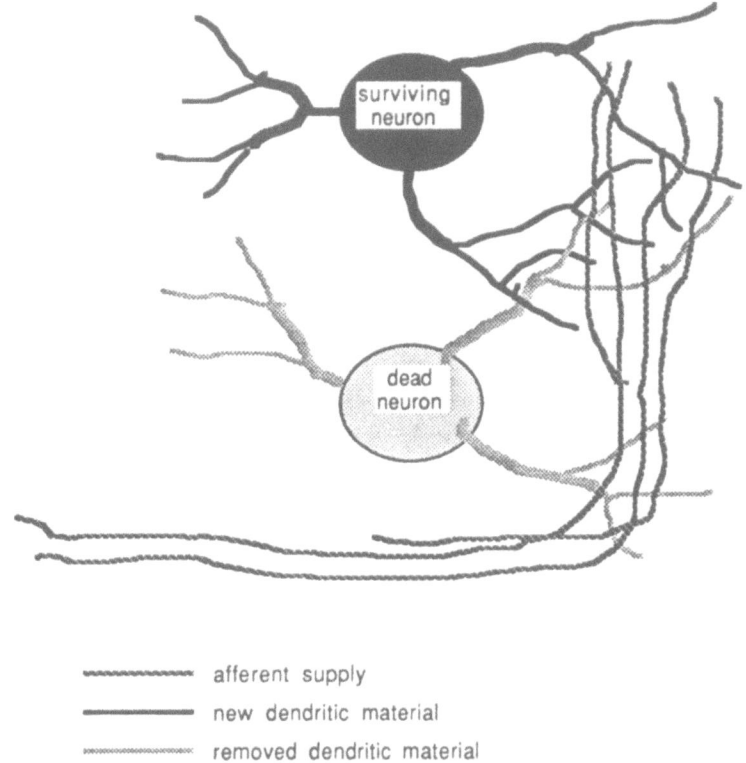

afferent supply

new dendritic material

removed dendritic material

Figure 3. A model of dendritic proliferation as a response to reduced competition for afferent supply consequent to the death of neighbor neurons. This model assumes that afferent supply is not reduced *pari passu* by the death of a portion of the neurons in the target zone.

Their results show that the surviving ganglion cells extend their dendrites to occupy the afferent zone vacated by the dead neurons. A model of reduced competition for afferent supply is shown in Fig. 3.

3. REGRESSIVE INFLUENCES

The emphasis thus far has been on discussion of the proliferative influences on the dendritic tree in aging brain. However, regressive influences are also clearly present. Qualitative observation of Golgi–Cox-stained material from aging human brain is sufficient to demonstrate the presence of obviously regressing neurons as well as flourishing neurons. We presume that in regions that are losing neurons with age, some of these obviously regressing neurons may be in the process of dying. Thus, at any given instant a neuron may travel one of two paths: regression and death or proliferation and survival. The two paths are illustrated in Fig. 4.

It is important to note that in our human cortical material the obviously regressing neurons were greatly outnumbered by qualitatively intact neurons, so that the net effect was determined by the surviving, flourishing neurons. Although it would have been interesting to quantify separately the extent of the regressing and the flourishing dendritic trees, this was not possible since there are significant numbers of neurons whose status (regressing or flourishing) is not obvious. It should also be emphasized that the numbers of apparently regressing neurons did not vary as a function of age over the age range usually studied (~50–100 years).

In addition to dendritic regression antecedent to neuron death, there also may be dendritic regression independent of impending death of the neuron. For example, the hypothalamic supraoptic nucleus (SON) has been shown not to lose neurons in the aging Sprague–Dawley rat (Hsu and Peng, 1978; Peng and Hsu, 1982). Flood and Coleman (1983) have extended this observation to the SON of the aging F344 male rat. However, data show that there is reduction in dendritic extent of 34% between the ages of 20 and 27 months (Flood and Coleman, 1983). It is notable that the dendrites of these SON neurons are directed ventrally, toward the base of the brain. A possibly important contribution to the

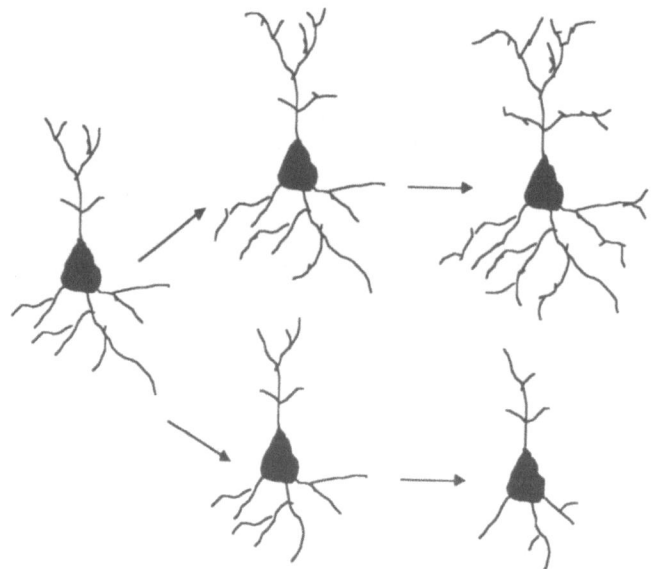

Figure 4. At any instant, a healthy neuron may start down one of two paths: (1) to continue to survive, flourish, and proliferate additional dendritic surface or (2) to regress.

understanding of this dendritic regression may come from work showing that the noradrenergic innervation of the SON progressively moves away from the ventral portion of this nucleus with advancing age, producing a partial NE denervation of the distal portions of the dendritic tree with increasing age (Sladek *et al.*, 1980). Denervation-induced dendritic regression has previously been described in other systems (e.g., Jones and Thomas, 1962; Matthews and Powell, 1962; Rubel *et al.*, 1981; Caceres and Steward, 1983). It remains to be determined whether the role of the NE denervation in the SON is equivalent to the denervations by other afferent sources produced in other systems or whether the NE denervation plays a special role related to the proposed noradrenergic regulation of plasticity.

4. BALANCE OF INFLUENCES

The considerations outlined above suggest a model of multiple influences regulating the extent of the receptive dendritic surface in the aging brain that may be classed as proliferative or regressive influences. The proliferative influences may be derived through the death of close neighbor neurons and mediated through trophic factors (including neurotransmitters) or through reduced competition for afferent supply. The regressive influences may be either the impending death of the neuron or loss of afferent supply through the death of afferent neurons or the pruning or retraction of afferent axons. The net extent of the dendritic material in a region at any given moment would then be a function of the balance between the proliferative and regressive forces.

The ability of surviving neurons to proliferate additional neuritic material in conjunction with the death of their neighbors serves the nervous system well throughout the developmental continuum. Study of a number of different neuronal populations has demonstrated that during early development there is an overproduction of neurons that is followed by the death of "excess" neurons as connections are established. In general, the explanations advanced for the phenomenon of neuron death during early development revolve around appropriateness of connections. Thus, a neuron that fails to make appropriate input and output connections becomes a candidate for elimination. It has been suggested that not only the qualitative but also the quantitative appropriateness of the connections may modulate neuron death (see Cunningham, 1982, for review and hypotheses).

Just as there is excess production of neurons during early development, there is also a period of overproduction of neuronal processes that is followed by the degeneration of "inappropriate" connections and the expansion of other connections destined to form the mature nervous system. Systems in which such phenomena have been demonstrated (in whole or in part) include motor neurons, geniculocortical fibers, climbing fibers, and callosal fibers (see Oppenheim,

1981, for review). Thus, the early development of the nervous system depends on the elimination of selected subsets of neurons and processes (both of neurons that die and of neurons that will survive) with the proliferation of those processes that are related to appropriate connections.

An extensive literature has demonstrated that the adult brain retains the mechanisms of response to manipulation of connections that were present in the brain during early development. Neurons cut off from their targets may die (e.g., Tong *et al.*, 1982; Weller *et al.*, 1979), and processes contract or expand in response to pressures exerted on them (e.g., Standler and Bernstein, 1982; Caceres and Steward, 1983; Linden and Perry, 1982). We argue that the work we have described represents a continuation into the aged brain of these same mechanisms. Thus, certain of the mechanisms developed toward optimizing the early development of the nervous system seem to remain throughout the life span. In later life, these mechanisms are useful to recovery from injury, cerebrovascular incidents, and age-related deficits such as neuron loss and decrements in transmitter systems.

If responses to appropriateness of connections are directed at the formation of connections that are more optimal from a functional point of view, we may ask whether the continuation of these same mechanisms during later life may not mean that as we get older we get better. In other words, are the neurons and connections that are eliminated largely the ones that are inappropriate? Although it may be comforting to some of us to adopt this position, there can be no argument regarding age-related declines in at least certain aspects of behavioral functioning of the healthy nervous system. Determining whether other aspects are improving sufficiently to produce a net increase in functional ability requires major value judgments and remains open to question.

ACKNOWLEDGMENT. The authors were supported by Grants AG 01121 and AG 03644 from the National Institute on Aging.

REFERENCES

Amaral, D. G., Avendano, C., and Cowan, W. M., 1980, The effects of neonatal 6-hydroxydopamine treatment on morphological plasticity in the dentate gyrus of the rat following entorhinal lesions, *J. Comp. Neurol.* **194**:171–192.

Banker, G. A., 1980, Trophic interactions between astroglial cells and hippocampal neurons in culture, *Science* **209**:809–810.

Black, I. B., Adler, J. E., Dreyfus, C., Jonakait, G., Katz, D., LaGamma, E., and Markey, K., 1984, Neurotransmitter plasticity at the molecular level, *Science* **225**:1266–1270.

Buell, S. J., and Coleman, P. D., 1979, Dendritic growth in the aged human brain and failure of growth in senile dementia, *Science* **206**:854–856.

Buell, S. J., and Coleman, P. D., 1981, Quantitative evidence for selective dendritic growth in normal human aging but not in senile dementia, *Brain Res.* **214**:23–41.

Caceres, A., and Steward, O., 1983, Dendritic reorganization in the denervated dentate gyrus of the rat following entorhinal cortical lesions: A Golgi and electron microscopic analysis, *J. Comp. Neurol.* **214**:387–403.

Coleman, P. D., Buell, S. J., Magagna, L., Flood, D. G., and Curcio, C. A., 1986, Stability of dendrites in cortical barrels of C57B1/6N mice between 4 and 45 months, *Neurobiol. Aging* **7**:101–105.

Cowan, W. M., Stanfield, B. B., and Kishi, K., 1980, The development of the dentate gyrus, *Curr. Top. Dev. Biol.* **15**:103–157.

Coyle, J. T., and Molliver, M. E., 1977, Major innervation of newborn rat cortex by monoaminergic neurons, *Science* **196**:444–447.

Cunningham, T. J., 1982, Naturally occurring neuron death and its regulation by developing neural pathways, *Rev. Cyto.* **74**:163–186.

Cupp, C. J., and Uemura, E., 1980, Age-related changes in prefrontal cortex of *Macaca mulatta:* Quantitative analysis of dendritic branching patterns, *Exp. Neurol.* **69**:143–163.

Curcio, C. A., and Coleman, P. D., 1982, Stability of neuron number in cortical barrels of aging mice, *J. Comp. Neurol.* **212**:158–172.

Curcio, C. A., Buell, S. J., and Coleman, P. D., 1982, Morphology of the aging central nervous system: Not all downhill, in: *Advances in Neurogerontology,* Volume 3, *The Aging Motor System* (J. A. Mortimer, F. J. Pirozzolo and G. J. Maletta, eds.), Praeger, New York, pp. 7–35.

Davis, L. G., Arentzen, R., Reid, J. M., Manning, R. W., Wolfson, B., Lawrence, K., and Baldino, F., Jr., 1986. Glucocorticoid sensitivity of vasopressin mRNA levels in the paraventricular nucleus of the rat, *Proc. Natl. Acad. Sci. U.S.A.* **83**:1145–1149.

Daw, N. W., Videen, T., Rader, R., Robertson, T., and Coscia, C., 1985, Substantial reduction of noradrenaline in kitten visual cortex by intraventricular injections of 6-hydroxydopamine does not always prevent ocular dominance shifts after monocular deprivation, *Exp. Brain Res.* **59**:30–35.

Ebersole, P., Parnavelas, J. G., and Blue, M. E., 1981, Development of the visual cortex of rats treated with 6-hydroxydopamine in early life, *Anat. Embryol.* **162**:489–492.

Felten, D. L., Hallman, H., and Jonsson, G., 1982, Evidence for a neurotrophic role of noradrenaline neurons in the postnatal development of rat cerebral cortex, *J. Neurocytol.* **11**:119–135.

Flood, D. G., and Coleman, P. D., 1983, Age-related changes in dendritic extent of neurons in supraoptic nucleus of F344 rats, *Neurosci. Abstr.* **9**:930.

Flood, D. G., Buell, S. J., DeFiore, C. H., Horwitz, G. J., and Coleman, P. D., 1985, Age-related dendritic growth in dentate gyrus of human brain is followed by regression in the "oldest old," *Brain Res.* **345**:366–368.

Flood, D. G., Buell, S. J., Horwitz, G. J., and Coleman, P. D., 1987, Dendritic extent in human dentate gyrus granule cells in normal aging and senile dementia, *Brain Res.* **402**:205–216.

Goldman-Rakic, P., and Brown, R. M., 1981, Regional changes of monoamines in cerebral cortex and subcortical structures of aging rhesus monkeys, *Neuroscience* **6**:177–187.

Hinds, J. W., and McNelly, N. A., 1977, Aging of the rat olfactory bulb: Growth and atrophy of constituent layers and changes in size and number of mitral cells, *J. Comp. Neurol.* **171**:345–368.

Hsu, H. K., and Peng, M. T., 1978, Hypothalamic neuron number of old female rats, *Gerontology* **24**:434–440.

Jones, W. H., and Thomas, D. B., 1962, Changes in the dendritic organization of neurons in the cerebral cortex following deafferentation, *J. Anat.* **96**:375–381.

Kasamatsu, T., and Pettigrew, J. D., 1976, Depletion of brain catecholamines: Failure of ocular dominance shift after monocular occlusion in kittens, *Science* **194**:206–209.

Kasamatsu, T., Pettigrew, J. D., and Ary, M., 1981, Cortical recovery from effects of monocular deprivation: Acceleration with norepinephrine and suppression with 6-hydroxydopamine, *J. Neurophysiol.* **45**:254–266.

Lauder, J. M., and Bloom, F. E., 1974, Ontogeny of monoamine neurons in the locus coeruleus, raphe nuclei and substantia nigra of the rat. I. Cell differentiation, *J. Comp. Neurol.* **155**:469–482.

Lidov, H. G., and Molliver, M. E., 1982, The structure of cerebral cortex in the rat following prenatal administration of 6-hydroxydopamine, *Dev. Brain Res.* **3**:81–108.

Linden, R., and Perry, V. H., 1982, Ganglion cell death within the developing retina: A regulatory role for retinal dendrites? *Neuroscience* **7**:2813–2827.

Lovell, K. L., 1982, Effects of 6-hydroxydopamine-induced norepinephrine depletion on cerebellar development, *Dev. Neurosci.* **5**:359–368.

Maeda, T., Tohyama, M., and Shimizu, N., 1974, Modification of postnatal development of neocortex in rat brain with experimental deprivation of locus coeruleus, *Brain Res.* **70**:515–520.

Matthews, M. R., and Powell, T. P. S., 1962, Some observations on transneuronal cell degeneration in the olfactory bulb of the rabbit, *J. Anat.* **96**:89–102.

McGeer, E., and McGeer, P. L., 1976, Neurotransmitter metabolism in the aging brain, in *Neurobiology of Aging* (R. D. Terry and S. Gershon, eds.), Raven Press, New York, pp. 389–403.

Mouritzen Dam, A., 1979, The density of neurons in the human hippocampus, *Neuropathol. Appl. Neurobiol.* **5**:249–264.

Müller, H. W., and Seifert, W., 1982, A neurotrophic factor (NTF) released from primary glial cultures supports survival and fiber outgrowth of cultured hippocampal neurons, *J. Neurosci. Res.* **8**:195–204.

Oppenheim, R. W., 1981, Neuronal cell death and some related regressive phenomena during neurogenesis: A selective historical review and progress report, in: *Studies in Developmental Biology* (W. Cowan, ed.), Oxford University Press, New York, pp. 74–133.

Parnavelas, J. G., and Blue, M. E., 1982, The role of the noradrenergic system on the formation of synapses in the visual cortex of the rat, *Dev. Brain Res.* **3**:140–144.

Peng, M. T., and Hsu, H. K., 1982, No neuron loss from hypothalamic nuclei of old male rats, *Gerontology* **28**:19–22.

Perry, V. H., and Linden, R., 1982, Evidence for dendritic competition in the developing retina, *Nature* **297**:683–685.

Pettigrew, J. D., and Kasamatsu, T., 1978, Local perfusion of noradrenaline maintains visual cortical plasticity, *Nature* **271**:761–763.

Purves, D., and Hadley, R. D., 1985, Changes in the dendritic branching of adult mammalian neurons revealed by repeated imaging *in situ*, *Nature* **315**:404–406.

Robain, O., Lanfumey, L., Adrien, J., and Farkas, E., 1985, Developmental changes in the cerebellar cortex after locus ceruleus lesion with 6-hydroxydopamine in the rat, *Exp. Neurol.* **88**:150–164.

Rogers, J., Silver, M. A., Shoemaker, W. J., and Bloom, F. E., 1980, Senescent changes in a neurobiological model system: Cerebellar Purkinje cell electrophysiology and correlative anatomy, *Neurobiol. Aging* **1**:3–11.

Rogers, J., Zornetzer, S. F., and Bloom, F. E., 1981, Senescent pathology of cerebellum: Purkinje neurons and their parallel fiber afferents, *Neurobiol. Aging* **2**:15–25.

Rogers, J., Zornetzer, S. F., Bloom, F. E., and Mervis, R. E., 1984, Senescent microstructural changes in rat cerebellum, *Brain Res.* **292**:23–32.

Rubel, E. W., Smith, Z. D. J., and Steward, O., 1981, Sprouting in the avian brainstem auditory pathway: Dependence on dendritic integrity, *J. Comp. Neurol.* **202**:397–414.

Schade, J. P., and Baxter, C. F., 1960, Changes during growth in the volume and surface area of cortical neurons in the rabbit, *Exp. Neurol.* **2**:158–178.

Sievers, J., Berry, M., and Baumgarten, H., 1981, The role of noradrenergic fibers in the control of postnatal cerebellar development, *Brain Res.* **207**:200–208.

Sladek, J. R., Jr., Khachaturian, H., Hoffman, G. E., and Scholer, J., 1980, Aging of central endocrine neurons and their aminergic afferents, *Peptides* **1**(Suppl. 1):141–157.

Standler, N. A., and Bernstein, J. J., 1982, Degeneration and regeneration of motoneuron dendrites after ventral root crush: Computer reconstruction of dendritic fields, *Exp. Neurol.* **75**:600–615.

Sumner, B. E. H., and Watson, W. E., 1971, Retraction and expansion of the dendritic tree of motor neurons of adult rats induced *in vivo*, *Nature* **233**:273–275.

Tong, L., Spear, P., Kalil, R., and Callahan, E., 1982, Loss of retinal X-cells in cats with neonatal or adult visual cortical damage, *Science* **217**:72–75.

Uemura, E., 1985, Age-related changes in the subiculum of *Macaca mulatta:* Dendritic branching pattern, *Exp. Neurol.* **87**:412–427.

Weller, R. E., Kaas, J. H., and Wetzel, A. B., 1979, Evidence for loss of X-cells of the retina after long-term ablation of visual cortex in monkeys, *Brain Res.* **160**:134–138.

Wendlandt, S., Crow, T. J., and Stirling, R. V., 1977, The involvement of the noradrenergic system arising from the locus coeruleus in the postnatal development of the cortex in rat brain, *Brain Res.* **125**:1–9.

Woolsey, T. A., and Van der Loos, H., 1970, The structural organization of layer IV in the somatosensory region (SI) of the mouse cerebral cortex. The description of a cortical field composed of discrete cytoarchitectonic units, *Brain Res.* **17**:205–242.

17

Practical and Theoretical Issues in the Uses of Fetal Brain Tissue Transplants to Promote Recovery from Brain Injury

DONALD G. STEIN

1. INTRODUCTION

In this chapter I discuss some of the practical and theoretical problems associated with the use of fetal brain tissue transplants to promote functional recovery from brain injuries. Readers familiar with the research on neuroplasticity and recovery from brain damage know that within the last few years, there has been a veritable explosion of literature demonstrating that embryonic brain tissue is capable of being transplanted directly into the host brain of an adult animal. In the correct environment, there is now no longer any doubt that the fetal tissue grows substantially, becomes integrated with the host brain, and, under the right conditions, is capable of mediating significant recovery from central nervous system (CNS) injury (Sladek and Gash, 1984; Bjorklund and Stenevi, 1985; Gash *et al.*, 1985).

Virtually any methodology employing fetal brain tissue appears to yield successful results. For example, suspensions of dissociated neurons and glia injected into the host CNS (e.g., ventricles or brain parenchyma) are as effective in becoming integrated into the host brain as are blocks of tissue transected from the embryo and grafted into the site of a previous injury or inserted into the cerebral ventricles. For some types of analyses fetal brain tissue has even been placed into the optic chamber (Olson *et al.*, 1983) or used to study potential regeneration in the severed peripheral nerves (Doering and Aguayo, 1984).

DONALD G. STEIN • Dean of the Graduate School and Associate Provost for Research, Rutgers University, Newark, New Jersey 07102.

The hundreds of recently published research papers in this exciting research area attest dramatically to the fact that transplants of neural tissue grow and become integrated with the host brain. However, the issue that I wish to discuss in this chapter is not the various methodologies employed in placing transplants into the brain; there are already numerous and detailed reviews and books on this subject that accomplish this goal (e.g., Sladek and Gash, 1984; Bjorklund and Stenevi, 1985). Instead, I would like to focus on the question of how transplants might act to promote functional recovery in the damaged adult CNS. The issue here is whether reciprocal axonal connections formed between host and transplant tissue are necessary to promote functional recovery following traumatic injury to the brain.

2. SPECIFICITY OF NEURAL CONNECTIONS BETWEEN HOST AND TRANSPLANT TISSUE

In the context of developing treatment strategies for brain injury, it is important to gain insight into the question of how transplants might mediate their beneficial effects and whether there are side effects that could prove to be detrimental to the recipient. Suppose, for example, that brain tissue transplants work to promote or enhance recovery by establishing specific neuronal connections with the host brain. If this is one of the critical circumstances required for functional recovery, then research should focus on how to maximize neuronal growth and the reestablishment of specific synaptic connections. One could assert, then, that the damaged portions of the brain recover by using the newly formed pathways provided by the "installation" of new neurons to replace those that were lost as a result of the injury. Accordingly, the source of transplantable material as well as the specificity of the embryonic tissue (e.g., could frontal cortex be used to replace occipital?) would be among the critical factors to consider in the development of appropriate treatments for various kinds of brain and spinal cord injuries.

Although there are a wealth of examples from which to choose, a study by Sunde et al. (1984) characterizes the "specificity" approach to how transplants might work. These researchers irradiated rat pups (600 rad) on the first day of birth to reduce the number of granule cells in the dentate gyrus of the hippocampus and to mimic some of the naturally occurring abnormal development or degenerative diseases that can affect this structure. Immediately after the irradiation, healthy dentate tissue taken from newborn rat donors was grafted into one side of the bilaterally irradiated hippocampus. About 5 months later, the rats sustained lesions of the entorhinal cortex or hippocampus to trace host fibers that might have penetrated into the transplants. The recipients were then killed, and their brains were processed to determine whether cholinergic projections had developed.

Sunde and colleagues observed that "the transplants, when appropriately located in the hippocampal region, established specific and *highly ordered* (my emphasis) afferent and efferent connections with the host brain. Misplacement of the transplants resulted in abnormal connections." These authors were able to demonstrate very specific and highly ordered host–transplant interconnections in neonatal transplant recipients. Given such specificity, Sunde *et al.* speculated that the transplants of the same neuronal cell types could play an important role in "the structural repair of the (x-irradiation) damaged neuronal circuits."

Freed *et al.* (1984) also provide evidence to support the notion that functional recovery from brain injury is dependent on transplant–target specificity. More importantly, such specificity would also be required in adults. Freed's team first created unilateral 6-hydroxydopamine lesions of the substantia nigra and then screened the animals for depletions of striatal dopamine by examining apomorphine- or amphetamine-induced stereotypic rotations (this rotational behavior is thought to result from a lesion-induced imbalance in the levels of striatal dopamine). Once they had induced consistent stereotyped rotation, fetal substantia nigra (E17) was implanted into the right lateral ventricle in one group of rats, and other animals received transplants of either frontal cortex or tectum.

The group that was given transplants of fetal substantia nigra had significantly reduced rotation in response to apomorphine or amphetamine challenges. In fact, only the substantia nigra grafts caused the animals to turn to the other side, indicating that these grafts actually provided a surplus of dopamine and thus a reversal of the asymmetry. All three types of grafts survived and grew substantially in the host brains, but just the nigral grafts induced dopaminergic binding sites in the striatum (as measured by [^{3}H]spiroperidol binding) to return to normal levels. Like the findings of Sunde *et al.* (1984), the work of Freed and his colleagues can be taken to indicate that the specificity of connections between the transplant and the host tissue can be an important factor in determining the success of the fetal tissue to mediate functional recovery. As mentioned earlier, this view has been supported by the large number of anatomic studies that have recently been conducted to characterize the detailed morphology of transplant–host interactions [see also Fine *et al.* (1985), who studied anatomic reorganization following transplants of embryonic ventral forebrain neurons in rats with lesions of the nucleus basalis].

Nonetheless, the question remains as to whether the specific connectivity is essential for recovery to be observed. In the Freed *et al.* experiments, specific connections were inferred from the fact that there were increased receptor binding sites for spiroperidol, which is a dopamine receptor agonist that binds to postsynaptic membranes in the striatum. If functional synaptic connections are formed, how could they integrate neuronal signals from other neurons with their cell bodies residing in the ventricles? What would be the appropriate, physiological signal necessary to initiate the release (or decrease) of dopamine? The neurons would be outside the "neuronal circuits" processing information in the nigrostriatal system. (In this context, increased receptor binding could also be interpreted as indicating that DA-producing transplant tissue reduces postsynaptic supersensitivity induced by the initial lesion.) If the ventricular implants work by providing adequate levels of transmitter substances, then one could ask whether specific fetal cells from the substantia nigra are essential, or could any dopamine-containing cells (or adequate dopamine analogue) produce the same effects on receptor binding or dopamine histofluorescence? This issue has not yet been completely resolved, although it is an important question in determining the most effective and safe way to deliver neurotransmitter replacement to the damaged or diseased brain.

In a review of his work on recovery from lesion-induced stereotyped rotation, Freed *et al.* (1984) addressed the question of whether specificity of cell types was critical in mediating functional recovery. He was able to demonstrate a more remarkable "plasticity" by employing isografts of adult adrenal chromaffin cells into the lateral ventricles of rats with unilateral lesions of the substantia nigra. Chromaffin cells are capable of changing their phenotypes under the right environmental conditions (e.g., cultured in the presence of nerve growth factor) to express a more neuronal phenotype, and they can release large concentrations of dopamine when sufficiently stimulated.

By transplanting the chromaffin cells into the lateral ventricles, Freed was able to reduce lesion-induced rotations by 43%, whereas similar implants of sciatic nerve reduced rotations by only 11%. Histofluorescence studies showed relatively high levels of catecholamine fluorescence in the striatum of animals with chromaffin cell allografts. But, as Freed pointed out, the chromaffin cells produced functional recovery without specific reinnervation of the striatum. These cells enhanced the recovery by "passive diffusion of the catecholamine into the striatum."*

I believe that this is an important issue from a treatment perspective, because if specific neuronal connections are not essential, then it might be possible to develop therapies that would not require implantation of additional tissue taken from embryos or neonates to obtain good functional recovery. From the clinical perspective, treatments that are less traumatic than those required by fetal brain tissue transplants are desirable.

First, in order to do transplants following brain injury, additional surgery is dictated regardless of whether it is performed immediately after injury or at some later time. Second, where does one obtain adequate amounts of tissue of appro-

---

*Recently, neurosurgeons in Mexico and the United States have reported "spectacular" results to the public media following autologous transplants of adrenal gland chromaffin cells taken from relatively young patients with severe parkinsonism. The adrenal medulla was excised from the patient and was then placed either into the cerebral ventricle adjacent to the corpus striatum or into a "pocket" in the striatum created by surgically removing some of the caudate itself. Medical reports given to the press suggested almost complete elimination of tremors within 3 days of surgery in patients who were otherwise unresponsive to conventional drug therapy.

priate type or age of embryonic tissue for transplants into patients? Third, if functional recovery is obtained, is it long lasting? Would the risks to the patient be worthwhile if it were determined that fetal transplants might have to be made on a repeated basis in order to sustain functional recovery?

2.1. Some New Experimental Tests of Transplant Specificity

Because these and similar questions were of interest to us, we decided to examine whether the specificity of neuronal connections between transplant and host brain is essential to obtain functional recovery from brain damage. Prevailing conceptions of brain function emphasize this specificity. Inherent in this perspective is the idea that the transplant functions to "replace" the properties held by the damaged area. It is also assumed that the fetal tissue facilitates recovery only if there is a mutual exchange of innervation between the transplant and the host and that the fetal tissue is homologous to the host tissue. We reasoned that if the transplants could mediate behavioral recovery from brain injuries without specificity (e.g., using nonhomologous tissue or tissue that did not form reciprocal axonal connections with the host), it might be possible to develop alternative treatments that would not require the direct implantation of embryonic tissue into the damaged host brain.

In part, we began to consider the possibility that fetal tissue transplants were mediating recovery by releasing as yet unspecified trophic factors into the host's central nervous system rather than by establishing new connections with the host brain. If trophic factors are involved in the process of recovery (Hart *et al.*, 1978; Nieto-Sampedro *et al.*, 1982; Manthorpe *et al.*, 1983; Nieto-Sampedro and Cotman, 1985), then one could initiate treatments for brain injury by direct systemic or intracerebral injections of the substances themselves. As a result, the practical and ethical problems concerning the use of transplants as therapeutic tools can be avoided.

Our interest in using trophic factors to promote recovery developed out of our own unexpected findings that embryonic brain tissue transplants facilitate recovery of learned spatial discrimination performance after bilateral frontal cortex injuries in adult rats. In our first experiment (Labbe *et al.*, 1983), we removed the medial frontal cortex bilaterally in three groups of male adult rats and used one group as intact controls. One group was given transplants of embryonic frontal cortex taken from animals in the 19th day of gestation (E19); another received E19 implants of cerebellar tissue, and the remaining group had brain lesions and no further treatments. Animals with medial frontal cortex lesions typically demonstrate severe and long-lasting impairments in their ability to learn spatial alternation problems, especially when there is a brief delay between training trials.

In this experiment we began testing our rats only 4 days after the transplants were made. Under these conditions, we were able to show that the rats that had been given implants of embryonic frontal cortex learned the spatial alternation problem much more rapidly than animals with implants of cerebellar tissue or lesions alone (Fig. 1). Despite the improved performance, however, the animals with transplants never learned the alternation task as well as their intact counterparts. In this experiment, the recovery was observed very soon after the transplants were made. Although some sprouting of axon terminals can occur within 24 hr after deafferentation, it is unlikely that within a week after the implant, specific neuronal connections between host and transplant could grow sufficiently to affect behavior directly [recently, however, Tsukuhara (1985) did show limited lesion-induced sprouting in the feline red nucleus when the afferents from the interpositus nucleus were damaged, but fibers from the sensory motor cortex were already present and simply extended presynaptic terminals]. In our model of brain injury, the lesions are always made bilaterally so that there is little possibility of reafferentation from the same fibers that would innervate

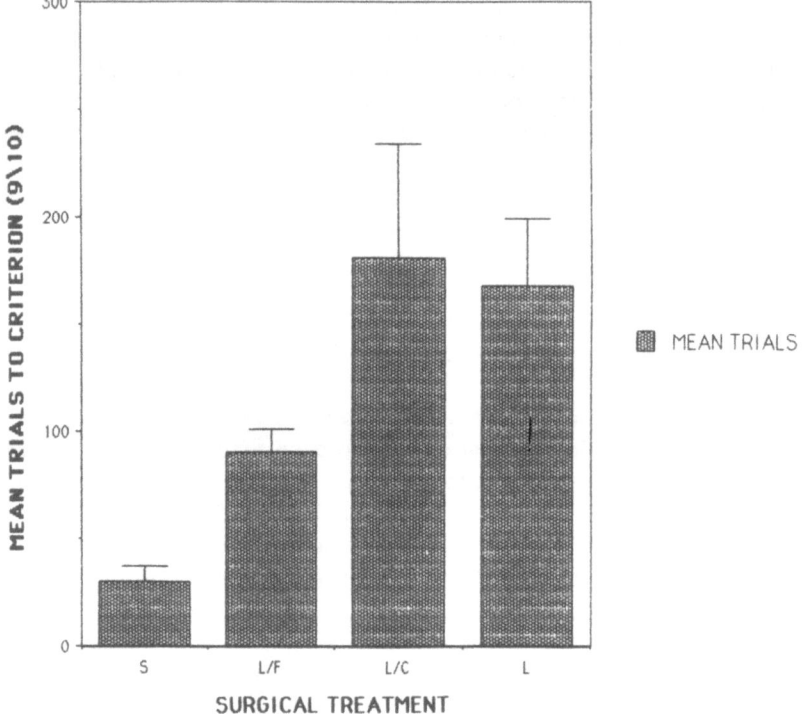

Figure 1. Average trials to criterion and standard errors on spatial alternation learning. Adult rats received sham operations (S) or bilateral frontal cortex lesions followed 7 days later by transplants of E19 frontal cortex (L/F) or E19 cerebellar tissue (L/C). The L/F group performed the learned alternation significantly better than the L/C group or those with lesions alone (L).

the area in a normal brain. If normal neuronal connections do not form, then what mediates the behavioral recovery? Examination of the brain tissue from all of the groups several months later revealed that the embryonic frontal tissue had grown substantially and had, in fact, become integrated with the host frontal cortex (Fig. 2).

Using retrograde transport techniques, we also demonstrated that the transplants had received only very sparse projections from the dorsomedial thalamic nuclei that normally project to the medial frontal cortex (Beckstead, 1979). None of the rats that had received embryonic cerebellar tissue had transplants that survived. This latter finding first led us to believe that transplants had to be homologous to the area in which they were placed, but later experiments caused us to reject this hypothesis.

2.2. Are Transplants Morphologically "Normal"?

In a subsequent study, Elliott Mufson, Randy Labbe, and I (Mufson *et al.*, 1987) examined the morphology of transplants (E19) placed into the frontal cortex. After a survival period of about 3 months, the animals were killed, and their brains were prepared for examination. Although there were many individual neurons within the transplant that appeared normal, the transplants never developed any of the characteristics of the intact cortex. For example, the laminae of the intact frontal cortex were completely absent, and the pattern of myelinated fibers did not resemble normal tissue. We did, however, see cholinergic cells within the transplants, a complex pattern of blood vessels indicating integration into the blood supply of the host animal, and some penetration of cholinergic fibers between the host brain and the transplant.

2.3. Is Homologous Embryonic Tissue Required to Obtain Recovery?

In general, the frontal transplants were very similar to those we had observed in our behavioral studies. They grew substantially in the host brain but never came close to approximating the characteristics of normal, healthy cortex. Despite this fact, the frontal transplants have been shown to enhance recovery from cognitive deficits caused by bilateral removal of the medial frontal cortex. The results of these experiments suggested to us that the formation of specific neuronal connections between the transplant and host brain is not necessarily a requirement for behavioral recovery to occur. If our hypothesis was correct, we reasoned that it would be possible to place nonhomologous tissue into a lesion site and observe functional recovery. For example, could implants of frontal cortex tissue improve performance if placed in an area of the brain with markedly different functions?

To answer this question, we created bilateral lesions of the occipital cortex in adult male rats and tested the animals on both brightness and pattern discrimi-

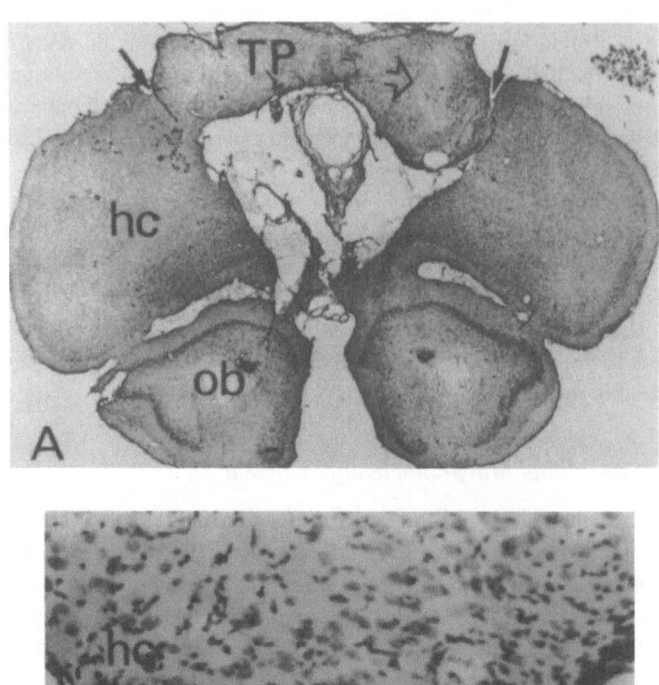

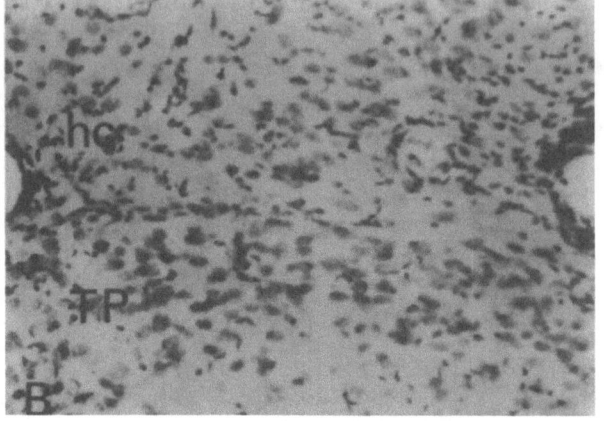

Figure 2. Photomicrographs of embryonic brain tissue transplants into the frontal cortex. (A) The transplant in coronal section (10×). The arrows indicate the locations where the host and fetal tissue were attached. (B) Micrograph (100×) showing integration of host cortex (hc) with the transplant (TP). This tissue was stained with cresyl-echt violet. (C) Low magnification (10×) showing trans-

nation problems (Stein *et al.*, 1985). The animals were given E19 transplants of frontal or occipital cortex directly into the zone of injury. As can be seen in Fig. 3, the animals that had received frontal transplants directly into the damaged occipital cortex were able to solve the brightness discrimination task better than lesion-only counterparts or animals with E19 occipital cortex implants. Note, however, that the recovered performance was not as good as that seen in the intact animals; the frontal transplants had only a partial beneficial effect, and the occipital transplants no effect at all. As was the case in Freed's experiments cited above, both types of transplants survived and grew in the host brain (with few of

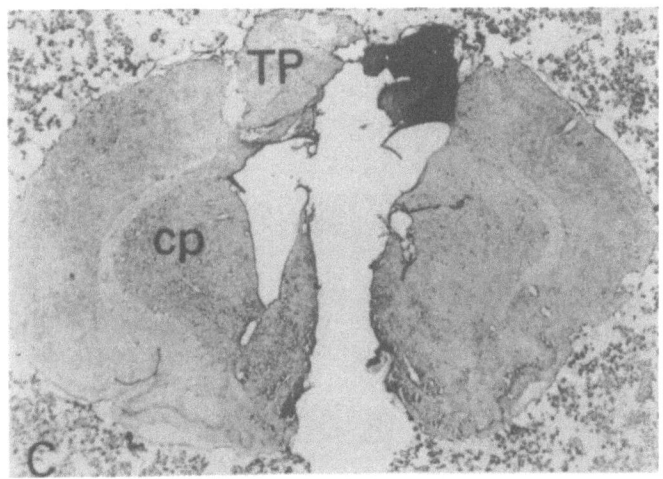

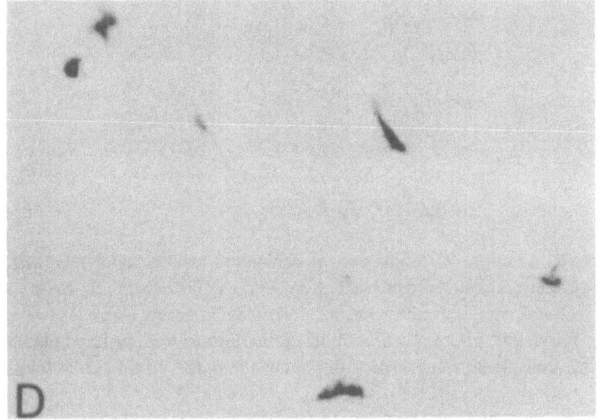

plant injected unilaterally with horseradish peroxidase (HRP). The injection site is black, and there is no leakage into the host cortex. (D) A few labeled neurons (100×) in the unstained lateral dorsomedial nucleus of the thalamus labeled by injection of HRP into the transplant. (Reproduced at 70%.)

the morphological characteristics of normal visual cortex), yet only the frontal cortex mediated the partial recovery we observed. Our histological evaluation of the transplants did not reveal any reciprocal neuronal contacts between host and transplant tissue (as measured by retrograde tracer techniques in which HRP was injected either into the transplant or into adjacent host cortex).*

*However, when HRP was injected into healthy, adjacent visual cortex or when this substance leaked from the transplant into the host brain, neurons in the dorsal lateral geniculate nucleus were clearly labeled. This suggested that our failure to see reciprocal connections was due to the fact that there were none formed, even though there was limited behavioral recovery.

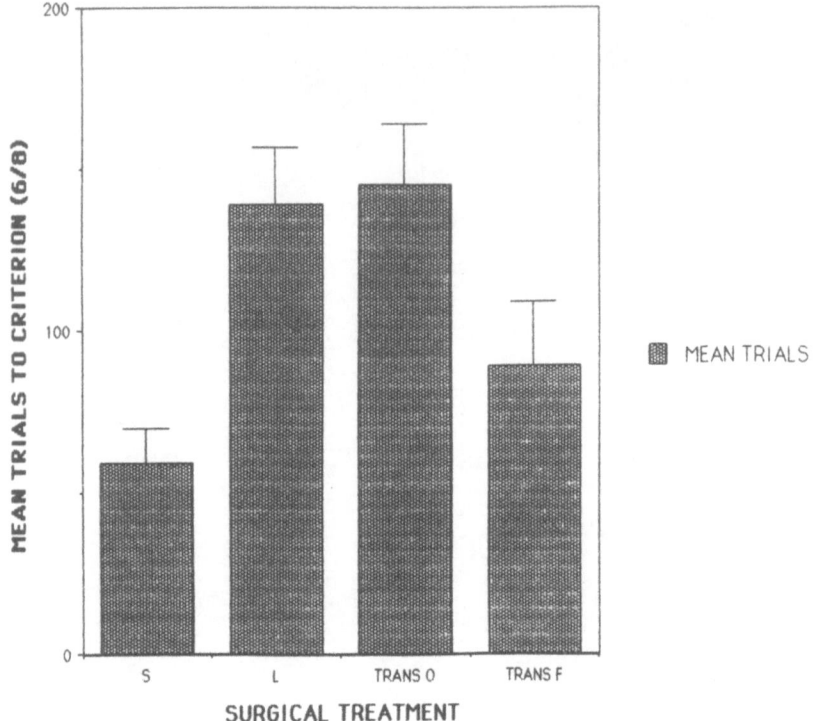

Figure 3. Trials to criterion on brightness discrimination task in adult rats with lesions of the occipital cortex given transplants of embryonic frontal or occipital cortex. S, sham-operated controls; L, group with bilateral occipital cortex lesions; TRANS O, group given lesions plus embryonic occipital cortex; TRANS F, group given occipital cortex lesions followed by implants of embryonic frontal cortex. All transplants were made 7 days after the initial visual cortex injuries.

Although relative differences in the rate of maturation may have accounted for the failure of the more developed E19 occipital cortex to affect recovery, the main point here is that frontal transplants were successful in enhancing visual performance after only 1 week of residence in the damaged host brain, and they apparently did so without forming specific reciprocal neuronal connections.

3. IS THERE A CRITICAL POSTOPERATIVE PERIOD FOR TRANSPLANT EFFECTIVENESS?

The surprising results of several recently completed experiments led us to question even more strongly whether the long-term presence of transplants is even a necessary condition for recovery of function in brain-damaged subjects. In the first study, we sought to determine whether the interval between initial brain injury and subsequent implantation of fetal tissue would play a role in mediating behavioral recovery from frontal cortex injuries. From a clinical perspective, this could be an important problem because it might not always be possible to perform the implantation at "just the right time." In other words, we were concerned with whether there was a critical period during which implants of fetal tissue would be more effective. Some evidence to this effect had already been provided by Nieto-Sampedro and colleagues (1982, 1985), who had shown that if transplants are made earlier than 7 days postoperatively, the viability and size of the transplant material are much less than if the same implants are made between 7 and 10 days after injury. Nieto-Sampedro and colleagues argued that the 7- to 10-day waiting period was necessary for the endogenous production of trophic factors whose presence appeared to insure the viability and growth of the embryonic tissue (see Nieto-Sampedro, 1987).

In our study we were more concerned with what would happen if transplants were made after even longer postoperative delays of 14, 30, and 60 days after the initial frontal cortex injuries. Accordingly, we replicated our first study by using a 7-day postoperative implant group as well as groups of adult rats that had received the implants of E19 frontal cortex directly into the damaged medial–frontal cortex at 14, 30, and 60 days postinjury. After the appropriate interval had lapsed, the transplants were made, and 4 days later, the animals began testing on the same spatial alternation task employed in previous experiments.

The behavioral results are summarized in Fig. 4. We observed that animals that were given implants 7 and 14 days after the medial frontal cortex removals performed significantly better than age- and postoperative-interval-matched lesion-only controls. These two groups also learned the problem better than groups that had received their implants 30 or 60 days after the injury; these latter animals were as impaired as those with lesions and no further treatment; the late transplants conferred no beneficial consequences whatsoever. The data are important because they demonstrate that there is no "spontaneous" recovery *per se,* because even after a very long postoperative delay, the animals with the late transplants were worse than those that received transplants at an earlier time.

As with all of our studies, the rats were killed for histological examination on completion of behavioral testing. Of course, those animals in the groups that failed to show any evidence of recovery (i.e., the 30- and 60-day delay groups) had almost no evidence of successful transplant "takes." Thus, of the 24 animals in the two groups that had received late transplants, small grafts were found in only two cases. In contrast, every animal in the 7-day group had a successful transplant that grew and integrated with the host brain.

Since the rats receiving transplants 14 days after injury also showed significant recovery, we expected to find healthy and viable transplants similar to those seen consistently in the 7-day group. To our surprise, only two animals had transplants that survived, and they were smaller and appeared less viable than those seen in the 7-day group we had been studying.

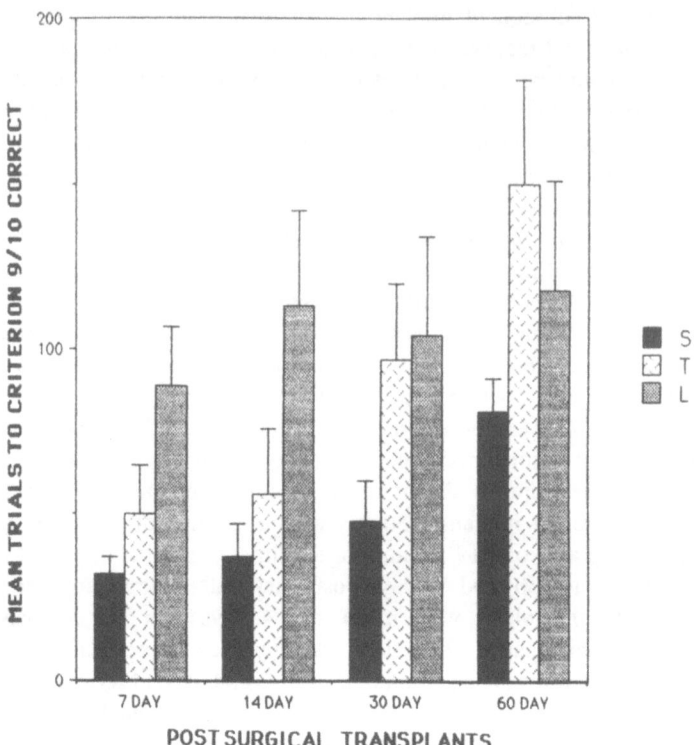

Figure 4. Trials to criterion on a spatial alternation task in rats with bilateral frontal cortex lesions given sham operations or transplants 7, 14, 30, or 60 days after the initial surgery. The black bars represent sham-operated animals; the hatched bars represent the performance of rats with E19 frontal tissue transplants; and the gray bars are the rats with lesions only.

Thus, it appears from these data that transplants do not need to be present throughout the entire testing period in order to mediate functional recovery. These data, along with our previous findings, also suggest that specific transplant–host neuronal interconnections are not essential for behavioral recovery to occur.

4. DO TROPHIC FACTORS PLAY A ROLE IN TRANSPLANT-INDUCED RECOVERY?

The alternative explanation that transplants release substances that provide trophic support to damaged neural tissue appears viable. Yet, despite the rather strong evidence I have discussed supporting this contention, it is not completely conclusive. Obviously, the best test of the notion that transplants mediate recovery by the release of neurotransmitters, peptides, or trophic factors during the initial stages of their presence in the injured brain would be to remove them at some point after they had had their effects and then retest the subjects on another task to determine whether the deficits would reappear. Under these circumstances, it would be hard to argue that axonal connections were mediating recovery, because they would have been completely disrupted by the removal of the grafts prior to behavioral retesting.

4.1. Recovery Seems to Persist when Transplants Are Removed

We have recently completed the behavioral aspects of this experiment. First, we created medial frontal cortex lesions in adult rats that were given implants of E19 frontal cortex and tested on spatial alternation (the same task used in our previous experiments with frontal cortex grafts). As usual, the animals were compared to lesion-alone controls and intact, age-matched counterparts. After the original acquisition learning had been tested for all animals, one group of rats had their transplants removed by aspiration, one kept their transplants, and the other (lesion-alone, intact controls) remained the same except for the fact that they were given "sham surgery" at the time of the transplant removals.

Under these conditions the animals were then tested on a spatial navigation task in a circular water maze. Briefly stated, this task requires the rats to find a small, visually obscured platform submerged just under the murky water in the tank. This behavioral measure requires the rats to use extramaze, room cues (e.g., cabinets on the wall, posters, doors) to locate the platform in a fixed period of time and then remember the location of the platform on subsequent trials with or without the platform present. The dependent variable is the distance swum per unit time and the time taken to find the platform per trial. As Fig. 5 shows, after the first session (in which none of the rats know the task), the animals with transplants demonstrate a marked decrease in swimming distance within the first few sessions in the water maze in comparison to rats with lesions alone. The figure also shows that animals with transplants removed performed the spatial navigation task as well as those with the transplants intact. Both groups learned more quickly than the lesion-alone rats, but once again, they were not as effective in mastering the water maze as their normal counterparts.

After the animals had completed the water maze, they were then retested on the spatial alternation task in the T maze. The retention test showed that the group with transplants removed retained the original learning as well as those animals with transplants intact during the retesting sessions. Here again, the removal of the transplants after recovery had occurred did not work against the animals.

These data must be taken as rather strong evidence that the transplants themselves need not be present at all times for recovery to occur or to be

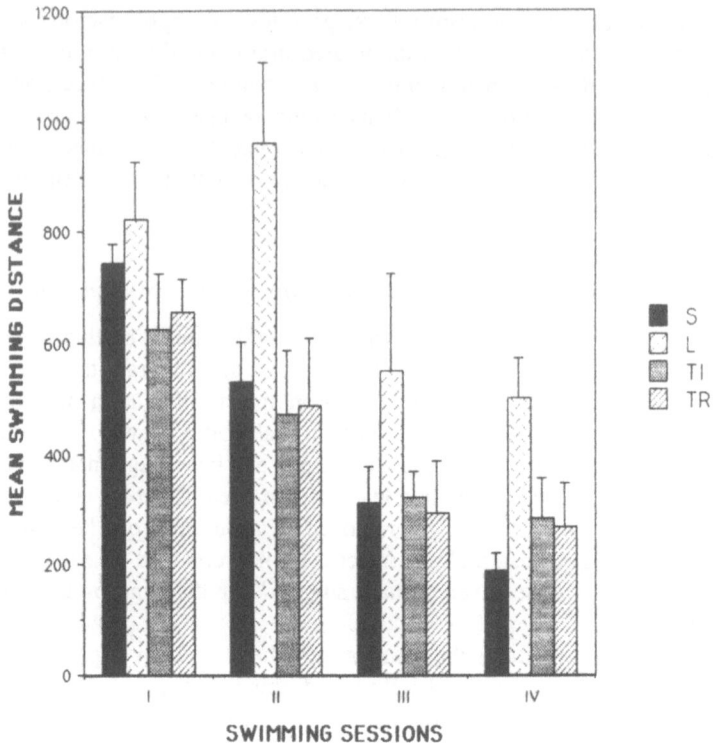

Figure 5. Spatial navigation performance in the water maze. Mean swimming distance was recorded over four sessions (represented by Roman numerals) for animals with sham operations (black bars), lesions alone (hatched bars), transplants intact at the time of testing (gray bars), or transplants removed (diagonally striped bars) prior to testing. These data show that animals with lesions alone are impaired during the training sessions, but those that had transplants for several months prior to testing were able to locate the submerged platform in the tank of water. More importantly, even those animals with their transplants removed performed the task as well as those with transplants intact at the time of testing, and both groups were better than their lesion-alone counterparts.

maintained. If this is the case, several points must be addressed. First, it becomes difficult to argue that anatomic reinnervation is a necessary and critical component of transplant-induced recovery of function. Second, our findings that grafts are capable of mediating functional recovery within 1 week after implantation make it unlikely that the effects are related to specific neuronal connections. Third, the fact that frontal cortical tissue implants can mediate partial restoration of visual function when placed into damaged occipital cortex also argues against the notion of specificity of neuronal connectivity as a critical factor in recovery.

4.2. Do Transplants Release or Stimulate the Production of Trophic Substances?

With the above points in mind, the hypothesis that transplants work by affecting the release of trophic factors that enhance the survival and function of neurons in the damaged host brain becomes more attractive. This hypothesis implies that the transplants operate as "minipumps" that manufacture and release diffusible factors into the host brain. The diffusion of these factors would not be dependent on the establishment of specific neuronal connections with the host brain. In fact, the hypothesis does not even require that the transplanted neurons themselves necessarily participate in the recovery process, although they might do so in certain circumstances not yet discovered. An alternative but equally plausible hypothesis is that the embryonic tissue transplants could serve to stimulate the host brain tissue to produce or enhance its output of neurotrophic factors that are released in response to brain injury (Nieto-Sampedro, 1987). In either case, the functional recovery would be the result of primarily neurochemical rather than morphological reorganization in the CNS.

What might be some of the diffusible substances produced and released by fetal brain tissue grafted into a damaged CNS? In a recent experiment, Ole Steen-Jorgenson of the National Hospital of Copenhagen and I examined the role of transplants in the synthesis and release of synaptic marker proteins (NCAMs) in the damaged central nervous system.

In order to investigate the putative role of trophic substances in the recovery process, we implanted frontal cortex taken from 19-day-old rats directly into a lesion site created 7 days previously by aspiration of the left medial frontal cortex in adult male rats; the procedures we employed were essentially identical to those used in our earlier experiments except that we killed the rats after a recovery period of 27 days for the biochemical assays. As markers for recovery we used the concentrations of neuronal antigens measured both in the cortical area immediately adjacent to the lesion and in the intact contralateral frontal cortex. To control for the specificity of the proteins we examined, we also measured the concentration of glutamine synthetase (a marker for glial cells) in the same tissue samples. The markers for neurons were NCAM (synonymous with D_2), D_3 protein, and D_1 protein, and the marker for astrocytes was glutamine synthetase. These markers have all been previously characterized. The neuronal cell adhesion molecule, NCAM, is involved in interneuronal synaptic adhesion and can be taken as a marker for newly formed neuronal membranes and synapses; NCAM is also found in glial cells, but only in low concentrations. All three markers are integral membrane proteins and do not display selectivity for any neurotransmitter system.

In this experiment we used 18 adult Sprague–Dawley rats as subjects ($n = 6$/group). One group of rats received a unilateral lesion of the frontal medial cortex followed 7 days later by implantation of frontal cortex taken from embryos in the 19th day of gestation. One group received only a frontal cortex lesion and no further treatment. The third group of rats served as unoperated controls.

At the end of the 27-day survival period, all of the rats were killed, and their brains were exposed for removal of brain, which consisted of tissue taken from the transplants, brain tissue adjacent to the lesion, and tissue from homologous sites in the contralateral intact hemisphere. The samples were homogenized and analyzed for the neuronal markers by immunoelectrophoresis and for glutamine synthetase by immunodiffusion.

The results of our analyses showed that the rats with medial frontal cortex lesions differed from both the intact controls and lesions-plus-transplant subjects. The individual results for each marker and area of brain examined are shown in Fig. 6. In the rats without transplants, the samples that were taken adjacent to the lesion showed a very significant decline in NCAM levels such that the concentration of this marker was decreased to 50% of control levels next to the site of the injury. In contrast, there was only a 27% decrease in those rats that had received fetal brain tissue transplants into the damaged frontal cortex.

In the lesion-only group, the contralateral, intact cortex also showed a notable effect on D_3 protein levels. Here, the concentration of D_3 was 71% of control value and was also significantly lower than the concentration in rats with lesions plus transplants. Based on these findings it seems that a frontal cortex injury results in a decreased concentration of neuronal markers both at the site of the lesion and in intact contralateral cortex. Our experiment shows that transplants of embryonic brain tissue can arrest the lesion-induced decline of NCAM and D_3 completely in the contralateral cortex and partially (but significantly) in the cortex immediately adjacent to the injury. When the cortical samples from controls were directly compared to those from rats with lesions alone, the NCAM concentration was lowest in the tissue close to the lesion. A similar evaluation of the rats with lesions plus transplants showed that both the NCAM and the D_3 protein concentrations were lower in tissue adjacent to the injury than in the contralateral tissue.

Glutamine synthetase clearly behaved differently compared to the neuronal markers, but the concentrations of this marker were not significantly affected by the presence of the lesions or the transplants. In four of the rats, samples from the transplants themselves showed that the concentration of neuronal markers was below corresponding values in adult frontal cortex, whereas the concentration of glutamine synthetase was nearly double the concentration of the host frontal cortex. This finding simply suggests that astrocytic proliferation in the transplant and its vicinity had occurred.

Based on our analysis of neuronal protein markers, we believe that there is a lesion-induced loss of neuronal membrane material, possibly synapses, in cortical tissue adjacent to the damaged frontal cortex and to a lesser degree in the

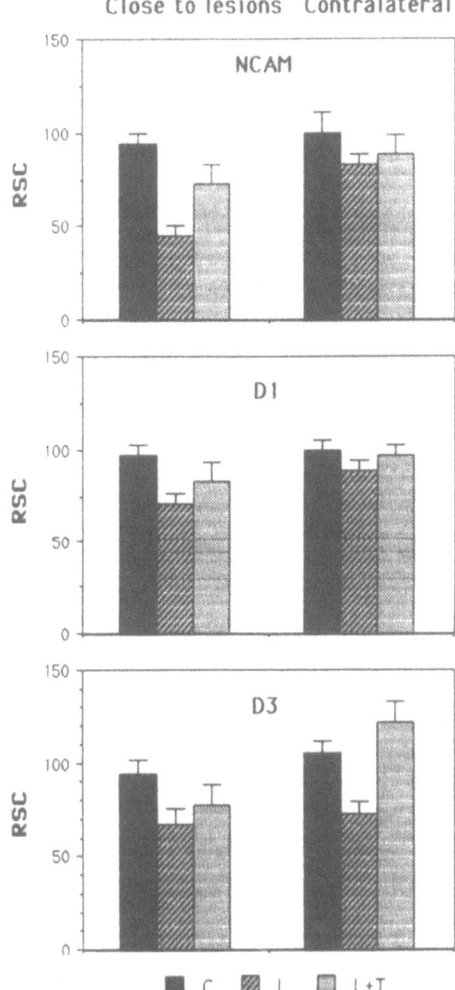

Figure 6. The relative specific concentrations (RSC) of neuronal markers in the area adjacent to the frontal cortex lesions and in the contralateral frontal cortex. Tissue samples were taken 28 days after the unilateral E19 frontal tissue transplants were placed into the lesion area. The transplants themselves were grafted onto the host brain 7 days after the initial injury. C, intact control rats; L, rats with lesions alone; L + T, rats with lesions plus transplants. NCAM is the neuronal cell adhesion molecule; D_3 and D_1 proteins are both markers for mature synapses.

contralateral intact frontal area. Loss of neuronal membrane material in the two brain areas so far studied can be explained as a consequence of the removal of the neuronal input caused by the aspiration of the medial frontal cortex.

We have now shown that transplants of embryonic brain tissue can attenuate the decreases in neuronal marker concentrations induced by a previous lesion. The capability of transplants to counteract the effect of lesions in the intact, contralateral hemisphere appears to rule out explanations based on the growth of direct innervation from neurons in the transplant. Such growth would demand that axons traverse the dorsal part of the corpus callosum and terminate in the appropriate area of the opposite cortex. It is more likely, however, that trophic factors produced by the transplant could reach the tissue adjacent to the lesion and the contralateral hemisphere by diffusion. These trophic factors might reduce the second wave of synaptic degeneration that often accompanies traumatic injury. Such diffusible factors could be produced by proliferating or developing neurons or glial cells in the transplant or the zone of injury or even by neurons or glial cells in the contralateral, intact cortex that respond to the injury by becoming more metabolically active (Whitaker-Azmitia 1987).

4.3. Glial Cells May Play an Important Role in Transplant-Mediated Functional Recovery

The specific substances that reestablish or maintain synaptic and dendritic contacts are not yet completely defined, but there is growing evidence that such factors need not be produced by neurons themselves. In this context, Kesslak *et al.* (1986) have recently presented evidence for an astroycytic contribution to transplant-induced behavioral recovery after frontal cortex ablations. These investigators used two types of preparations to obtain injury-induced endogenous trophic factors from neonatal rats.

In the first preparation they damaged the occipital cortex and filled the lesion cavity with saline-soaked gelfoam. After 5 days, the gelfoam fragments were removed, and cell-free extracts were prepared from this medium and from brain tissue taken from the tissue adjacent to the lesions. The solution obtained in this manner was used to soak fresh, sterile gelfoam, which was then placed into the damaged frontal cortex of adult rats immediately after the injury (the lesions and testing procedures were actually a replication of our work described above).

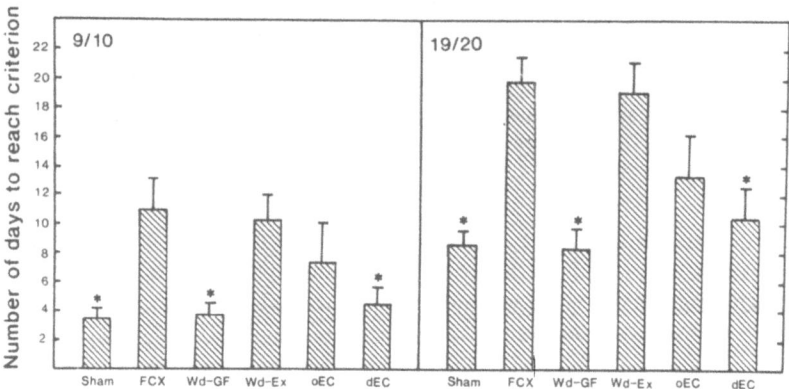

Figure 7. Days to criterion for two different performance criteria in rats with frontal cortex lesions given wound extract or implants of gelfoam soaked in wound extract. The specific groups are: Sham, sham-operated rats; FCX, frontal cortex ablations; Wd-GF, wound-gelfoam fragment implants; Wd-Ex, wound fluid extracts; oEC, no-delay embryonic cortex transplants; dEC, delayed embryonic cortex transplants (with permission of the authors, see Kesslak *et al.*, 1986).

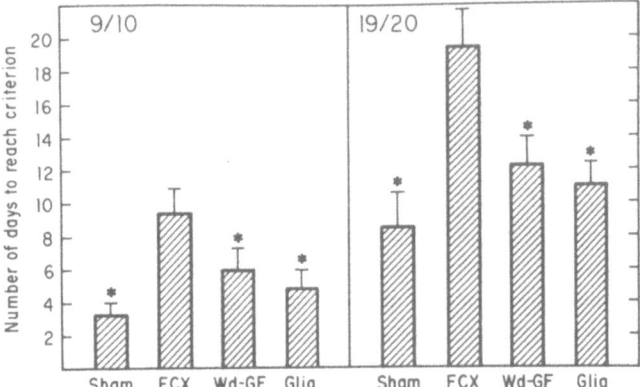

Figure 8. Days required to attain two different performance criteria on spatial alternation after frontal cortex lesions and implants of reactive astrocytes. The groups are: Sham, no lesions; FCX, frontal cortex lesions; Wd-GF, gelfoam fragments taken from the initial wound; Glia, astrocytes in a gelfoam matrix (with permission of the authors, see Kesslak *et al.*, 1986).

For the second preparation of astrocyte cultures, purified rat brain astrocytes were first incubated in trypsin solutions, washed in growth medium, and suspended at 10^8 cells/ml. This solution was then applied to fresh gelfoam, and the fresh cells were allowed to grow in culture medium for 4–10 days prior to being placed into the damaged frontal cortex of adult recipients.

Ten days after inserting the gelfoam into the wound cavity, the animals began testing on a reinforced spatial alternation task similar to the one we employed in our own work. Fifty days after the surgery, the animals were killed for histology and reconstruction of the lesions and transplants.

As Fig. 7 shows, wound extracts placed directly into the area of a frontal cortex lesion promote significant functional recovery rapidly and without the presence of neurons from the transplant.

Figure 8 shows that purified astrocytes promote behavioral recovery to the same extent as wound gelfoam, and, by reference to our own work, as effectively as transplants of embryonic neuronal tissue itself. The results of Keslak *et al.* support the hypothesis that functional recovery from brain injury is mediated by trophic substances rather than by the assumed critical reinnervation of neurons from transplanted embryonic brain tissue.

5. SYSTEMIC INJECTIONS OF TROPHIC FACTORS CAN ALSO PROMOTE FUNCTIONAL RECOVERY

Taking all of the findings together, it seems increasingly evident that specific neuronal connections between transplants and host brain tissue may not be

required for the mediation of functional recovery after brain damage. If neurotrophic substances produced by nonneural cells such as glia (Kesslak *et al.*,
1986; Whitaker-Azmitia, 1987; Nieto-Sampedro, 1988) can facilitate recovery,
it then becomes possible to ask whether direct systemic or intracerebral injections
of trophic substances would be just as effective in promoting or enhancing
functional recovery.

In fact, there is an increasing body of literature showing that, indeed,
chronic injections of substances that promote neuronal growth can be effective in
reducing the severity of brain damage. For example, it is now well documented
that single or multiple intracerebral injections of nerve growth factor (NGF) can
enhance behavioral recovery from different types of brain injury. Quite some
time ago we showed that intracaudate injections of NGF could attenuate the
perseverative deficits that accompany bilateral lesions of the caudate nucleus
(Hart *et al.*, 1978; Stein and Will, 1983). Interestingly, the NGF treatments
seemed to affect the size and number of reactive astrocytes in the area of the
injury rather than sparing neurons (Stein, 1981). At the time we did not have an
adequate explanation of the findings, but in the light of the work of Nieto-
Sampedro and Kesslak and colleagues cited above, it is likely that the glial cells
were secreting trophic factors that sustained or helped to maintain metabolic
functions in cells that would ordinarily be impaired as a result of the CNS injury.

More recently, Will, *et al.* (1987) have demonstrated that NGF injections
can facilitate recovery of spatial memory in adult rats with lesions of the septum
or hippocampus. These workers have argued that the NGF injections stimulate
neurons to maintain levels of ACh that would ordinarily decline after damage to
the limbic-system cholinergic pathways that innervate the hippocampus.

In a similar approach, our laboratory has shown that chronic intraperitoneal
injections of GM_1 gangliosides can also facilitate recovery following damage to
the caudate nucleus, the frontal cortex, and nigrostriatal bundle (Sabel *et al.*,
1984). Ganglioside effects on recovery from brain injury are very rapid, and
although sparing of neurons has been observed (Sabel *et al.*, 1984), it is also
possible that this class of substances works by reducing lesion-induced edema
that occurs during the initial stages of the injury (Karpiak and Mahadik, 1984).

One final possibility that has not received much attention is that trophic
factors released by transplants or provided by direct injections may serve to
activate alternative and ordinarily silent pathways. There is some literature demonstrating that such pathways do exist and that they can become functionally
active immediately after the onset of an injury (Wall, 1977; Merzenich *et al.*,
1983; Wall and Kaas, 1985). Such rapid recovery would be consistent with what
has been observed in our transplant experiments (see also Kesslak *et al.*, 1986) as
well as with our injections of neurotrophic substances.

6. PROBLEMS AND RISKS IN USING EMBRYONIC BRAIN TISSUE GRAFTS FOR THE TREATMENT OF BRAIN INJURY

Although I cannot review all of the relevant literature here, there are sufficient grounds for optimism that effective pharmacological treatments for brain and spinal cord injuries may soon be available. Such treatments have certain advantages over transplants. Very recent data suggest that transplants of embryonic brain tissue may never develop a normal blood–brain barrier to macromolecules circulating in the blood (Rosenstein, 1987). The failure of grafted tissue to establish a blood–brain barrier even as long as 1 year after implantation means that systemic substances could enter the brain via the grafts. For example, blood-borne antibodies could potentially cause an immune rejection reaction or inflammation of CNS tissue (e.g., Kay and Baker, 1979). Hormones or steroids could also penetrate the brain more freely and accelerate degenerative processes (Cotman and Scheff, 1979).

There is also some evidence to suggest that brain grafts may eventually be rejected after long periods of time in the brain (S. B. Dunnett, personal communication). Transplants can also lead to severe disruption of recovery when exposed to high levels of endogenous estrogen (Kolb, 1987) because the fetal tissue grows so extensively that it acts as a space-occupying tumor.

Under such circumstances, one should exercise extreme caution in using embryonic tissue grafts to facilitate recovery from brain injury in human patients. In this context, it will be particularly appropriate to monitor the long-term effectiveness of the autologous adrenal grafts that were recently performed on Parkinson's patients in the United States and Mexico.

Injections can be made with less risk to subjects than the repeated surgeries that might be required to obtain complete beneficial consequences with grafts. Gangliosides and other factors (NGF, GMF, EGF) can be injected without much concern for the problem of immune rejection that might occur with grafting of fetal brain tissue in higher mammals (see Brundin *et al.*, 1985, for a discussion of this problem). Injectable substances are easier to employ. Finding adequate types of neurotrophic substances is not as much of a problem as finding adequate fetal tissue for grafting into patients. In this context, the age of the donor, viability of donor tissue, and availability, amount, and immunocompatibility of donor tissue would have to be considered in planning a course of treatment using grafted tissue. In the case of treatment for degenerative or metabolic diseases of the CNS, both trophic factors and transplants may have important roles to play; however, the disease process that attacks the patient's nervous system might very well affect the transplanted tissue as well. Injectable neurotrophic substances could be given repeatedly without less risk, should conditions require it.

7. CONCLUSIONS

It is now becoming increasingly clear that there are no simple answers to the question of how functional recovery occurs after brain injury. Research on the uses of fetal brain tissue grafts to promote recovery have certainly captured the interest of both the scientific community and the public and have provided a great stimulus to our efforts to find an effective treatment for brain and spinal cord injuries. Where only a few years ago, there were only a handful of researchers concerned with rehabilitation of damaged brains, there are now hundreds of scientists attacking the problem. I believe that without transplant research we would not have made as many advances in our understanding of how growth and guidance factors operate to establish and maintain normal connections between neurons.

The full story on the uses of transplants to promote recovery can hardly be considered complete. As the studies reviewed here show, the specificity of neuronal connections between graft and host tissue is not a necessary precondition for behavioral recovery. Nevertheless, there may very well be circumstances in which the specific circuitry provided by implants of neural tissue may be essential for functional recovery to occur. However, in order to ascertain which functions may or may not rely on the specific reinnervation of transplants, a more balanced research approach is necessary. Most of the attention in the neuroscience community has focused on describing the anatomy and the morphology of transplants. There is clearly less interest in how transplanted tissue might operate to promote behavioral recovery. Yet, without a better understanding of the relationship among neural/glial grafts, trophic substances, and behavior, progress toward the development of effective treatments for brain and spinal cord injuries may be long delayed.

ACKNOWLEDGMENTS. The research conducted at Clark University and described in this chapter was made possible through a research contract with the American Paralysis Association and a grant from the National Institute of Mental Health (ADAMHA).

REFERENCES

Beckstead, R. M., 1979, An autoradiographic examination of corticocortical and subcortical projections of the mediodorsal projection (prefrontal) cortex in the rat, *J. Comp. Neurol.* **184**:43–62.

Bjorklund, A., and Stenevi, U. (eds.), 1985, *Neural Grafting in the Mammalian CNS,* Elsevier, Amsterdam.

Brundin, P., Nilsson, O. G., Gage, F. H., and Bjorklund, A., 1985, Cyclosporin A increases survival of cross-species intrastriatal grafts of embryonic dopamine containing neurons, *Exp. Brain Res.* **126**:1–5.

Cotman, C. W., and Scheff, S., 1979, Synaptic growth in aged animals, in: *Aging: Physiology and*

Cell Biology of Aging (A. Cherkin, C. W. Finch, N. Kharasch, T. Makinodan, F. L. Scott, and B. L. Strehler, eds.), Raven Press, New York, pp. 109–120.

Doering, L., and Aguayo, A. J., 1984, *In vitro* preparations of embryonic brain survive and differentiate after transplantation into peripheral nerves of adult rats, *Soc. Neurosci. Abstr.* **10:**49.

Fine, A., Dunnett, S. B., Bjorklund, A., Clarke, D., and Iverson, S. D., 1985, Transplantation of embryonic ventral forebrain neurons to the neocortex of rats with lesions of the nucleus basalis magnocellularis I. Biochemical and anatomical observations, *Neuroscience* **16:**769–786.

Freed, W. J., Hoffer, B. J., Olson, L., and Wyatt, R. J., 1984, Transplantation of catecholamine-containing tissues to restore functional capacity to the damaged nigrostriatal system, in: *Neural Transplants: Development and Function* (J. Sladek, Jr. and D. M. Gash, eds.), Plenum Press, New York, pp. 373–406.

Gash, D. M., Collier, T. J., and Sladek, J. R., Jr., 1985, Neural transportation: A review of recent developments and potential applications to the aged brain, *Neurobiol. Aging* **6:**131–150.

Hart, T., Chaimas, N., Moore, R. Y., and Stein, D. G., 1978, Effects of nerve growth factor on behavioral recovery following caudate nucleus lesions in rats, *Brain Res. Bull.* **3:**245–250.

Karpiak, S., and Mahadik, S. P., 1984, Reduction of cerebral edema with G_{M1} ganglioside, *J. Neurosci. Res.* **12:**485–492.

Kay, M. B., and Baker, L. S., 1979, Cell changes associated with declining immune function, in: *Aging: Physiology and Cell Biology of Aging* (A. Cherkin, C. W. Finch, N. Karasch, T. Makinoda, F. L. Scott, and B. L. Strehler, eds.), Raven Press, New York, pp. 27–50.

Kesslak, J. P., Nieto-Sampedro, M., Globus, J., and Cotman, C. W., 1986, Transplants of purified ostrocytes promote behavioral recovery after frontal cortex ablation, *Exp. Neurol.* **92:**377–390.

Kolb, B., 1987, Behavioral recovery of function following brain grafts, in: *20th Annual Winter Conference on Brain Research, Vail, CO.*

Labbe, R., Firl, A. C., Jr., Mufson, E. J., and Stein, D. G., 1983, Fetal brain transplants: Reduction deficits in rats with frontal cortex lesions, *Science* **217:**470–472.

Manthorpe, M., Nieto-Sampedro, M., Skaper, S. D., Lewis, E. R., Barbin, G., Longo, F. M., Cotman, C. W., and Varon, S., 1983, Neuronotrophic activity in brain wounds of the developing rat. Correlation with implant survival in the wound cavity, *Brain Res.* **267:**47–56.

Merzenich, M. M., Kaas, J. H., Wall, J., Nelson, R. J., Sur, M., and Felleman, D. J., 1983, Topographic reorganization of somatosensory cortical areas 3b and 1 in adult monkeys following restricted deafferentation, *Neuroscience* **8:**33–35.

Mufson, E. J., Labbe, R., and Stein, D. G., 1987, Morphologic features of embryonic neocortex grafts in adult rats following frontal cortical ablation, *Brain Res.* **401:**162–167.

Nieto-Sampedro, M., 1988, Growth factor induction and order of events in CNS repair, in: *Pharmacological Approaches to the Treatment of Brain and Spinal Cord Injuries* (D. G. Stein and B. Sabel, Plenum Press, New York.

Nieto-Sampedro, M., and Cotman, C. W., 1985, Growth factor induction and temporal order in central nervous system repair, in: *Synaptic Plasticity* (C. W. Cotman, ed.), Guilford Press, New York, pp. 407–456.

Nieto-Sampedro, M., Lewis, E. R., Cotman, C. W., Manthorpe, M., Skaper, S. D., Barbin, G., Longo, F. M., and Varon, S., 1982, Brain injury causes a time-dependent increase in neuronotrophic activity at the lesion site, *Science* **221:**860–861.

Olson, L., Seiger, A., and Stromberg, I., 1983, Intraocular transplantation in rodents: A detailed account of the procedure and examples of its use in neurobiology with special reference to brain tissue grafting, in: *Advances in Cellular Neurobiology* (S. Federoff and L. Hertz, eds.), Academic Press, New York, pp. 407–442.

Rosenstein, J. M., 1987, Neocortical transplants in the mammalian brain lack a blood–brain barrier to macromolecules, *Science* **235:**772–773.

Sabel, B., Dunbar, G., and Stein, D. G., 1984, Gangliosides minimize behavioral deficits and enhance structural repair after brain injury, *J. Neurosci. Res.* **12**:429–443.

Sladek, J. R., Jr., and Gash, D. M. (eds.), 1984, *Neural Transplants: Development and Function*, Plenum Press, New York.

Stein, D. G., 1981, Functional recovery from brain damage following treatment with nerve growth factor, in: *Functional Recovery from Brain Damage* (M. W. van Hof and G. Mohn, eds.), Elsevier/North Holland Biomedical Press, Amsterdam, pp. 423–444.

Stein, D. G., and Will, B., 1983, Nerve growth factor produces a temporary facilitation of recovery from entorhinal cortex lesions, *Brain Res.* **261**:127–131.

Stein, D. G., Labbe, R., Attella, M. J., and Rakowsky, H. A., 1985, Fetal brain tissue transplants reduce visual deficits in adult rats with bilateral lesions of the occipital cortex, *Behav. Neural Biol.* **44**:266–277.

Sunde, N., Laurberg, S., and Zimmer, J., 1984, Brain grafts can restore damaged connections in newborn rats, *Nature* **310**:51–53.

Tsukuhara, N., 1985, Synaptic plasticity in the rat nucleus and its possible behavioral correlates, in: *Synaptic Plasticity* (C. W. Cotman, ed.), Guilford Press, New York, pp. 201–230.

Wall, J. T., and Kaas, J. H., 1985, Cortical reorganization and sensory recovery following nerve damage and regeneration, in: *Synaptic Plasticity* (C. W. Cotman, ed.), Guilford Press, New York, pp. 231–260.

Wall, P. D., 1977, The presence of ineffective synapses and the circumstances which unmask them, *Phil. Trans. Soc. Lond. [Biol.]* **278**:361–372.

Whitaker-Azmitia, P. M., Ramirez, A., Noreika, L., Gannon, P. J., and Azmitia, E. C., 1987, Onset and duration of astrocytic response to cells transplanted into the adult mammalian brain, in: *Cell and Tissue Transplantation into the Adult Brain* (E. C. Azmitia and A. Bjorklund, eds.), Ann. NY Acad. Sci. **495**, pp. 10–23.

Will, B., Hefti, F., Pallage, V., and Toniolo, G., 1987, Nerve growth factor: Effects on CNS neurons and on behavioral recovery from brain damage, *Exp. Brain Res.* (in press).

18

Functional Electrical Stimulation and Its Application in the Rehabilitation of Neurologically Injured Individuals

JERROLD S. PETROFSKY

1. EARLY HISTORY OF ELECTRICAL STIMULATION IN MEDICINE

The concept of using electrical stimulation in medical therapy is not new, predating even the birth of Christ. A number of references are found in early writings concerning the use of electricity for therapy, including electric eels and other species with electric organs. Ancient physicians used electric discharges from the black torpedo fish for the treatment of arthritis, headaches, and other types of disorders. However, these early therapies were not very well understood.

It was really not until the writings of Kruger in 1744 that the first clear concept of the use of electricity for paralysis was proposed:

> But what is the usefulness of electricity, for all things must have a usefulness, that is certain. We have seen it cannot be looked for either in theology or in jurisprudence, and therefore nothing is left but medicine. The best effect would be found in paralyzed limbs to restore sensation and reestablish the power of motion.

Although Kruger certainly could not have fully understood what electricity was, early discoveries with static electricity had led him to believe that electricity could be used as a therapeutic modality for stroke and brain injury. Following his initial observation, a number of papers appeared in the literature supporting Kruger's ideas.

In 1746, Kratzenstein first introduced muscle contraction by static elec-

JERROLD S. PETROFSKY • Department of Psychiatry, University of California at Irvine, Irvine, California 92717.

tricity using static discharges from a leiden jar. This was followed by the now classic experiments of Luigi Galvani in 1791. Galvani reasoned that the nerves, like copper wire, conducted a form of electricity that caused the muscles to move. The nerves would resemble electrical cables with insulation around their surface. In an attempt to prove or disprove this fact, he performed an experiment in which he surgically connected a frog's spinal cord to its lower leg muscles with a piece of conductive metal. The muscle contracted.

Galvani's experiments, although correct in nature, were disproven later. The metal that Galvani used in these experiments was poisonous to the body, causing an oxidation–reduction reaction that generated enough voltage to cause the muscle to contract. After his experiments and their publication in the *Proceedings of the Bologna Academy,* the idea of using electricity in medical therapy took on new meaning.

When the experiments were repeated by Volta with inert wires in 1816 (some 25 years later), the true influence of electricity in the spinal cord was discovered.

The distinction between motor and sensory nerves was first made by Magendie in 1822. He found that he could insert needles into different nerves and that following electrical stimulation of some nerves, muscles contracted, whereas after stimulation of other nerves, pain was felt. In 1825, Sarlandiere first introduced electroacupuncture for stimulation of muscle. In 1829, Marianini found that stimulation with negative current was much more effective than stimulating with positive current for generating muscle contraction.

Duchene de Boulogne found that a muscle could be stimulated from cloth-covered electrodes lying on the surface of the skin. This was not only the first use of transcutaneous electrodes but also the first use of faradic stimulation of muscle. Although stimulation had been tried for nearly 2000 years, it was not until 1851 that Dubois-Remonde recorded the first action potentials in muscle by identifying them in the contracting arm muscles. This was the beginning of modern electromyography. The fact that nerves conduct electricity more slowly than copper wire was first observed in 1850 by the physicist Helmholtz.

From the early experiments of Galvani and Volta, the use of electricity as a therapeutic modality increased rapidly as the years progressed. In the East, the Chinese began using electricity for electroacupuncture, which involved inserting needles in the motor points of muscles and then passing electricity through the needles. Recent studies indicate that this type of electrical stimulation may have some therapeutic benefit.

Unfortunately, while people in the East were treating electricity seriously, people in the West were misusing it by claiming a wide variety of therapeutic benefits that could be derived with electrical stimulation. Electrical stimlation was used in the 1800s as a cure for heart disease, rheumatism, arthritis, and a wide variety of other clinical disorders. Obviously, many examples of the use and abuse of electricity for therapy resulted in less than enthusiastic acceptance of the potential use of electrical stimulation by the medical community. Such charlatanism caused electrical stimulation to disappear almost totally as a therapeutic modality until the 1900s.

Throughout the 1900s, a revival in interest in electrical stimulation as a therapeutic modality developed. As was the case in the 1800s, electrical stimulation was used initially as a potential cure for clinical disorders and as an aid to promoting health and well-being.

2. RECENT HISTORY OF FUNCTIONAL ELECTRICAL STIMULATION FOR PATIENT THERAPY IN SPINAL CORD INJURY

There have been a number of attempts to restore motor function in paralyzed patients, but these have not been notably successful. Goldsmith et al. (1982) described attempts to restore stroke- and spinal cord injury (SCI)-impaired function by replacing lost circulation with omental transplants. Naftchi (1982) and Faden et al. (1982) have tried to achieve similar results with medication. Perkins et al. (1981) and Kao et al. (1982) have attempted restoration by nerve tissue transplants. Thus far, such attempts have been only partially successful in selective patients whose conditions are specifically amenable to such treatment.

Greater success has been achieved in what has come to be called ''rehabilitation engineering'' with the application of engineering principles to the design and development of artificial systems that can substitute, in part, for the lost central command function. This effort has resulted in significant progress toward producing useful contractions of stroke, head trauma, and SCI patients' paralyzed muscles. Such movement can be valuable to the patient.

If the SCI is high enough in the spinal column, α motor neurons supplying the paralyzed muscle of the lower part of the body are not damaged. Therefore, the affected muscles do not undergo denervation atrophy; loss of muscle mass and strength result from simple disuse atrophy, and the muscle retains the ability to contract in response to electrical stimulation. For stroke or brain injury patients, all α motor neurons are left spared, making electrical stimulation even more applicable. Such functional electrical stimulation (FES) has been used in the research environment for many years. Trnkoczy et al. (1976) and Zealer and Dedo (1977) have described and discussed the use of FES in detail.

Functional electrical stimulation can be applied to the body in a number of different ways. These have included intramuscular wire electrodes (Peckham et al., 1976; Vodovnik et al., 1967; Marsolais and Kobetic, 1982), sleeve electrodes (Solomonow et al., 1978; Petrofsky, 1978; Petrofsky et al., 1976), and surface electrodes (Milner et al., 1970; Scott, 1968; Kralj and Jaeger, 1982).

Scott (1968) pointed out that surface electrodes require high current densities and that this could result in skin burns. Consequently, completely implantable systems have also been developed (Brindley *et al.*, 1978; Holle *et al.*, 1984).

Reswick and Vodovnik (1967) brought attention to the application of engineering principles and locomotor disabilities. The same group also studied stimulation of skeletal muscle in special situations (Crochetiere *et al.*, 1967). In order to produce rudimentary locomotions, Lieberson *et al.* (1961) used FES with open-loop control for the first time, followed by Long and Masciacrelli (1963). Petrofsky and Phillips (1983c) extended the use of FES to closed-loop control (stimulation modified by electronic sensors placed on the limbs) in order to produce leg movement in paralyzed persons. Closed-loop control (Petrofsky and Phillips, 1979a,b, 1983a; Petrofsky, 1979) had previously allowed the restoration of smooth and coordinated force and movement in cat skeletal muscle.

3. FUNCTIONAL ELECTRICAL STIMULATION

Functional electrical stimulation (FES) is the use of artificially generated electrical stimuli to a paralyzed muscle system to produce contractions that result in meaningful movement. These movements may be either functional (produce useful body movements) or therapeutic (physical condition enhancement). In recent years, the most effective FES systems are those in which the stimulation application is computer controlled. Stimulation is applied through a pair of electrodes above the involved muscle with either constant-voltage or constant-current control. There are a number of disadvantages and advantages to both types of stimulation; e.g., constant-current stimulation allows the power developed by muscle to be controlled better, but it is more painful for someone with partial or complete feeling in the skin. Therefore, the only real advantage of constant-current stimulation appears to be better controllability of the tension developed in the muscle. In either case, the current could be applied with a monophasic or biphasic waveform. However, there is a distinct advantage in balanced biphasic stimulation over either voltage-controlled or current-controlled monophasic stimulation in the lessening of skin erythemia.

Standard therapeutic stimulation usually involves either current-controlled or voltage-controlled stimulation at frequencies ranging from approximately 5 to 200 Hz. Average currents range from a few microamperes for subliminal stimulation to 500–600 mA peak current for strong therapeutic stimulation. These currents can be divided into two different categories: low stimulation currents (usually less than 1 mA) are used to assist healing of wounds, whereas stimulation currents of less than 5 mA are used for transcutaneous electrical neuromuscular stimulation (TENS).

The final variable for electrical stimulation is pulse width. Therapeutic pulse widths vary from 10 μsec to several milliseconds. Narrow pulse widths are used to stimulate deep muscles and have the additional advantage of not stimulating pain receptors in the skin. Therefore, stimulation with narrow pulse widths results in contraction of skeletal muscle deep below the surface of the skin with very little sensation (except a pinprick) under the area of the electrodes. However, to stimulate muscle with narrow pulse widths requires very large voltages and currents. To minimize such current, wider pulse widths can be used. There is more sensation to the skin but less peak current delivery in each pulse. Wider pulse widths also can be used much more readily to stimulate partially denervated muscle. They allow contraction of muscle in the face of a long absolute refractory period on the sarcolemma associated with denervation. In our own experience, partially denervated or fully denervated muscle seems to respond to stimulation only at frequencies below 20 Hz and at pulse widths in excess of 1 msec (unless the muscle has been totally destroyed).

The electrodes that can be used for electrical stimulation also vary. The oldest type of electrode used is metal, and platinum electrodes have been used from time to time. Silver electrodes. although popular, can create difficulties because silver can precipitate from the electrodes and cause carcinomas of the skin and/or permanent skin damage. Therefore, the most widely used silver electrode in use today is a silver/silver chloride electrode. In recent years, low-resistance carbonized rubber electrodes have come into use.

Another type of electrode that has been popular in recent years is an electrode designed into garments such as those designed by Biostimu-Trend Corporation in Opa Locka, Florida, called the Transcutaneous Transducer Garment. This offers the advantage of being easy to use throughout a day and embedding the wires in the garments to avoid wires pulling loose from the electrodes. These particular garments (Granek and Granek, 1985) offer a large number of electrode sites, which can be used for scoliosis, TENS treatment, or for muscle strengthening. By using conductive thread to connect the electrodes to a single connector, the garments can be made very practical with single lead-wire attachment to an electrical stimulator; however, the costs are very high.

In summary, therapeutic variables for electrical stimulation in the past have varied greatly. The obvious variables are stimulation pulse width, pulse amplitude, stimulation frequency, the type of electrode being used, and monophasic versus biphasic stimulation. Following the initial experiments by Rack and Westbury (1969), it has become increasingly popular in recent years to use sequential stimulation, rather than synchronous, motor unit stimulation. With sequential stimulation, a number of electrodes are placed around the muscle and alternately stimulated. This causes different parts of the muscle to contract out of phase with one another, resulting in much smoother contractions in muscle at lower stimulation frequencies. This has the advantage of avoiding the neuromuscular blockade (caused by depletion of ACh) that is commonly found with stimulation at frequencies above 50 Hz (Brown and Burns, 1949). With sequential stimulation, the muscle tetanizes into a smooth contraction at frequencies of 20 to 30 Hz.

4. AN ISOKINETIC MUSCLE EXERCISER FOR STRENGTH TRAINING

With the above rationale, work was recently done to develop two types of devices for training paralyzed muscle using electrical stimulation. Muscle trains as the result of a specific metabolic demand on nonparalyzed muscle. This demand has never been imposed with conventional electrical stimulation. However, by using computer technology, metabolic load on tissue can be controlled through electrical stimulation. The first of these devices was a progressive resistance exerciser. In a manner similar to voluntary progressive resistance exercise, this device allows paralyzed muscle to be exercised slowly while lifting a heavy load (Petrofsky et al., 1984a; Petrofsky and Phillips, 1983b). This type of isokinetic exercise has been universally shown to increase muscle strength rapidly in nonparalyzed muscle (Astrand and Rodahl, 1977). To cause this type

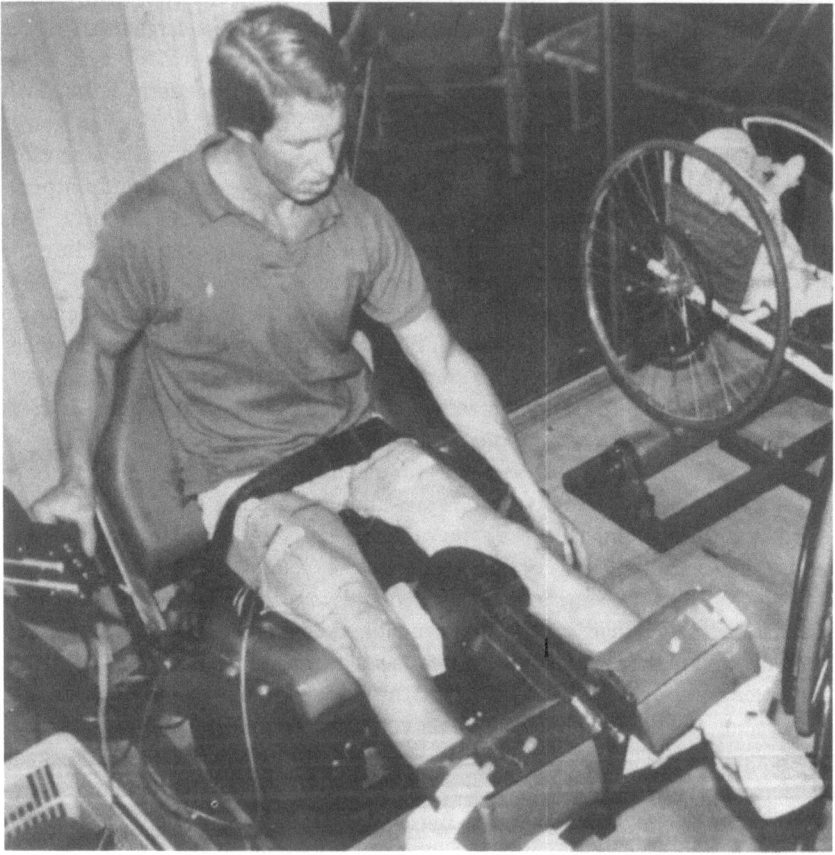

Figure 1. An isokinetic exerciser used for exercising muscles on paralyzed individuals.

of smooth but slow movement, feedback control was used (closed-loop control).

To exercise the quadriceps muscles as an example, a device was built as shown in Fig. 1. Sequential electrical stimulation of the quadriceps muscle was accomplished by three electrodes applied diagonally across the muscle over three motor points. Stimulation was then alternated between opposite pairs of electrodes at a frequency of 30 Hz with a pulse width of 300 μsec. The stimulation itself was current controlled and balanced biphasic (Petrofsky et al., 1983b).

To measure the movement of the muscles, an ankle support was applied around the ankle and then connected to weights. An appropriate load would be placed on the muscle by a computer, and when the muscle contracted, the leg would be lifted up and down to allow the muscles of the subject to engage in progressive resistance exercise. A sensor (potentiometer) in series with the chain was used as input for the computer controller to determine the position of the leg. A computer then used control equations to develop smooth muscle contractions for this muscle or other muscles in the body with other devices.

As described above, weight lifting can be accomplished using closed-loop control of skeletal muscle. In earlier experiments, simple closed-loop control was used with no adaptive control. These experiments (to which most of the data reported in this chapter pertain) still work quite well in terms of controlling movement in paralyzed muscle for exercise (Petrofsky et al., 1984a). However, use of adaptive control improved the control of muscle and reduced the possibility of errors such as end-loop oscillations, which could potentially injure a subject by hyperextending the knee. Exercise devices are now in use for all major muscle groups in the body.

5. AN AEROBIC EXERCISE BICYCLE FOR ENDURANCE
   TRAINING

To accomplish aerobic exercise, bicycling has been used. Aerobic exercise involves the lifting of weights for movement of the muscle against low loads but at high velocity and up to a point at which oxygen can be supplied to the muscle in sufficient quantities to provide the vast majority of the metabolism. This type of exercise is particularly important in that anaerobic types of exercise are a poor means of building the heart and circulatory system compared to aerobic training. To accomplish aerobic exercise, a special bicycle with postural support, knee stabilizers, breakaway shoes, and a leg position sensor in the pedals was developed (Petrofsky and Phillips, 1983b; Petrofsky et al., 1984c) (Fig. 2).

In previous studies, bicycling was induced by turning off and on a variety of different muscles, generally only the quadriceps and iliopsoas. In later studies, stimulation of the quadriceps, hamstrings, and gluteus maximus muscles was used. These three muscles compose the major power groups for the leg and are used during pedaling in voluntary exercise in nonparalyzed man. A review of methods used to induce aerobic exercise through computer-controlled bicycling

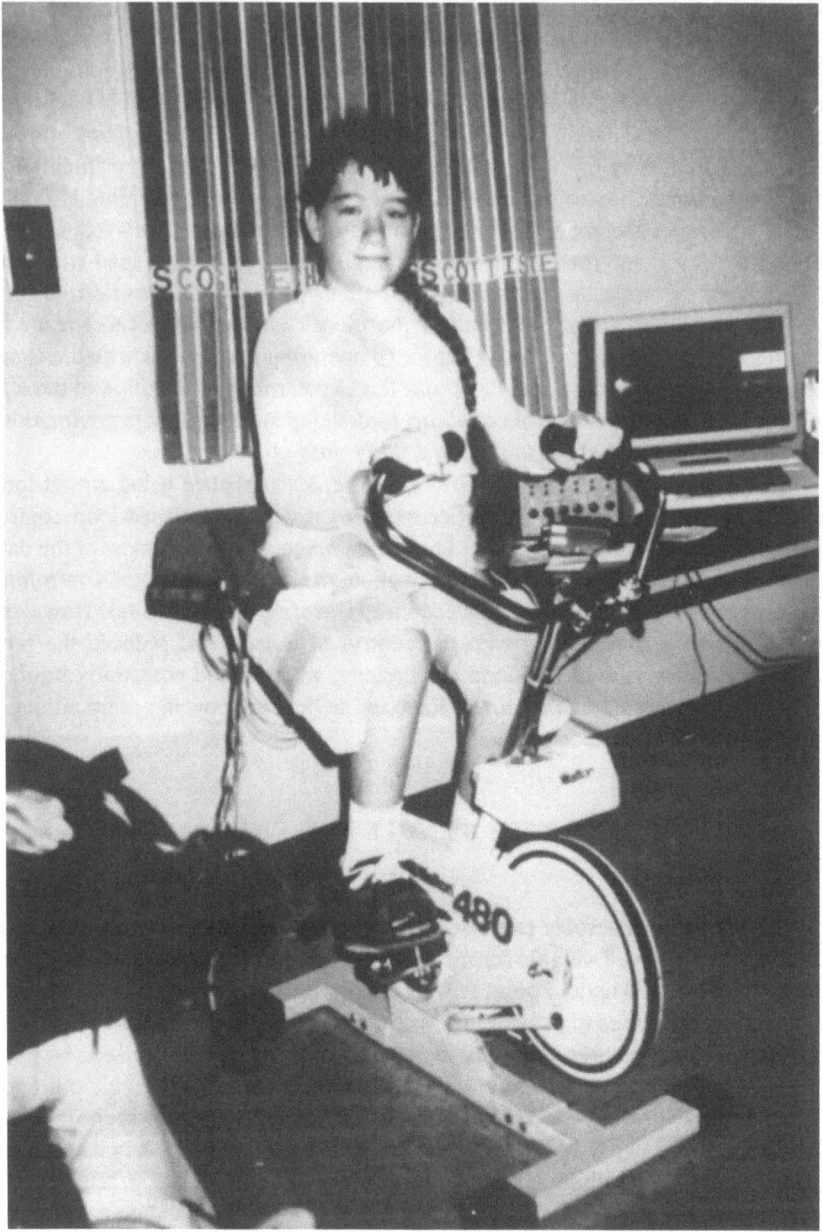

Figure 2. Modified ergometer used for exercising paralyzed individuals.

with both indoor and outdoor bicycles (three-wheeled bicycles used to induce exercise pedaling outdoors with an on-board computer) is given elsewhere (Petrofsky et al., 1983a, 1984c). These devices are now in use in hundreds of rehabilitation centers in the United States (Therapeutic Technologies Inc.).

One significant addition to the bicycle system in the last two years has been a cartridge controller (Petrofsky et al., 1985e), which, in many ways, resembles a game cartridge. In this particular case, the cartridge can contain either a read-only memory (ROM) or electronically erasable programmable read-only memory (EEPROM). The cartridge is used to store the program for each individual subject. The program itself (as described above) is not different for each individual subject, but essential data parameters such as blood pressure, heart rate, EKG, and body temperatures as well as the workload are different. In a clinical or hospital setting, one problem associated with electrically induced exercise is that if the workload is overdone, injury can result. For paraplegics and quadraplegics with impaired autonomic nervous systems, there is also the possibility of compromise to the cardiorespiratory system from overload. For example, recent studies have shown that paraplegics generally do not increase their blood pressure during isokinetic leg exercise because of autonomic nervous damage in afferent pathways in the lower spinal cord (Hendershot et al., 1985). Because of patients' impaired cardiorespiratory systems (quadriplegics, for example, cannot, in many cases, control their heart rate), physiological monitoring was believed to be essential for active physical therapy in a clinical setting. The cartridge system was developed to provide such monitoring.

The cartridge can be programmed by the physician regarding the extent of the workload, the number of times per week the exercise could be done, and the maximum resting and exercising body temperature, heart rate, and blood pressure. When the cartridge is then plugged into the computer, the bicycle itself becomes a physiological monitoring station, measuring resting and exercising physiological variables, and is able to shut the exercise off if the physiological parameters (set by a physician) are exceeded. With the EEPROM cartridge, the exercise can be recorded on the cartridge so that significant physiological variables and work endurance are both stored on the cartridge and reviewed by the physician at a later date. In this respect, the cartridge is an "electronic physicians' prescription system" to allow the setting of safe limits and monitoring of exercise response of paraplegic and quadriplegic patients in either a clinical or home environment.

6. PHYSIOLOGICAL CHANGES AND PHYSICAL CONDITIONING RESPONSES TO FES-INDUCED ACTIVE PHYSICAL THERAPY

6.1. Background Information

There are numerous studies indicating that FES-induced exercise can affect atrophied paralyzed muscle. Vrbova (1963), Brown (1973), Dubowitz and

Brooke (1974), and Van der Meulen *et al.* (1974) have found that electrical stimulation of skeletal muscle increases the speed of contraction and oxidative metabolism. Vrbova (1963) and Cherepakhin *et al.* (1977) studied muscles microscopically and found that continued FES-induced exercise on muscles restored the normal distribution of fast and slow twitch fibers in the muscle. Salmons and Vrbova (1969), Peckham *et al.* (1976), Hudlicka *et al.* (1977), and Brown *et al.* (1976) all studied the effect of FES on fast and slow twitch muscle elements. Cooper *et al.* (1973) implanted sciatic nerve stimulators and found that after only 10 weeks of stimulation, weak and paralyzed muscles increased in strength up to 25%. Kralj and Vodovnik (1977) published a comprehensive history of the use of FES with emphasis on its use in paralyzed patients.

Unfortunately, these studies have varied with respect to stimulus parameters, patient condition, and conditions of study and were done without control of the motion produced. In most cases, these studies neglected to measure the effects of electrical stimulation on the cardiovascular system and such simple parameters as blood pressure and heart rate. Therefore, it is difficult to interpret their results. Additional, carefully designed, qualitative studies on the effects of FES-induced exercise have therefore been done over the last few years.

The relationship between exercise and body fitness is well known. Alam and Smirk (1938) observed a blood pressure increase with voluntary isometric skeletal muscle contractions. Lind *et al.* (1968), Petrofsky and Lind (1980a), Krayenbuehl *et al.* (1973), and Helfant *et al.* (1971) have extended these observations and conducted studies on the effects of lack of exercise on the body condition of paralyzed persons.

McCloskey and Mitchell (1972) and Coote *et al.* (1971) worked with electrical stimulation of cat muscles and observed an increase in arterial pressure with such stimulation. Lind *et al.* (1964) and Petrofsky *et al.* (1981) made the same observation in the work in humans and animals. Corbett *et al.* (1970) reported an increase in arterial pressure with electrical stimulation of paralyzed muscles of the quadriplegic patient.

There have been numerous studies indicating that the forced inactivity of the paralytics results in a reduction of cardiovascular function. This work has been reviewed by Davis *et al.* (1981) as well as Nilsson *et al.* (1975) and has described general cardiovascular deconditioning of paralytic persons. Hjeltnes (1977) observed a large change in AV oxygen difference in lower-limb paralytics and considered this an indication of impaired circulation associated with lower-body blood pooling. Clausen (1977), in reviewing the cardiovascular effects of exercise, noted that the literature supported the concept of reduced circulatory function in paralytic patients. Shephard (1977) pointed out that normal methods of assessment of physical fitness cannot be used in paraplegics. Davis *et al.* (1981) cited a number of researchers who have studied the use of armchair ergometers for upper-body exercise to enhance fitness in paralyzed persons. Shephard (1965) observed increases in fitness estimated at 20 to 30% in exercising humans with upper-body exercisers. Knutsson *et al.* (1973) observed an increase in work capacity and a decrease in heart rate response in paraplegics when they carried out an exercise program. Coutts *et al.* (1983), in well-designed experiments, studied cardiopulmonary function and oxygen uptake in paraplegics and quadriplegics using wheelchair ergometry and noted distinct improvement with the exercise. Huang *et al.* (1983) compared paraplegics with normal, able-bodied individuals and reported significant differences in fitness. All these workers used wheelchair ergometry as the exercise medium.

Bergofsky (1964) observed the diminished thoracic compliance in patients with high SCI injuries and hypothesized that this may be related to a loss of intracostal function. Axen (1984) studied the responses of paralyzed patients to imposed pressure loading of the lungs and did not observe any significant difference between normal subjects and paraplegics.

6.2. *Functional Electrical Stimulation as a Therapeutic Modality*

Obviously, with any type of technology, especially technology to be used in rehabilitation, the real proof of that technology is not in the ability to control movement or the smoothness of that control but in the benefits that can be derived for the patient. Active physical therapy, like any other medical technology, must have proven medical benefits far superior to simple therapeutic electrical stimulation with a single pair of electrodes to be accepted as a well-used rehabilitation modality. Also, active physical therapy must be practical in a clinical setting in terms of the time required for a standard exercise session. Therefore, all protocols described below only deal with 15 min of exercise accomplished a maximum of once per day.

6.3. *Cardiovascular Responses*

As in voluntary exercise, blood pressure and heart rate increased throughout the duration of the work in both paraplegics and quadriplegics. However, blood pressure and heart rate changes associated with exercise (normalized in terms of percentage duration of the work over the 15-min period) were dramatically different between these groups of subjects during isokinetic leg exercise that fatigued the test muscle in 15 min. Although differences in individual response were found (Petrofsky *et al.*, 1985e), the average response, presented in Fig. 3 (from Phillips *et al.*, 1984), shows a significant difference in the response of paraplegic and quadriplegic subjects. Paraplegic subjects showed a small increase in blood pressure (both systolic and diastolic) throughout the duration of the exercise, with a large increase in heart rate. In contrast, quadriplegic subjects showed a sharp increase in blood pressure but a small to variable increase in heart rate. The differences in heart rate and blood pressure associated with work in paraplegic and quadriplegic subjects have been attributed to the level of the

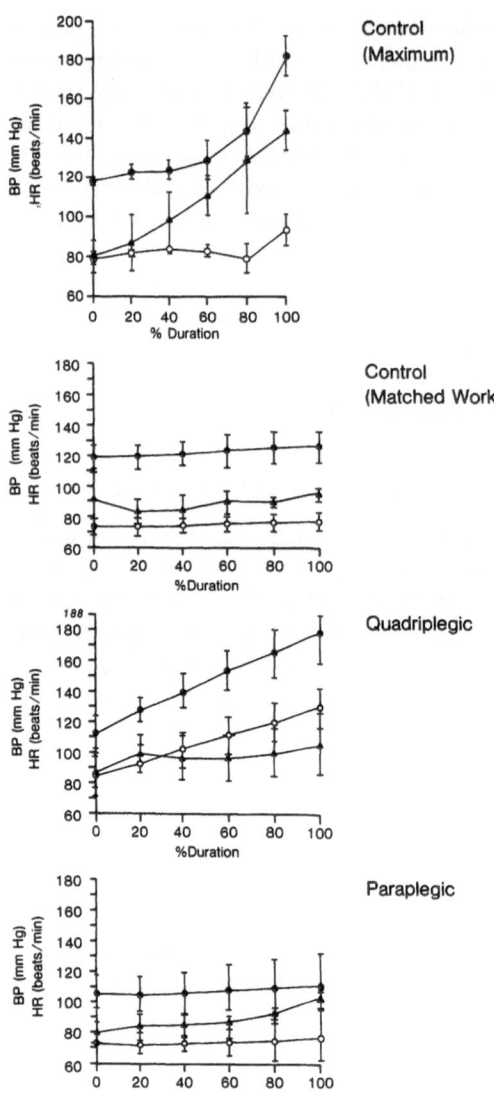

Control
(Maximum)

Control
(Matched Work)

Quadriplegic

Paraplegic

Figure 3. The average blood pressure
and heart rate responses of paraplegic,
quadriplegic, and control subjects associated with isokinetic exercise set to
fatigue the subject in a 15-min period,
and also matched work for control
subjects. (●) Systolic blood pressure;
(○) diastolic blood pressure; (▲)
heart rate.

injury and its effect on the normal reflex pathways associated with the blood
pressure response that occurs during exercise.

The increase in blood pressure and heart rate throughout the duration of the
exercise could be modified by varying the rest interval between the contractions
(Hendershot *et al.*, 1985). When the intercontraction interval was less than 6 sec,
blood pressure and heart rate both increased throughout the duration of the

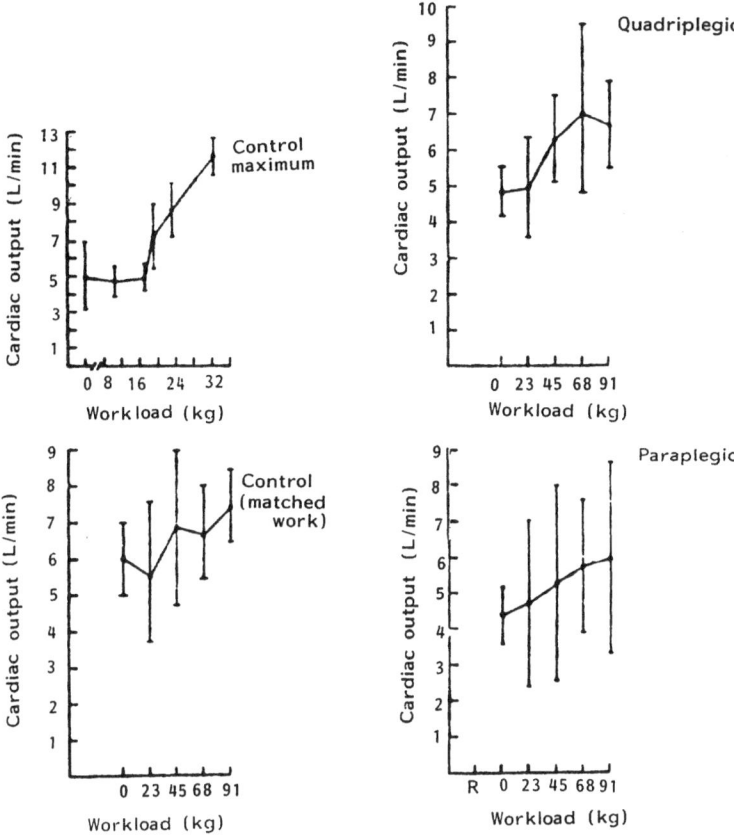

Figure 4. Cardiac output in quadriplegic, paraplegic, and control subjects during isokinetic exercise.

exercise to their maximum values. With intercontraction intervals greater than 6 sec, little change was seen in heart rate and blood pressure.

As in other types of isokinetic exercise, the increased metabolic demand from exercising muscle resulted in an increase in cardiac output (Fig. 4), ventilation (Fig. 5), and increased stress on the heart as gauged by the echocardiogram as shown in Fig. 6. However, as can be seen by these figures, the demand on the cardiorespiratory system was small. This is not surprising, since the size of the muscle group and the fact that the work was largely anaerobic dictates only a small increase in oxygen requirement in exercising muscle. It is not surprising that when training studies were accomplished with months of leg training (Fournier *et al.*, 1984), only small long-term effects were seen on conditioning of the blood pressure and heart rate responses either at rest or during the exercise. However, these forms of exercise (i.e., isokinetic and isometric) have never been

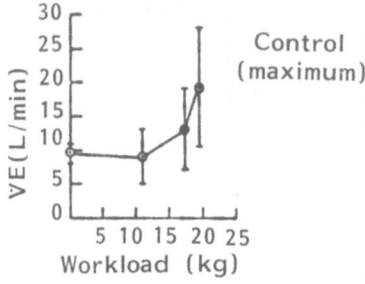

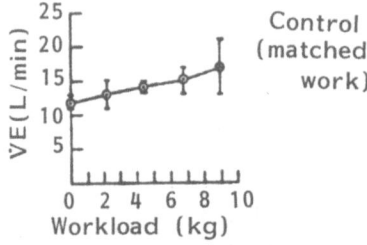

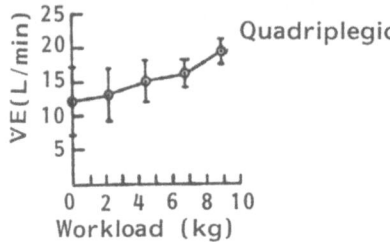

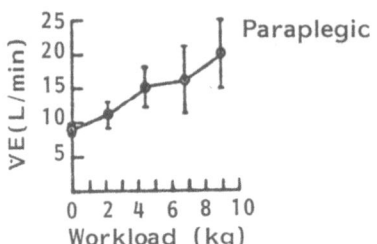

Figure 5. Ventilation in paraplegic, quadriplegic, and control subjects during isokinetic exercise associated with extension of the quadriceps muscle.

associated with cardiorespiratory conditioning, even in nonparalyzed men (Astrand and Rodahl, 1977). Therefore, to train muscular endurance as well as the cardiorespiratory system, as described above, bicycling was either induced on an indoor bicycle, such as the special bicycle described above, or on an outdoor bicycle as shown in Fig. 7.

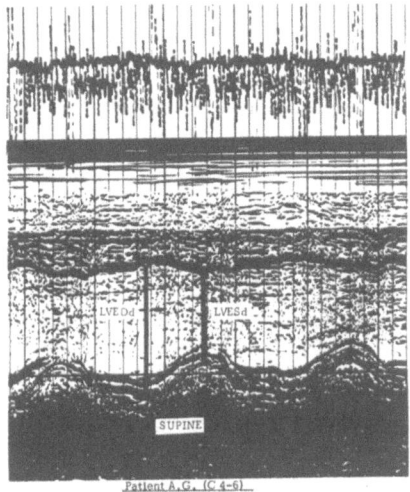 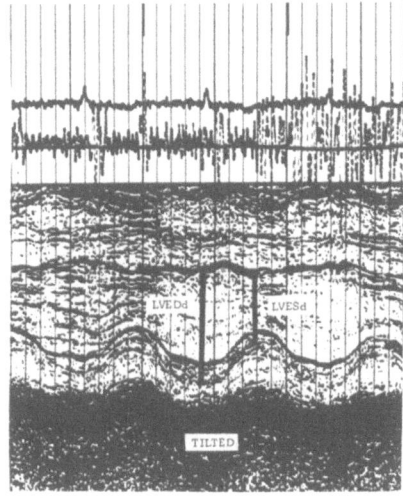

Patient A.G. (C 4-6)
M-Mode Echocardiography in Supine Position
Left ventricular end diastolic dimension (LVEDd) = 45 mm
Left ventricular end systolic dimension (LVESd) = 30 mm
Cardiac Index (CI) = 2.27 L/Min./M$_2$
Ejection Fraction (EF) = 70%

Patient A.G. (C 4-6)
M-Mode Echocardiography in Tilted Position (40°)
LVEDd = 38 mm   LVESd = 24 mm
CI = 1.7 L/Min./M$_2$ EF = 74%

Figure 6. Echocardiogram showing performance of the heart during tilting in a typical quadriplegic subject.

Like the leg trainer studies described above, the bicycle studies were set to optimize their potential use in a clinical setting. Exercise was set for 15 min maximum duration. In the case of the bicycles, for the experiments described below in which training was accomplished, subjects initially started at a workload that they could pedal between 10 and 15 min. If the workload could not be continued for 10 min, then the workload was reduced on the bicycle before the next experiment. If the subjects could successfully pedal 15 min, the workload was increased in $\frac{1}{8}$-kp steps for the next experiment.

As was the case with leg training, the blood pressure and heart rate increased in both paraplegic and quadriplegic subjects. This was in contrast to isokinetic or isometric exercise, where paraplegics showed only a modest increase in blood pressure throughout the exercise, possibly because the drive for increasing cardiac output, and hence blood pressure, during dynamic exercise is typically different from that associated with isometric and isokinetic exercise (Petrofsky and Phillips, 1984).

During dynamic exercise, the drive associated with increasing cardiac output comes from a large increase in venous return. Therefore, even with the absence of normal neural pathways associated with heart rate control as may occur in quadriplegics, the increase in cardiac output could be driven by a Frank–Starling mechanism because of a large increase in venous return from the size of the exercising muscle masses in the lower part of the body. This increase in

Figure 7. The outdoor bicycle used by paraplegic and quadriplegic subjects with an onboard computer to elicit computer-controlled bicycling.

cardiac output in the face of even a constant peripheral resistance would result in increases in both systolic and diastolic blood pressures. Similar phenomena occur with upper-body exercise.

With lower-body exercise in nonparalyzed man, mean blood pressure usually changes little if at all throughout the duration of even the most fatiguing work (Astrand and Rodahl, 1977). In contrast, both systolic and diastolic blood pressure can increase throughout the duration of fatiguing work resulting from the association with upper-body exercise because of the small muscle mass and large increase in cardiac output per change in peripheral resistance (Astrand and Rodahl, 1977).

These data point to the complexity of interpretation of heart rate and blood pressure responses in paralyzed man. Typically, in nonparalyzed man, the level of exercise stress is set by monitoring the blood pressure and heart rate during exercise. If certain predetermined values are exceeded, the exercise is terminated as being too stressful. However, in paralyzed man, the complex interactions among peripheral resistance,'cardiac output, and the disrupted autonomic nerve pathways make normal assessment of cardiopulmonary stress difficult at best. Further, the wide range of individual differences in response associated with different levels of autonomic nervous system damage certainly point to the unreliability of blood pressure and heart rate as an accurate assessment of cardiovascular stress (Petrofsky *et al.*, 1985e).

Although heart rate and blood pressure certainly showed a variable response associated with dynamic exercise, there was also a pronounced training effect in quadriplegic subjects. Quadriplegic subjects generally have a low tolerance for change in body position because of the disruption of the autonomic pathways. This orthostatic intolerance is most pronounced when tested on a tilt table. Rapid change in body position can cause subjects to black out and blood pressure to become erratic. After as little as 6 weeks of physical training on the bicycle ergometer, the cardiovascular system was changed markedly. Tolerance for tilt was increased (Danopulos *et al.*, 1985), and there was an increase in both resting systolic and diastolic blood pressures as shown in Fig. 8. Illustrated here are the resting and exercising mean blood pressures in four quadriplegic subjects before and after 6 weeks of training. During the course of this training regimen, resting blood pressure increased, and exercising blood pressure associated with training on the bicycle ergometer was reduced.

It must be noted that this dramatic training effect and increase in orthostatic tolerance was associated with just 15 min of exercise, three days per week, with only the quadriceps and iliacus muscles contracting. This is significant since the

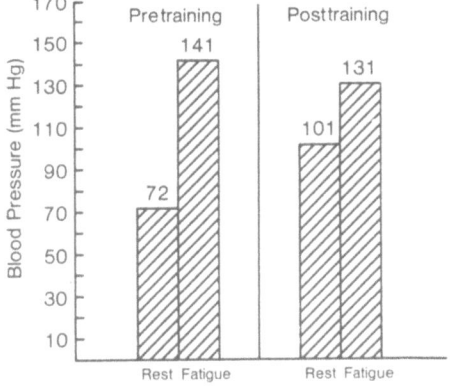

Figure 8. The changes in blood pressure and heart rate associated with training over 6-week period both at rest and during exercise (mean blood pressure only shown) in quadriplegic subjects when exercising on the bicycle ergometer (see text).

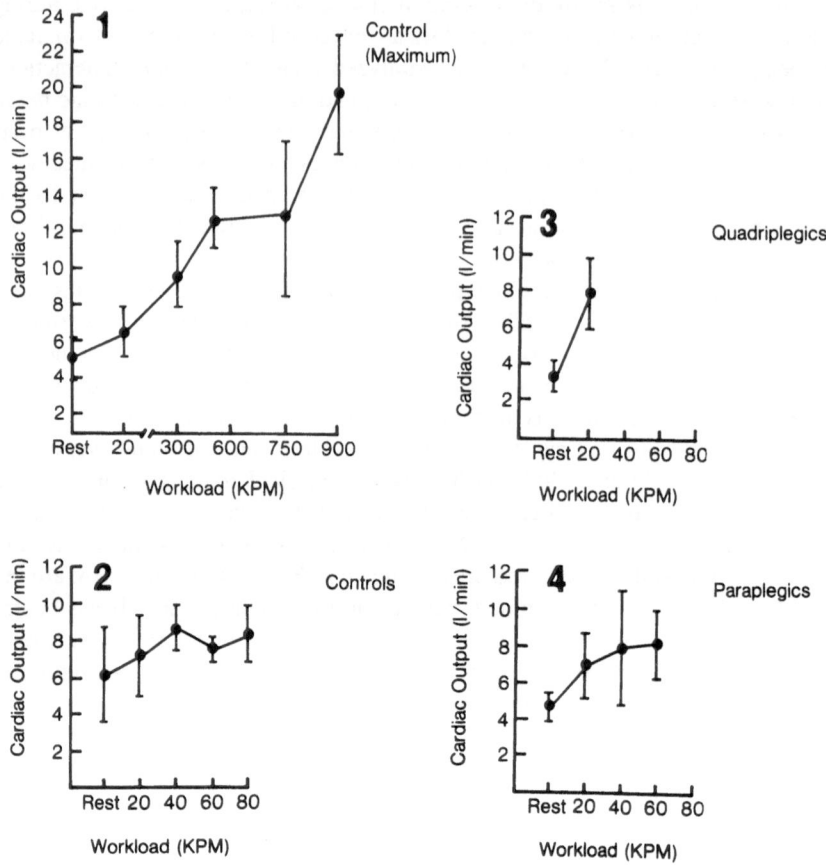

Figure 9. The cardiac output in paraplegic, quadriplegic, and control subjects when exercising on the bicycle ergometer. In these studies, only the iliapsoas and the quadriceps muscles were used.

muscle groups doing the exercise here are certainly small, as were the resulting changes in cardiac output (Fig. 9) associated with aerobic exercise. Recent experiments described above involved bicycling using the hamstring, quadriceps, and gluteus maximus muscles (Petrofsky *et al.*, 1985c). Using these three groups of muscles can increase cardiac output severalfold higher than that shown in Fig. 9, with corresponding increases in oxygen uptake.

6.4. Thermoregulatory Responses

In addition to dysfunction of the autonomic nervous system associated with paraplegia and quadriplegia, there is dysfunction of the thermoregulatory system as well. Claus-Walker and Halstead (1981) have extensively reviewed the literature on partial decentralization of the autonomic nervous system including thermoregulatory effects. Downey *et al.* (1967) showed that shivering occurred in quadriplegics when the core temperature was lowered, indicating that the deep thermal sensors are still operable. In 1969, the same workers showed that the stimulation of skin temperature receptors by cold increased the temperature at which shivering occurred in paralytics, and in 1973, they extended these studies to show that central cooling increases hand blood flow in paralyzed persons. There has been little work in assessment of thermoregulatory function in patients with autonomic impairment. In our own work (Petrofsky and Phillips, 1984), it was shown that people who are paralyzed from spinal cord injury (both paraplegics and quadriplegics) had reduced tolerance to work in the heat.

For example, Fig. 10 shows the central core temperature in paraplegic and quadriplegic subjects during rest in a climatic chamber for 30 min at temperatures of 30, 35, and 40°C. In these cases, the relative humidity was kept constant at 50%. Subjects at rest were able to tolerate the heat load quite well. However, when exercise was added with an armcrank ergometer at levels of work of 25 W for the entire 30-min period, control subjects were still able to maintain good body temperature control. In contrast, paraplegic subjects showed a sharp increase in body temperature associated with increasing environmental temperature, and quadriplegics could not maintain work at all in a 40°C environmental temperature and barely maintained work in a 35°C environmental temperature for the 30-min period (Petrofsky and Phillips, 1984).

Bicycling exercise is considered to be one of the most efficient forms of exercise. Here, however, nonparalyzed man's (Petrofsky and Lind, 1980b) and

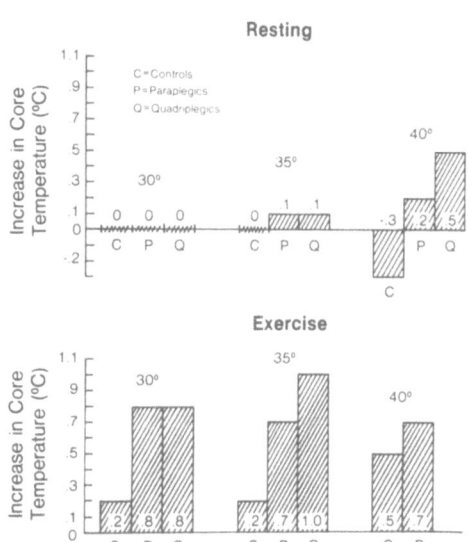

Figure 10. Increase in body temperature of a typical group of paraplegic and quadriplegic subjects during exposure to heat. In this particular case, subjects were exposed to temperatures of 30, 35, and 40°C at 50% relative humidity. During this exposure, a tympanic temperature probe was inserted near the eardrum to measure central core temperature. Central core temperature was measured under two conditions. In one condition, subjects sat in the environmental chamber for 30 min at rest, whereas in the other, subjects performed 25 W of work on an armcrank ergometer while sitting in the exercise chamber.

paralyzed man's (Petrofsky *et al.*, 1985c) efficiency for bicycling are still only approximately 25%. In other words, during bicycling, three-quarters of the energy associated with the breakdown of ATP is converted into heat instead of energy for muscular contraction. Given the above, then, it is not surprising that heat buildup would be significant during exercise in paralyzed individuals.

This was believed to occur because the sweat glands are also paralyzed below the level of injury in many paraplegic and quadriplegic patients. For this reason, the mean body sweat rates were significantly lower in paralyzed individuals than controls. These lower mean body sweat rates resulted in an inability of quadriplegics to thermoregulate properly at higher environmental temperatures. Heat stress must be carefully monitored when quadriplegics are involved, especially when exercising in warm environmental temperatures.

6.5. Muscular Response

Reports in the exercise literature (Astrand and Rodahl, 1977; Simonson, 1971) cite that training of skeletal muscle is specific for the type of exercise being accomplished. For example, simply lifting a weight up and down a few times a day does not increase muscle strength. Increasing strength in a muscle requires that progressive resistance exercise be accomplished. Mueller (1932) demonstrated that if weights were lifted at approximately half to two-thirds of the muscle's maximum strength very slowly up and very slowly down, muscle would develop strength rapidly. In contrast, activities during which the muscle works rapidly but at low loads, such as running, are associated with only small increases in muscular strength (Astrand and Rodahl, 1977). The type of metabolic load induced in the muscle can therefore be very instrumental in determining the changes that occur biochemically in skeletal muscle with long-term physical training.

As seen from the above studies, a number of investigators have examined the use of electrical stimulation as a means of restoring strength in skeletal muscle. In at least the last 30 years, electrical stimulation has generally been used to train muscle by applying two electrodes (synchronous stimulation) above the muscle and then stimulating it with different patterns of electrical current. These studies have ranged from altering the amplitude of stimulation to altering the frequency and pulse width. Muscles have been stimulated at tetanic frequencies rather than being turned off and on for various periods of time during a given day (Salmons and Vrbova, 1969; Peckham *et al.*, 1973; Pette *et al.*, 1973, 1975; Hudlicka *et al.*, 1977). The results of these experiments have been interesting but not very encouraging. Generally, they seem to show that although some strength can be restored to paralyzed muscle by FES, the increase in strength is small compared with that which can be obtained with voluntary physical training. Although endurance also increases, even with months to years of electrically induced exercise such as that described above, muscle still fatigues fairly rapidly. Further, the fatigue lasts for extended periods of time. This is in direct contrast to voluntary activity, where the muscle fatigues rapidly but then recovers quickly (Astrand and Rodahl, 1977).

One interesting phenomenon associated with this increase in strength, however, is spillover to other muscles. To determine the cross-sectional area of muscles, an anatomic point was chosen on the leg, and a CT scan was done. These CT scans showed pre- and posttraining data in terms of the size of not only the quadriceps muscles but other muscles in the upper and lower leg as well. As shown in Fig. 11, exercise 3 days a week over a 4-month period by a typical paraplegic subject resulted in a sharp increase in the mass of the muscles in the exercising leg (right leg) but also in a small increase in the mass of the left leg. During such exercise experiments, it is not uncommon to see some contralateral muscle activity in one leg when the other is stimulated. For example, at high stimulation levels, it is not uncommon to stimulate not only the quadriceps but the hamstring muscles as well because of some current spread to the peroneal nerve located on the dorsal aspect of the femur. With this type of electrical stimulation, muscle biopsy studies have shown that electrical stimulation during

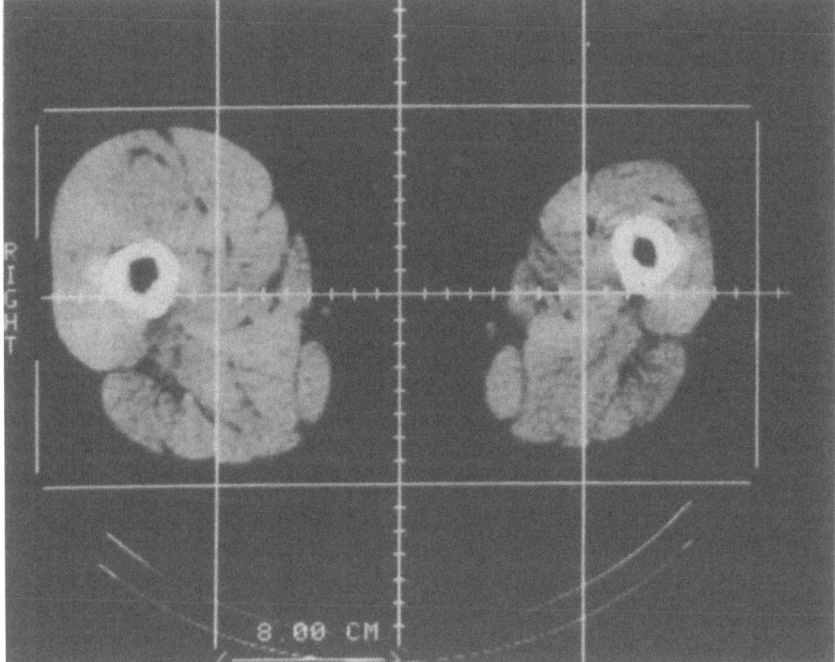

Figure 11. A CT scanner cross section of the thigh in a paralyzed individual. The left leg was used as the control, whereas the right leg was exercised 15 min a day, 3 days per week, using closed-loop control isokinetic exercise (see text).

leg lifting is associated predominantly with hypertrophy of fast twitch (type I) muscle fibers. Electrically induced weightlifting has clearly resulted in an increase in muscle fiber size and blood flow to the muscle fibers with no apparent increase in muscle fiber sprouting or nerve sprouting.

After years of paralysis, muscles are inherently very weak. In the past, in most cases, bicycling was impossible as an exercise starting point. Muscular strength has been shown to be less than 10% of the average voluntary strength of nonparalyzed individuals in paraplegics and quadriplegics (Petrofsky and Phillips, 1983b). For this reason, then, in most previous studies, subjects were first trained with weights on the leg trainer (described in Section 4 of this chapter) to build up muscular strength prior to any type of endurance training.

Although subjects could lift in excess of 20 lb for a period of 15 min with their leg muscles before they engaged in bicycling, initial endurance for bicycling was very short (Petrofsky and Phillips, 1983b). When patients were initially placed on the bicycle ergometer with stimulation of the iliacus and quadriceps muscles to induce pedaling at a speed of 50 rpm, endurance was approximately only 30 sec for a typical subject. However, as the exercise was continued on a 3-day-per-week basis, endurance increased rapidly to about 30 min after seven sessions. As with weightlifting, when exercise was continued on a 1-day-per-week basis, endurance training was much less. On a 5-day-per-week basis, endurance increased much more rapidly. Some individuals display a very rapid increase in endurance on the bicycle, whereas others show a very slow increase in endurance when exercising on a 3-times-per-week basis for a maximum of 15 min on the bicycle ergometer.

7. FUNCTIONAL ELECTRICAL STIMULATION AND WALKING

A final form of exercise that has been developed recently is walking. A number of different laboratories around the world (Kralj et al., 1980; Petrofsky and Phillips, 1983c, 1985; Petrofsky et al., 1984b, 1985b; Marsolais and Kobetic, 1982; Holle et al., 1984) have been working in recent years with computer-controlled electrical stimulation for walking. Walking, like any other form of muscular exertion, is more than simply ambulation. Walking in itself can be used for physical training. In our own studies, computer-controlled walking was originally initiated with closed-loop control movement with sensors at the knee and other experiments at the hips and on the foot. Recent experiments in this area by our own group have allowed fairly smooth, coordinated walking with surface transcutaneous electrical stimulation. By using electrodes above the quadriceps, gluteus maximus, and hamstring muscles as well as the pectineus and tensor fasciae latae muscles, it has been possible to allow both paraplegic and quadriplegic subjects to stand with some balance and to walk limited distances. Initial studies, however, show that fatigue would occur within a few hundred feet of walking and that a great deal of stress was placed not only on the cardiorespiratory system but on the legs. Therefore, recent studies, as shown in Fig. 12, have used a combination of lightweight braces (Louisiana State University reciprocating gait orthosis) with electrical stimulation.

In these studies, electrical stimulation was applied only to the major power muscle groups of the legs, these being the quadriceps, hamstring, and gluteus maximus muscles. Because of the reciprocating action of the legs associated with the cables connecting the two sides of the braces, stimulation to tuck the pelvis on one side of the body caused ipsilateral extension of the hip. This extension of the hip then initiated steplike walking movements. For this reason, the timing for walking can be kept fairly simple, as shown by the control algorithm in Fig. 13. Standing and sitting, because of the complexity of the knee joint and the balance

Figure 12. A paraplegic research associate walking with the combined RGO–FES walking system.

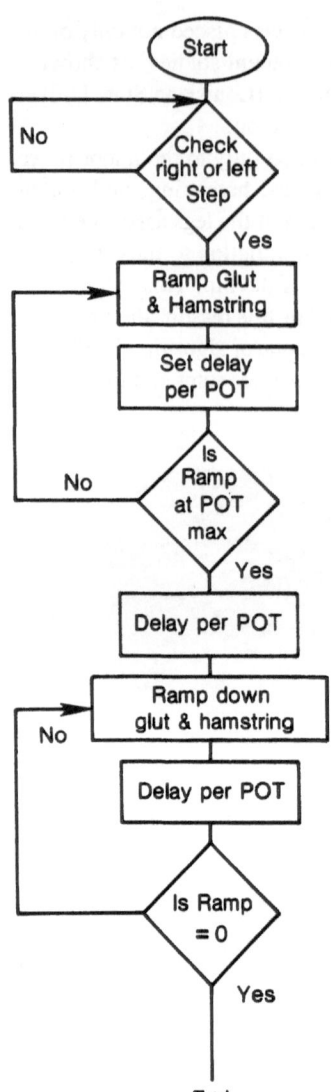

Figure 13. The algorithm used to induce walking in the combined RGO–FES walking system.

process, still required sensors to be placed on the knee and on the hips to detect hip position, hip and knee lock, and knee position to coordinate movement for a smooth, coordinated stand and for smooth coordinated sitting. A complete description of standing and sitting as well as walking is given elsewhere (Petrofsky *et al.,* 1985b).

For walking, the hamstring and gluteus maximus muscles were pulsed on the opposite side of the body from that desired to place a step in the forward position. This forward positioning of the leg by ipsilateral stimulation was coordinated through the anterior and posterior cables associated with the LSU reciprocating gait orthosis (RGO) (Douglas *et al.*, 1983; McCall *et al.*, 1983). To take a step forward, weight was shifted to the left-hand side of the body, and stimulation was applied to the gluteus maximus and hamstring muscles. To take a step back, weight is shifted to the opposite leg, and with stimulation again to the same gluteus maximus and hamstring muscles, the leg moves backwards. In this manner, then, walking forwards, backwards, and turning by a combination of walking forwards and backwards, could be accomplished with the RGO through stimulation of only the hamstring and gluteus maximus muscle groups.

By using electrically conductive clothing (Transcutaneous Transducer Garment) (Granek and Granek, 1985; Petrofsky *et al.*, 1985b), it was possible to develop a garment that could be used to stimulate muscle. By simply putting on a special garment and plugging a single DB25 connector into a computer, a subject could be prepared to walk or pedal a bicycle with only a few minutes' preparation time. Physiological costs of this type of walking are shown in Figs. 14 and 15. Shown in Fig. 14 is the weight distribution in a typical subject associated with walking in the RGO with and without electrical stimulation. As can be seen here, from the impulse loading on the upper panels and average weight distribution on the lower panels, weight distribution is decidedly different with and without electrical stimulation. Weight distribution was measured by developing a load cell and putting the load cell on the bottom of Laufstrand canes. These canes then provided an electrical output that could be amplified and either fed directly to a Brush recorder or averaged through an RMS converter, the output of which was provided for the same Brush recorder. This figure then shows the change in weight distribution in the right and left arms (R and L) during walking at a speed of approximately 1 mile per hour. As can be seen from this figure, weight distribution was low in the RGO, with impulses only averaging as high as 30 lb on each arm. This amounted to approximately 30% of the body weight of this particular subject. However, with FES-induced walking, the power to walk was decidedly moved to the lower part of the body, and weight carried on the arms was reduced by approximately 50% with a doubling of speed. Therefore, with walking in a combined brace and FES system, impulse loading and average strength on the arm in this particular subject, for example, was only a few pounds, which amounted to about 3% of the subject's body weight as an average weight distribution (Petrofsky *et al.*, 1985d). Electromyographic studies of the key muscle groups in the upper part of the body confirm these data, as shown in Fig. 15. When FES is used for walking, the amount of activity in key muscle groups such as the pectoralis major and the rotator cuff muscles is dramatically reduced, showing less muscle activity in the upper part of the body (Petrofsky *et al.*, 1985d).

The FES-induced walking in itself uses most of the muscle power in the lower part of the body. As shown in Fig. 14, where the same paralyzed subject is

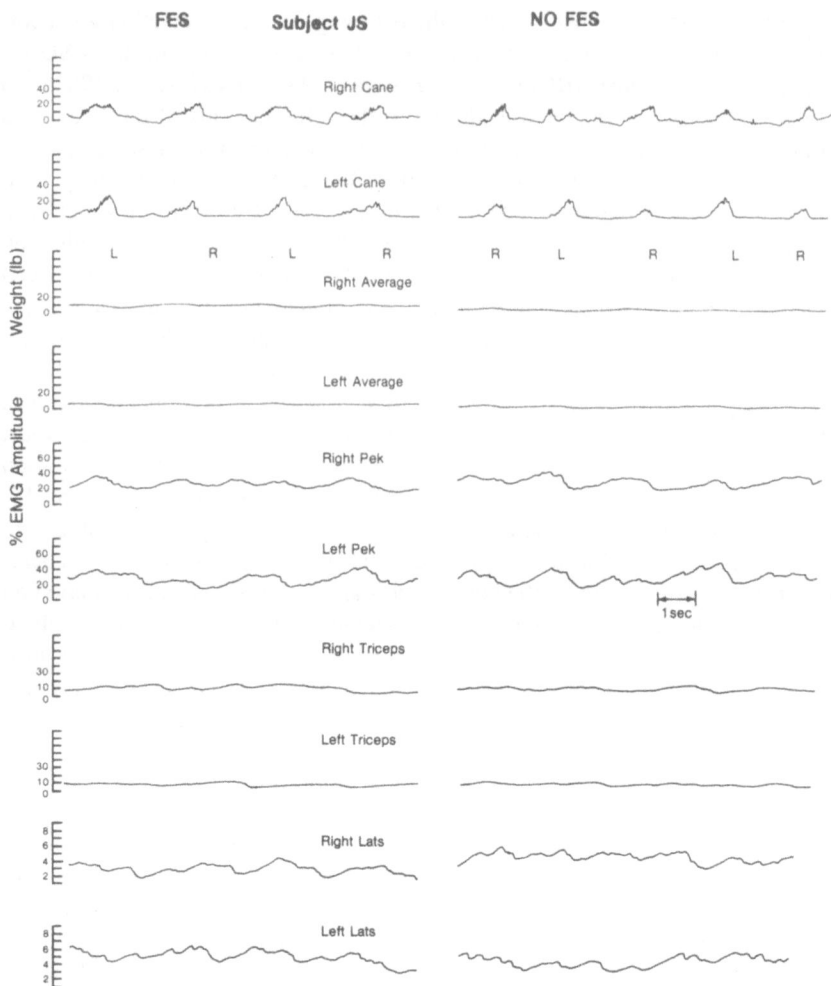

Figure 14. The weight distribution on the canes (Laufstrand canes, see text) associated with computer-controlled walking with a combination of RGO and FES technology (see text).

walking with and without electrical stimulation with movement being measured by lights on the shoulder, hips, and knees, it can be seen that walking is very smooth in an FES–RGO combined system. However, the cost of the work is not dramatically different. For example, after 15 min of walking in two paraplegic subjects, blood pressure was 130/84 and 127/68 mm Hg with and without electrical stimulation, respectively, and heart rate was 118 and 108 beats per minute with and without electrical stimulation, respectively.

Oxygen uptake, on the other hand, showed slight differences. The average

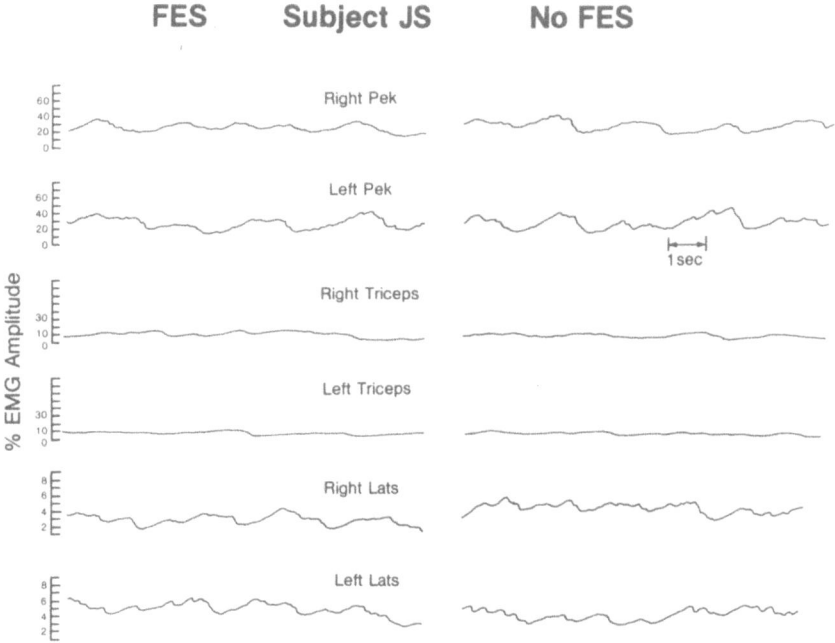

Figure 15. The electromyogram recorded on the pectoralis major muscles, the triceps muscles, and the rotator cuff muscles associated with walking with and without FES in the LSU RGO in a typical paraplegic subject (see text).

oxygen costs for walking in these same two subjects without electrical stimulation was 21.1 ml/kg per min, and with electrical stimulation was 15.6 ml/kg per min. On the surface it appears that the oxygen cost of walking with FES was lower than without FES (Petrofsky *et al.*, 1985a). In point of fact, the oxygen uptake of the exercise of the set was the same. However, when expressed in oxygen cost per meter of movement, because of the doubling in speed associated with walking with the lower part of the body, the oxygen cost per meter of work was lower, whereas the total oxygen cost of the exercise was the same. The absolute oxygen cost of walking was approximately 1.2 liters/min. in these subjects. This amounted to about 35% of their estimated maximum VO_2. Exercise at this level, with changes in blood pressure and heart rate as shown above, can be very therapeutic in terms of an aerobic exercise modality (Fig. 16), providing the same type of heart rate and blood pressure changes (Petrofsky *et al.*, 1985a).

However, for the full exercise effects of walking to be realized (especially for long distances), the subjects must be relieved of the necessity for continually looking at their feet. This occurs because of sensory anesthesia below the level of spinal cord injury (that often occurs with the motor paresis). In the absence of

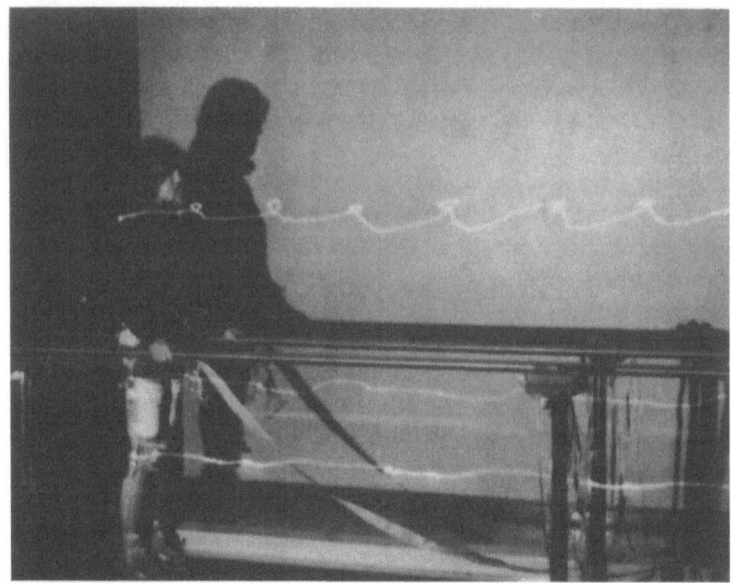

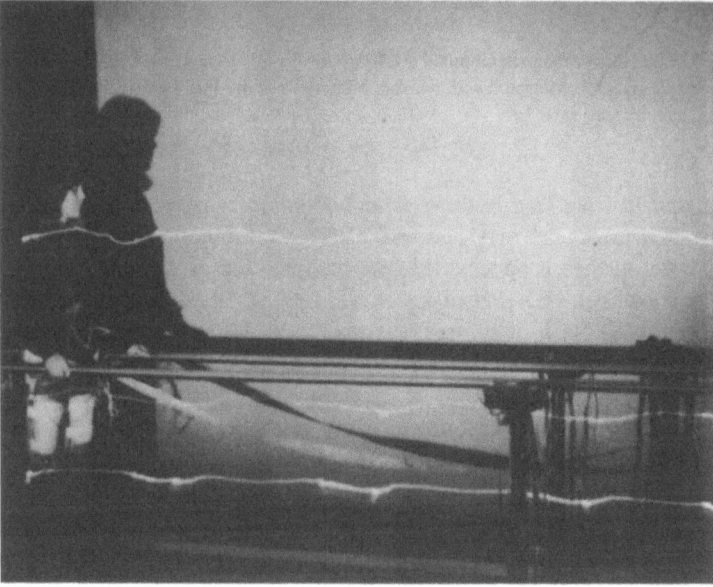

Figure 16. The locus of motion of the shoulder, hip, and knee joints measured in a typical paraplegic subject during walking with and without electrical stimulation. Locus of motion was measured by placing a grain-of-wheat light on the shoulders, knees, and hips and then using time-lapse photography to measure the extent and range of motion (see Petrofsky *et al.,* 1985a).

proprioceptive and tactile feedback, the subject substitutes visual feedback in order to maintain balance. We have developed a tactile feedback system that translates load (and position) information on the plantar surface of the feet into tactile (and spatial) information on the subject's chest wall above the level of injury (Phillips and Petrofsky, 1985a). This has been demonstrated in the laboratory environment to improve significantly the ability of a spinal-cord-injured individual to balance his/her body weight (with no visual cues) during prolonged standing (Phillips and Petrofsky, 1985b).

8. SUMMARY

For people who are fully paralyzed, a number of different therapeutic modalities have been developed to allow aerobic and anaerobic exercise for muscular conditioning. But more important than simple muscular conditioning is the enhanced conditioning of the cardiorespiratory axis that can occur with properly controlled and safe electrically induced exercise.

A number of factors are yet to be determined. For example, it becomes increasingly important to understand such factors as changes in skin blood flow, peripheral blood flow, and how they affect the formation of pressure sores, changes in renal and bladder function and how these are affected by exercise in paralyzed people. With different levels of autonomic nerve damage, it is possible that the response may be different in different individuals to these forms of exercise. However, the enhanced blood flow to the limbs during exercise and enhanced muscle training should, in itself, prevent the formation of pressure sores and, as is the case for diabetics, help whole-body function associated with electrically induced work. These matters need to be carefully investigated in the future, as well as more comprehensive studies about the effect of aging, body weight, length of time since injury, and male and female sex differences on the effect of electrically induced exercise on whole-body homeostasis.

ACKNOWLEDGMENTS. The author wishes to acknowledge all the volunteers who contributed their time to serve as research subjects.

REFERENCES

Alam, M., and Smirk, F. H., 1938, Unilateral loss of a blood pressure raising, pulse-accelerating, reflex from voluntary muscle due to a lesion of the spinal cord, *Clin Sci.* 3:247–252.

Astrand, P. O., and Rodahl, K., 1977, *Textbook of Work Physiology*, 2nd ed, McGraw Hill, New York.

Axen, K., 1984, Adaptations of quadriplegic men to consecutively loaded breaths, *J. Appl. Physiol.* 56(4):1099–1103.

Bergofsky, E. H., 1964, Mechanism for respiratory insufficiency after cervical cord injury. A source of alveolar hypoventilation, *Ann. Intern. Med.* **61**(3):435–446.

Brindley, G., Polkey, C. E., and Rushton, D. N., 1978, Electrical splinting of the knees in paraplegia, *Paraplegia* **16**:428.

Brown, G. L., and Burns, B. D., 1949, Fatigue and neuromuscular block in mammalian skeletal muscle, *Proc. R. Soc. Lond.* **136**:182–189.

Brown, M. D., Cotter, M. A., Hudlicka, O., and Vrbova, G., 1976, The effects of different patterns of muscle activity on capillary density, mechanical properties and structure of slow and fast rabbit muscles, *Pfluegers Arch.* **36**:241.

Brown, W. P., 1973, Functional compensation of human motor units in health and disease, *Neurol. Sci.* **2**(2):199–209.

Cherepakhin, M. D., Kakurin, L. I., Ilina-Kakueva, E., 1977, Evaluation of the effectiveness of electrostimulation of the muscles in preventing disorders relating to prolonged limited motor activity in man, *Kosm. Biol. Avaikosm. Med.* **11**(2):64–68.

Clausen, J. P., 1977, Effects of physical training on cardiovascular adjustments to exercise in man, *Physiol. Rev.* **57**:779–815.

Claus-Walker, J., and Halstead, L., 1981, Metabolic and endocrine changes in spinal cord injury: I. The nervous system before and after transection of the spinal cord, *Arch. Phys. Med. Rehabil.* **62**:595–601.

Cooper, E. B., Bunch, W. H., and Campa, J. F., 1973, Effects of chronic human neuromuscular stimulation, *Surg. Forum* **24**:477–479.

Coote, H. H., Hilton, S. M., and Perez-Gonzales, J. F., 1971, The reflex nature of the pressor response to muscular isometric exercise, *J. Physiol. (Lond.)* **215**:789–804.

Corbett, J. S., Frankel, H. L., and Harris, P. J., 1970, Cardiovascular changes associated with skeletal muscle spasm in tetraplegic men, *J. Physiol. (Lond.)* **215**:381–387.

Coutts, K. D., Rhodes, E. C., and McKenzie, D. C., 1983, Maximal exercise responses of tetraplegics and paraplegics, *J. Appl. Physiol.* **55**(2):479–482.

Crochetiere, W. T., Vodovnik, L., and Reswick, J. B., 1967, Electrical stimulation of skeletal muscle—a study of muscle as an actuator, *Med. Biol. Eng.* **5**:11–125.

Danapulos, D. M., Kezdi, P., Stanley, E. L., Petrofsky, J. S., Phillips, C. A., and Meyer, L. S., 1985, Cardiovascular circulatory dynamics in subjects with chronic cervical spinal cord injury in supine and head up tilt position, *J. Neurol. Orthop. Surg.* **6**:265–270.

Davis, G. M., Kofsky, P. R., Kelsey, J. C., and Shephard, R. J., 1981, Cardiorespiratory fitness and muscular strength of wheelchair users, *CMA J.* **125**:1317–1323.

Douglas, R., Larson, D., and D'Aubrosia, R., 1983, LSU reciprocating gait orthosis, *Orthopedics* **6**:834–839.

Downey, J. A., Chiodi, H. P., and Darling, R. C., 1967, Central temperature regulation in the spinal man, *J. Appl. Physiol.* **22**(1):91–94.

Dubowitz, V., and Brooke, M., 1974, *Muscle Biopsy, a Modern Approach*, W. B. Saunders, Philadelphia.

Faden, A. I., Jacobs, T. P., Feuerstein, G., and Holaday, J. W., 1981, Dopamine partially mediates the cardiovascular effects of noloxone after spinal injury, *Brain Res.* **213**:415–421.

Fournier, A., Goldberg, M., Green, B., and Petrofsky, J. S., 1984, A medical evaluation of the effects of computer assisted muscle stimulation in paraplegic patients, *Orthopaedics* **7**:1119–1133.

Goldsmith, H., Steward, E., Chen, W. F., and Duckett, S., 1982, Application of intact omentum to the normal and traumatized spinal cord, in: *Spinal Cord Reconstruction* (C. Kao, ed.), Raven Press, New York.

Granek, H., and Granek, M., 1985, Transcutaneous transducer garments: Current mammalian usage and future applications, *J. Neurol. Orthop. Surg.* **6**:271–278.

Helfant, R. H., DeVilla, M. A., and Meister, S. G., 1971, Effect of sustained isometric handgrip exercise on left ventricular performance, *Circulation* **44**:982–993.

Hendershot, D., Petrofsky, J. S., Phillips, C. A., and Moore, M., 1985, Blood pressure and heart responses in paralyzed and non-paralyzed man during isokinetic leg training, *J. Neurol. Orthop. Surg.* **6**:259–264.

Hjeltnes, N., 1977, Oxygen uptake and cardiac output in graded arm exercise in paraplegics with low level spinal lesions, *Scand. J. Rehabil. Med.* **9**:107–113.

Holle, J., Gruber, H., Frey, M., Kern, H., Stohn, H., and Thoma, H., 1984, Functional electrical stimulation in paraplegics, *Orthopedics* **7**:1145–1160.

Huang, C., McEachran, A., Kuhlemeier, K. V., and DeVivo, M. J., 1983, Prescriptive ergometry to optimize muscular endurance in acutely injured paraplegic patients, *Arch. Phys. Med. Rehabil.* **64**:518–582.

Hudlicka, O., Brown, M., Cotter, M., Smith, M., and Vrbova, G., 1977, The effect of long-term stimulation of fast muscles on their blood flow, metabolism and ability to withstand fatigue, *Pfluegers Arch.* **37**:141.

Kao, C., Bunge, R., and Reier, P., 1982, *Spinal Cord Reconstruction,* Raven Press, New York.

Knutsson, E., Lewenhaupt-Olsson, E., and Thorson, M., 1973, Physical work capacity and physical conditioning in paraplegic patients, *Paraplegia* **11**:205–216.

Kralj, A., and Jaeger, R. J., 1982, Posture switching enables prolonged standing in paraplegic patients functionally electrically stimulated, in: *Proceedings of the Fifth Annual Conference on Rehabilitation Engineering,* Houston. Pub. (IEEE 1982)

Kralj, A., and Vodovnik, L., 1977, Functional electrical stimulation of the extremities, *J. Med. Eng. Tech.* **15**:75–80.

Kralj, A., Bajd, T., and Turk, R., 1980, Electrical stimulation providing functional use of paraplegic patient muscles, *Med. Prog. Technol.* **7**:3–15.

Krayenbuehl, H. P., Rutishauser, W., Schoenbeck, M., and Amende, I., 1973, Evaluation of left ventricular function from isovolumic pressure measurements during isometric exercise, *Am. J. Cardiol.* **29**:323–330.

Lieberson, W. T., Homquest, H. J., Scott, D., and Dow, M., 1961, Functional electrotherapy: Stimulation of the peroneal nerve synchronized with the swing phase of the gait in hemiplegic patients, *Arch. Phys. Med. Rehabil.* **42**:101–105.

Lind, A. R., Taylor, S. H., Humphreys, P. W., Kennelly, B. M., and Donald, K. W., 1964, The circulatory effects of sustained voluntary muscle contraction, *Clin. Sci.* **27**:229.

Lind, A. R., McNicol, G. W., Bruce, R. A., MacDonald, H. R., and Donald, K. W., 1968, The cardiovascular responses to sustained contractions of a patient with unilateral syringomyelia, *Clin. Sci.* **35**:45–53.

Long, C. III, and Masciacrelli, V. D., 1963, An electrophysiologic splint for the hand, *Arch. Phys. Med. Rehabil.* **44**:499–503.

Marsolais, E. B., and Kobetic, R., 1982, Functional walking of paralyzed patients by means of electrical stimulation, in: *Proceedings of the Fifth Annual Conference on Rehabilitation Engineering,* Houston.

McCall, R. E., Douglas, R., and Nicholas, R., 1983, Surgical treatment in patients with myelodysplasia before using the LSU reciprocation system, *Orthopedics* **6**:843–848.

McCloskey, P. I., and Mitchell, J. H., 1972, Reflex cardiovascular and respiratory responses originating in exercising muscle, *J. Physiol. (Lond.)* **224**:173–186.

Milner, M., Quanbury, A. O., and Basmajian, J., 1970, Surface electric stimulation of lower limb, *Arch. Phys. Med. Rehabil.* **51**(9):540–545.

Mueller, E. A., 1932, Das Arbeitsmaximum bei statischer Haltearbeit, *Arbeitsphysiol.* **5**:605.

Naftchi, N. F., 1982, Functional restoration of the traumatically injured spinal cord in cuts by clonidine, *Science* **217**:1042–1044.

Nilsson, S., Staff, P., and Pruett, E., 1975, Physical work capacity and the effect of training on subjects with long standing paraplegia, *Scand. J. Rehabil. Med.* **7**:51–56.

Peckham, P. H., Mortimer, J. T., and Van der Meulen, J. P., 1973, Physiologic and metabolic changes in white muscle of cat following induced exercise, *Brain Res.* **50**:424.

Peckham, P. H., Mortimer, J. T., and Marsolais, E. B., 1976, Alteration in the force and fatigability of skeletal muscle in quadriplegic humans following exercise induced by chronic electrical stimulation, *Clin. Orthop.* **114**:326.

Perkins, C. S., Aguayo, A. J., and Bray, C. M., 1981, Behavior of Schwann cells from trembler mouse unmyelinated fibers transplanted into myelinated nerves, *Exp. Neurol.* **71**:515–526.

Petrofsky, J. S., 1978, Control of the recruitment and firing frequencies of motor units in electrically stimulated muscles in the cat, *Med. Biol. Eng. Comp.* **16**:302–308.

Petrofsky, J. S., 1979, Sequential motor unit stimulation through peripheral motor nerves in the cat, *Med. Biol. Eng. Comp.* **17**:87–93.

Petrofsky, J. S., and Lind, A. R., 1980a, The blood pressure response during isometric exercise in fast and slow twitch skeletal muscle in the cat, *Eur. J. Appl. Physiol.* **44**:223–230.

Petrofsky, J. S., and Lind, A. R., 1980b, Comparison of metabolic and ventilatory responses of men to various lifting tasks and bicycle ergometry, *J. Appl. Physiol.* **45**:60–63.

Petrofsky, J. S., and Phillips, C. A., 1979a, Constant velocity contractions in skeletal muscle by sequential stimulation of muscle efferents, *Med. Biol. Eng. Comp.* **17**:583–592.

Petrofsky, J. S., and Phillips, C. A., 1979b, Microprocessor controlled stimulation in paralyzed muscle, *IEEE NAECON Rec.* **79**:198–210.

Petrofsky, J. S., and Phillips, C. A., 1983a, Electrical stimulation of paralyzed muscle: A practical application of muscle mechanics, in: *Mechanics of Skeletal and Cardiac Muscle* (C. A. Phillips and J. S. Petrofsky, eds.), Charles C. Thomas, Springfield, IL, pp. 119–136.

Petrofsky, J. S., and Phillips, C. A., 1983b, Active physical therapy—a modern approach to rehabilitation therapy, *J. Neurol. Orthop. Surg.* **4**:165–173.

Petrofsky, J. S., and Phillips, C. A., 1983c, Computer controlled walking in the neurologically paralyzed individual, *J. Neurol. Orthop. Surg.* **4**:153–164.

Petrofsky, J. S., and Phillips, C. A., 1984, The use of functional electrical stimulation for rehabilitation of spinal cord injured patients, *CNS Trauma J.* **1**:57–72.

Petrofsky, J. S., and Phillips, C. A., 1985, Closed loop control of movement in skeletal muscle, *CRC Crit. Rev. Biomed. Eng.* **13**:35–96.

Petrofsky, J. S., Rinehart, J. S., and Lind, A. R., 1976, Isometric strength and endurance in slow and fast muscles in the cat, *Fed. Proc.* **35**:291.

Petrofsky, J. S., Phillips, C. A., and Lind, A. R., 1981, Interrelationships between recruitment order, muscle temperature and endurance for static exercise and the blood pressure response in the cat, *Circ. Res.* **48**:132–136.

Petrofsky, J. S., Heaton, H. H., and Phillips, C. A., 1983a, Outdoor bicycle for exercise in paraplegics and quadriplegics, *J. Bioeng.* **5**:287–296.

Petrofsky, J. S., Heaton, H. H., Phillips, C. A., and Glaser, R. M., 1983b, Application of the Apple as a microprocessor controlled stimulator, *Colleg. Microcomput.* **1**:97–104.

Petrofsky, J. S., Heaton, H. H., and Phillips, C. A., 1984a, Leg exercise for training of paralyzed muscle by closed loop control, *Med. Biol. Eng.* **22**:298–303.

Petrofsky, J. S., Phillips, C. A., and Heaton, H. H., 1984b, Feedback control system for walking in man, *Comp. Biol. Med.* **14**:135–149.

Petrofsky, J. S., Phillips, C. A., Heaton, H. H., and Glaser, R. M., 1984c, Bicycle ergometer for paralyzed muscle, *J. Clin. Eng.* **9**:13–19.

Petrofsky, J. S., Hendershot, D., and Phillips, C. A., 1985a, Cardiovascular stresses during computer controlled walking in the paralyzed, *Fed. Proc.* **44**:447.

Petrofsky, J. S., Phillips, C. A., Douglas, R., and Larson, P., 1985b, Computer synthesized walking: An application of orthosis and FES, *J. Neurol. Orthop. Surg.* **6**:219–230.

Petrofsky, J. S., Phillips, C. A., and Hendershot, D., 1985c, The cardiorespiratory stresses which occur during dynamic exercise in paraplegic and quadriplegics, *J. Neurol. Orthop. Surg.* **6**:252–258.

Petrofsky, J. S., Phillips, C. A., Hendershot, D. M., Douglas, R., and Larson, P., 1985d, Weight distribution and muscle utilization during computer controlled walking in the paralyzed, *Physiologist* **28**:276.

Petrofsky, J. S., Phillips, C. A., and Petrofsky, S. H., 1985e, Electronic physicians prescription system for functional electrical stimulation (FES) patients, *J. Neurol. Orthop. Surg.* **6**:239–246.

Pette, D., Smith, M. E., Staudte, H. W., and Vrbova, G., 1973, Effects of long-term electrical stimulation on some contractile and metabolic characteristics of fast rabbit muscle, *Pfluegers Arch.* **338**:257–272.

Pette, D., Ramirez, B., and Mueller, W., 1975, Influence of the intermittent long-term stimulation on contractile, histochemical and metabolic properties of fibre population in fast and slow rabbit muscles, *Pfluegers Arch.* **361**:1–7.

Phillips, C. A., and Petrofsky, J. S., 1985a, Cognitive feedback as a sensory adjunct to functional electrical stimulation (FES) neural prostheses, *J. Neurol. Orthop. Med. Surg.* **6**:231–238.

Phillips, C. A., and Petrofsky, J. S., 1985b, Preliminary report on balance/stance in paraplegic with and without a cognitive feedback system, *Physiologist* **28**:302.

Phillips, C. A., Petrofsky, J. S., Hendershot, D. M., and Stafford, D., 1984, Functional electrical exercise: A comprehensive approach for physical conditioning of the spinal cord paralyzed individual, *Orthopedics* **8**:1112–1123.

Rack, P. M. H., and Westbury, D. R., 1969, The effects of length and stimulus rate on tension in the isometric cat soleus muscle, *J. Physiol. (Lond.)* **204**:443–460.

Reswick, J. B., and Vodovnik, L., 1967, External power in prosthetics and orthotics and overview, *Artif. Limbs* **11**(2):5–21.

Salmons, S., and Vrbova, C., 1969, The influence of activity on some contractile characteristics of mammalian fast and slow muscles, *J. Physiol. (Lond.)* **201**:535.

Scott, R. N., 1968, Myoelectric control systems. *Advances in Biomedical Engineering and Medical Physics*, Volume 2, (S. N. Levine, ed.), John Wiley & Sons, New York, pp. 101–110.

Shephard, R. J., 1965, The development of cardiorespiratory fitness, *Med. Serv. J. Can.* **21**:533–544.

Shephard, R. J., 1977, *Endurance Fitness*, 2nd ed., University of Toronto Press, Toronto.

Simonsen, E., 1971, *Physiology of Work Capacity and Fatigue*, Charles C. Thomas, Springfield, IL.

Solomonow, M., Foster, J., Eldred, E., and Lyman, J., 1978, Proportional control of paralyzed muscle, *Fed. Proc.* **37**:215.

Trnkoczy, A., Bajd, R., and Malez, C. M., 1976, A dynamic model of the ankle joint under functional electrical stimulation in free movement and isometric conditions, *J. Biomechan.* **9**:509–519.

Van der Meulen, J. P., Peckham, P. H., and Mortimer, J. T., 1974, Trophic functions of the neuron. Mechanisms of neurotrophic interactions. Use and disuse of muscle, *Ann. N.Y. Acad. Sci.* **228**:117–189.

Vodovnik, L., Crochetiere, W. J., and Reswick, J. B., 1967, Control of a skeletal joint by electrical stimulation of antagonists, *Med. Biol. Eng.* **5**:97–109.

Vrbova, G., 1963, The effect of motorneuron activity on speed of contraction of striated muscle, *J. Physiol. (Lond.)* **169**:513–526.

Zealer, D. L., and Dedo, H. H., 1977, Control of paralyzed axial muscles by electrical stimulation, *Trans. Am. Acad. Ophthalmol. Otolaryngol.* **84**:310.

19

Recovery of Language Disorders
Homologous Contralateral or Connected Ipsilateral Compensation?

ANDREW KERTESZ

1. INTRODUCTION

The study of recovery from language deficit is more than just a practical, prognostic exercise for the clinician or a base line for therapy. It also provides an important theoretical framework for cerebral reorganization. Human brain damage caused by stroke or trauma produces deficits that recover in two stages. The first stage is related to recovery from the acute effect of membrane failure, ionic imbalance, hemorrhage, cellular reaction, and possibly to the reestablishment of the circulation in the ischemic penumbra (Astrup *et al.*, 1981). Our interest has focused on second-stage recovery that takes place months, even years, after injury, and its mechanisms remain largely a mystery. A significant amount of physiological and functional recovery is probably related to intact structures compensating for the functional loss. Axonal regrowth and collateral sprouting are important mechanisms in the peripheral and, in certain instances, in the central nervous system. However, in large lesions in man causing focal cognitive or language loss, compensation is likely affected by (1) ipsilateral physiologically and anatomically connected structures, (2) contralateral homologous cortical areas, or (3) subcortical systems hierarchically and physiologically related to the damaged structures or function in question.

ANDREW KERTESZ • Department of Clinical Neurosciences, Research Institute, St. Joseph's Hospital, University of Western Ontario, London, Ontario N6A 4V2, Canada.

2. RIGHT HEMISPHERE COMPENSATION

The fact that aphasic patients recover considerably has been known for a long time, even though aphasic syndromes have often been presented as if they were static. Wernicke (1886) postulated that the contralateral homologous cerebral cortex is responsible for the return of language function. Jackson (1873) considered the right hemisphere capable of automatic utterances and responsible for the residual output in global aphasia. The idea of contralateral hemispheric compensation has been reemphasized by Henschen (1920–22) in his monumental monograph of aphasic cases with localization. Subsequently this was called the "Henschen principle" by Nielsen (1946), who also believed that it is the right hemisphere that compensates for language deficits. Nielsen also thought that the right hemisphere was able to substitute mainly for comprehension but not for expressive function.

Initially, the principle of contralateral homologous cortical substitution was based on large left hemisphere lesions with good recovery where there was very little left hemisphere remaining to take over. More recently, CAT scan studies made the same point (Cummings et al., 1979; Landis et al., 1980). In addition, in some patients who became aphasic with a single left hemisphere stroke but recovered, a second right hemisphere stroke produced a language deficit again (Nielsen, 1946; Levine and Mohr, 1979; Cambier et al., 1983). These cases were rarely documented and may have represented bilateral language organization to begin with, rather than a commonly operating mechanism of functional transfer to the contralateral hemisphere.

The idea of compensation through right hemisphere function, even after partial left hemisphere damage, was supported by studies of sodium amobarbitol given to aphasics who had recovered (Kinsbourne, 1971; Czopf, 1972). The studies indicated that even though the aphasic disturbance occurred from a left hemisphere lesion, it was the right hemispheric injection that increased the language disturbance, indicating that the right hemisphere compensated for the previous deficit produced by the left side.

The residual language capacities of the right hemisphere are revealed by hemispherectomy studies, which indicated that a picture closely resembling global aphasia occurred after a left hemispherectomy. An exceptional case showed more recovery of repetition and automatic speech, but this patient had a serial removal of a glioma rather than a single operation (Burklund and Smith, 1977). In the reported cases of adult hemispherectomies, some comprehension recovered, but expressive language abilities were basically eliminated except for expletives, automatic phrases, a few purposeful words, repetition, and singing; altogether, it must be emphasized that fluent, propositional speech has not returned in any of these subjects.

There is also controversy about the extent of hemispherectomy. Many of the detailed psychological reports give little anatomic information, and at times some of the subcortical structures remain intact (hemidecortication). In view of the recent evidence for the role of subcortical structures in language function, this may be an important anatomic variable in discussing the effect of hemispherectomies.

The transfer of language to the right hemisphere occurs more easily in children, and this is thought to be related to the greater "plasticity" of the younger brain. Surgical removal of the left cerebral cortex for the treatment of infantile hemiplegia and epilepsy results in little, if any, language deficit (Krynauw, 1950; Gardner et al., 1955; Basser, 1962). If brain injury occurs early in life, it was thought that functional plasticity permits the right hemisphere to take over language functions completely. More detailed testing of linguistic abilities for hemidecortication indicated that although left and right hemispherectomy subjects were indistinguishable on tests of verbal intelligence, when complex syntactic tasks were used, the left-damaged patients performed less well even though their hemidecortication was in early infancy (Dennis and Kohn, 1975).

The effects of early injury in childhood are far from uniform. The severity of impairment changes with the function studied. Lansdell (1969) found that an inverse relationship existed between the age of onset of the first symptoms and verbal intelligence scores in cases of epileptogenic left hemisphere lesions. Another source of methodological controversy is that children and adults are not always comparable in recovery because the former usually become aphasic from trauma and infection and the latter from stroke. Recently, some children with stroke were shown to have more aphasic deficit, and the difference between children and adults in the degree of compensation has undergone some reevaluation (Woods and Teuber, 1978).

Somewhat against equipotentiality or the right hemisphere having language capacity in the beginning is the fact that aphasia very rarely occurs as the consequence of right hemisphere damage even in children (Woods and Teuber, 1978). One would expect the incidence of aphasia to be equal in both hemispheric injuries if the hemispheres were equipotential to begin with. Some of the studies of cerebral asymmetry indicate that the structural asymmetry for language may exist at birth, and therefore the hemispheres are not truly equipotential; maturation finalizes a process that begins before birth (Tezner et al., 1972; Wada et al., 1975; Witelson and Pallie, 1973).

The critical age of compensation for language loss remains the subject of controversy. This is variable and has not been accurately determined (Lenneberg, 1967). Some authors considered the critical age to be about 10 years (Perlstein and Sugar, 1954), but this may be related to the fact that hemispherectomies are usually done in infancy and only a few hemispherectomies were performed in teenagers for malignant glioma. In a case of dominant hemispherectomy in a 14-year-old boy (Hillier, 1954), comprehension was said to be retained, but the expressive speech level was not accurately documented. Reading was severely impaired. In a report of a 12-year-old girl (Gott, 1973), aphasia occurred after a tumor removal at the age of 8, and reappearance of the tumor at the age of 10 led to a left hemispherectomy. Following that she could sing and speak single words. Comprehension was one of the least impaired functions (Peabody Picture Vocabulary Test Score 70). She was able to follow one-step commands but required repetition for multistage commands. Metalinguistic tasks indicated that she understood prepositions and rhyming. Naming was about 70%, and she was able to repeat words and sentences, but initiation of speech was impaired. The patient's reading, writing, and signaling were worse than her verbal ability. Although this child performed better than all the adult hemispherectomies published, recovery was far from complete, and the linguistic deficit, resembling Broca's aphasia, persisted 2 years after dominant hemispherectomy. No observations were made concerning her sexual maturation.

Estimates from dichotic listening tests (Krashen and Harshman, 1972) suggested that the lateralization of language (the critical age of compensation) is completed by the age of 5 years. There is other evidence that indicates that the evolution of lateralization continues into later childhood (Bryden and Allard, 1978). It is likely that puberty or sexual maturation, with the biological effects of reproductive hormones on the brain, has a great deal to do with the loss of recovery. This explains why second language acquisition is imperfect after puberty and why most adults learning a second language cannot entirely master all the phonological and syntactic complexities. Although they may achieve fluency, they continue to have an accent and to make grammatical and idiomatic errors. This also confirms that phonology and syntax are basic left hemisphere operations and cannot easily be substituted or acquired after puberty, when the adult brain loses its plasticity, especially for these language functions in man.

Recent studies of callosal sections for epilepsy indicated that although the disconnected right hemisphere cannot produce propositional speech, language comprehension is surprisingly preserved in some examples (Sperry and Gazzaniga, 1967; Gazzaniga and Sperry, 1967). The right hemisphere appeared to understand fairly complex verbal instructions in these patients. It appeared to be able to retrieve named objects with the left hand or respond appropriately to left visual half-field presentation of words.

Zaidel (1978), after testing hemispherectomy and commissurotomy patients, concluded that the right hemisphere has considerable semantic capacity and can even store concepts to match pictures related semantically but not visually. However, the right hemisphere appeared deficient in phonemic discrimination and syntax. Zaidel compared right hemisphere scores on the picture vocabulary test with five diagnostic aphasia groups, concluding that the right hemisphere's performance came closest to that of Wernicke's aphasics. Performance on other subtests of auditory comprehension showed that they came closest to anomic or Broca's aphasics. The disconnected right hemispheres also show the same word frequency effect as aphasics, as well as the same relative order of deficits for specific semantic word categories such as object names, geometric forms, actions, colors, numbers, and letters. The right hemisphere seemed unable to resolve long sequential commands, possibly because of deficit short-term verbal memory, but was quite capable of decoding the pictorial references of single spoken words. These results suggest dissociable mental operations for various components of comprehension.

There is controversy whether or not the right hemisphere's deficiency in fully supporting language is related to its lack of syntactic ability or to a deficient short-term verbal memory. Phonetic decoding and representation require short-term capacity, and the right hemisphere appears to be deficient in this, as evidenced from dichotic listening tasks (Gordon, 1980; Milner *et al.*, 1968; Sparks and Geschwind, 1968). It is possible that the right hemisphere, rather than representing words phonetically, recognizes them by matching the sound from an acoustic template to a semantic association. Studies on right hemisphere reading also suggest the lack of internal phoneme–grapheme conversion mechanisms. Studies by Levy and Trevarthen (1976) showed that the right hemisphere could not rhyme or recognize phonemes. Phonemic decoding may be necessary for the production of language, which may be the major stumbling block in the output capacity of the right hemisphere. The issue remains controversial because some studies indeed show that rhyming capacity persists (Gott, 1973), and the right hemisphere can communicate by spelling words with anagram letters (Gazzaniga and LeDoux, 1978).

Recent studies of cerebral blood flow with xenon-133 also revealed a right hemisphere participation in recovery to a various degree (Knopman *et al.*, 1984). Positron emission tomography (PET) studies of cerebral metabolism showed a great deal of hypometabolism surrounding and even remote from cerebral infarcts, thus suggesting that homologous areas in the contralateral hemisphere play a role in comprehension (Metter *et al.*, 1981).

3. IPSILATERAL STRUCTURAL COMPENSATION

Von Monakow (1914), who established the principle of diaschisis to explain second-stage recovery, recognized the temporary nature of aphasic disturbance and, in fact, used aphasia as the most obvious model for studying recovery. His theory of diaschisis stipulated that acute damage to the nervous system, such as in stroke, deprives the surrounding, functionally connected tissue from innervation, and therefore they become inactivated, similar to the phenomenon of "spinal shock." As innervation is regained from somewhere else, so does function return to the otherwise undamaged structure. The phenomenon of diaschisis has been widely accepted and subsequently elaborated.

Even before von Monakow's diaschisis theory, it was noted that sudden lesions produced more deficit than slowly growing ones. Dax (1865) suggested that left hemisphere lesions may not result in aphasia if the lesion develops gradually. Modern animal experimentation confirmed this "serial-lesion" effect (Ades and Raab, 1946).

The reorganization theory postulated that the deficit eventually recovers not because of reinnervation of the connected area but rather because of reorganization or physiological rerouting, which takes place through functionally connected brain. This, of course, would allow both ipsilateral and contralateral structures to take part in the process.

The remote effect of stroke, seen on PET scan as hypometabolic areas, can also be construed as evidence for diaschisis. Decreased activity in the thalamus ipsilateral to the injury in both acute and chronic stages, in ipsilateral cortex adjacent and at a distance from the lesion, and in the visual cortex distal to retinal, optic-nerve, or optic radiation lesions have been described (Kuhl *et al.*, 1980; Metter *et al.*, 1981; Martin and Raichle, 1983). However, the lack of recovery of the hypometabolism in cases in which the aphasic deficit has recovered considerably throws some doubt on the significance of these distant hypometabolic areas in the interpretation of recovery (Black *et al.*, 1984; Demeurisse *et al.*, 1983; Nagata *et al.*, 1986).

Subsequent experiments indicated that subcortical structures also play an important role in compensation, thus revealing hierarchical representation (Jackson, 1873) through various evolutionary levels of the brain. When a higher level was impaired, a lower level representing the same function would take over. "Reverse ablations," however, demonstrated that certain areas considered instrumental in recovery could be removed with impunity; in other words, they are not used for normal function (Bucy, 1934).

4. FACTORS IN RECOVERY FROM APHASIA

In our studies of recovery from aphasia, we have measured the aphasic deficit with a standardized test, the Western Aphasia Battery (WAB), which is specifically designed for language function in an aphasic population, including severely and mildly affected patients (Kertesz, 1979, 1982). It is essential to use accurate deficit analysis with a standardized test battery as various factors in recovery are studied.

4.1. Initial Severity

Initial severity is one of the most important factors in recovery. Early investigators considered initial severity to have a highly predictive value (Godfrey and Douglass, 1959; Schuell *et al.*, 1964; Sands *et al.*, 1969; Sarno *et al.*, 1970a,b; Gloning *et al.*, 1976; Kertesz and McCabe, 1977). The severity of deficit at onset has a considerable effect on comparing recovery rates because mildly affected aphasics do not have much room for recovery (a "ceiling effect''), and severe aphasics often have more potential. Treated patients tend to be selected from the less severe groups and bias the results. Unless initial severity is considered a major factor to be controlled, studies of treatment should not be considered reliable. There are various methods of controlling for initial severity, such as analysis of covariance or using outcome measures instead of recovery rates or expressing the change as a percentage of initial severity.

4.2. Time from Onset

The time from onset when patients are studied is also an important factor. When patients are entered into recovery studies at various stages in their progress, comparison becomes very difficult. In our studies, we took care to start our evaluation within the acute period, usually between 10 and 45 days after a stroke. Since most of our patients were examined at exactly 14 days post-onset, this provides a rather homogeneous population. Only the more severely affected patients, who could not be examined at that time because of intercurrent medical illness or obtundation, were kept until the upper limit of the time period.

4.3. Etiology

Etiology is the third major factor that needs to be considered. Traumatic aphasia, for instance, recovers quickly if it is related to closed head injury. Persisting dysarthria, however, is common in severe trauma, and this often disrupts communication to such a degree that the extent of posttraumatic aphasia is difficult to determine. Penetrating head injury affects a different age group and behaves differently because of the variation in the speed and path of the missiles and the associated concussion. Therefore, posttraumatic aphasia is biologically different from the vascular type. There are many similarities, nonetheless, indicating that the recurring patterns of aphasic types are not necessarily related to the distribution of vascular lesions. A recent study by Ludlow et al. (1986) on Vietnam veterans showed that the lesions that produce a persisting asyntactic or Broca's aphasia are large, involving the subcortical structures and the parietal area in addition to Broca's area. Ludlow et al. (1986) reached very much the same conclusions that are reached studying stroke recovery.

4.4. Lesion Size

Lesion size and location have also been recognized as interrelated and complex factors. Until recently, clinicians relied on autopsy correlations, but modern neuroimaging has provided an opportunity to study lesion characteristics in vivo. We found, in our first study of lesion size measured on computerized tomography (CT) and recovery from aphasia, that the larger the lesion the poorer was the outcome; in other words, outcome correlated negatively with lesion size

(Kertesz *et al.*, 1979). This has been known to clinicians since Jackson's time and suggests, in principle, the so-called "mass effect" (Lashley, 1938). Recovery rates also showed a trend of negative correlation with one unexpected exception: the recovery rate of comprehension was found to be correlated positively with lesion size. This can best be understood if we look at another study of ours in which the best-recovered modality was found to be comprehension (Lomas and Kertesz, 1978). Patients with large lesions having global or severe Broca's aphasia often show greater improvement in comprehension. Patients with smaller lesions, such as anomics, already have good comprehension; therefore, they have less room for recovery (a ceiling effect). The large lesions with more recovery and small lesions with little change give rise to a paradoxically positive correlation unless the initial severity is covaried, as was done in our subsequent studies.

Since then, various other publications have emerged dealing with localization of the lesions in a somewhat different, symptom-oriented approach (Selnes *et al.*, 1983; Knopman *et al.*, 1983). Knopman *et al.* (1983) looked at lesion size and location and found that in the language area about 60 cm^2 was a critical mass, and larger lesions resulted in relatively less recovery. They concluded that speech has less redundancy, and if the surrounding areas are involved, less recovery will take place. The symptom-oriented approach generally leads to less focal localization of deficits, as some of these functions are widely distributed in the brain.

We recently studied lesion size and recovery in the 0- to 3-month and 0- to 12-month intervals in 82 patients in whom unilateral lesions were available on the CT scans for tracing and localization. Negative correlations were obtained throughout when initial severity was controlled for. These correlations were particularly high for fluency, and they were low in the comprehension tasks. Outcome and lesion size, as measured by the last evaluation at 12 months, correlated negatively in a highly significant fashion throughout all subtests.

The recovery of Broca's and Wernicke's groups was examined in further detail, as they represent important clinical syndromes with double dissociation of functional deficits of speech output, comprehension, and variable behavior on recovery. The recovery of Broca's aphasics was correlated with lesion size, age, the degree of atrophy, and initial severity. The most significant correlations were obtained in the outcome measures that showed a negative correlation for lesion size. Cerebral atrophy, as measured by the ratio of frontal horn width and brain diameter on the horizontal section at the level of the pineal, did not correlate significantly.

Lesion location was evaluated in Broca's aphasics and divided at the median for poor and good recovery. The structures with significant involvement (more than 50%) were the inferior frontal gyrus, especially the pars opercularis and triangularis, and the insula in both groups. Supramarginal, angular, and superior temporal gyri were not involved in those cases in which recovery was good. The subcortical regions showed significant differences in the involvement of the putamen and the caudate, which was twice as frequent in the persistent cases.

Wernicke's aphasics showed variable recovery. Six patients remained severely affected, and ten recovered to either anomic or conduction aphasia. The AQ changes were correlated with lesion size, age, initial severity, individual language parameters, and nonverbal control tests. The results showed a significant negative correlation between lesion size and recovery and less than significant correlation for age and outcome.

Lesion location was checklisted in 16 patients with Wernicke's aphasia. The most consistently involved structure was the superior temporal gyrus. In cases of poor recovery, the middle temporal gyrus and the supramarginal gyrus were significantly more frequently involved in addition to the postcentral gyrus and the insula, which was twice as frequently involved in the unrecovered cases. Subcortical structures did not seem to be as significant as for persisting Broca's aphasia.

5. VARIATIONS IN LANGUAGE LATERALITY AND ANATOMIC ASYMMETRY

The variable degree of recovery, which cannot entirely be explained by the extent and location of lesions, has been postulated to relate to individual differences in language laterality. The suggestion by Subirana (1969), Gloning *et al.* (1969), and Geschwind (1974) that left handers and right handers with a family history of left handedness recover better from aphasia because they have a more bilateral language distribution is based on anecdotal evidence. There is also a recently popularized but yet to be proven theory of more bilaterally distributed language in women (McGlone, 1980). Recent studies of anatomic asymmetry on CT scans, inspired by the demonstration of commonly larger planum temporale on the left by Geschwind and Levitsky (1968), have correlated better outcome with atypical asymmetry (Pieniadz *et al.*, 1983). We have studied the factor of anatomic asymmetry on CT, as measured by occipital width, frontal width, and protuberance (petalia), and could not confirm that it played a role in recovery in any of the aphasic groups. Neither did we find any sex differences in recovery (Kertesz, 1988).

6. CONCLUSIONS

The complex interaction of size and location of lesions, in addition to time from onset, etiology, and initial severity, is the main factor in the recovery of cognitive and language loss. Other biological factors, such as age, sex, and handedness, play a less significant role when a stroke population is followed.

Certain aspects of hemispheric specialization may vary according to individuals, even though anatomic asymmetries do not seem to play a role in recovery according to our preliminary findings. It could be that anatomic asymmetries relate more to handedness variables than language distribution, as suggested by some of our studies in normals (Kertesz et al., 1986); therefore, we are not seeing an effect on language recovery. The individual variations in the intra- and interhemispheric distribution of various functional components may contribute to an important extent to the ability of the mature brain to compensate after a single nonprogressive lesion. Other pathological variations, such as repeated stroke insults, cerebral atrophy, and intercurrent latent dementia, are factors to be considered or even examined directly, although they were controlled by exclusion in our studies.

Lesion size is undoubtedly a significant factor in the extent of recovery. However, the important exceptions are certain crucial areas in the left hemisphere that are more important for prognosis than others. Motor and premotor phonemic assembly mechanisms are elaborated by a cortical/subcortical network that can be damaged partially and followed by good recovery. However, if both cortical and subcortical components of the network are impaired, recovery is much less likely. Therefore, the role of hierarchical structural organization is important in the recovery of motor speech and Broca's aphasia. Contralateral, homologous cortical substitution is probably not a major mechanism for the recovery of speech output.

A complex network of various structures is also needed for the processing of language comprehension, although interhemispheric connections may be playing a larger role in comprehension than in motor output. It seems that when there is a restricted deficit in the dominant hemisphere auditory association area, the posterior superior temporal gyrus and the planum temporale can be compensated for by (1) the opposite or homologous hemispheric structures, (2) surrounding structures in the temporal and inferior parietal regions, or (3) the insula with a few exceptions of word deafness after a unilateral lesion. However, when most of these compensating structures are affected and the lesion is large, precluding right hemispheres access, recovery may not take place.

Since the time of von Monakow, the concepts of diaschisis have been further elaborated in physiological, anatomic, and pharmacological terms. Of course, axonal regrowth or collateral sprouting could be one of the mechanisms that contribute to recovery. It is not likely, however, that these regenerative processes can achieve the required specificity of connections for language processing. The only pathways in the mammalian CNS that have been shown to preserve some ability to regenerate are monoaminergic and cholinergic projections, which are diffuse and have a global, modulatory controlling function.

Functional reorganization, or compensation, is a behavioral term indicating new strategies developed by the brain-damaged individual. In addition to rerouting connections, the psychological processes of motivation, learning, and environmental enrichment are also considered. This model has, of course, very important implications for treatment. The classical rehabilitation studies of Franz and Oden (1917) indicated the role of motivation in the recovery of locomotion. Brain-lesioned animals and humans can substitute maneuvers and develop new solutions and strategies (Sperry and Gazzaniga, 1967).

The studies of right hemisphere language reveal that the substantial verbal ability, especially comprehension, of the right hemisphere is likely to play an important role in compensation for some left-hemisphere damage. It has been suggested that right-hemisphere difficulty with instructions and multistage commands is related to a short-term verbal memory constraint; subsequently, the right hemisphere cannot compensate when the left hemisphere is disconnected or damaged (Gott, 1973; Smith, 1966).

Several investigators (French et al., 1955; Kinsbourne, 1974; Moscovitch, 1981) have argued that the remaining left-hemisphere control or inhibition with partial lesions masked right-hemisphere language comprehension that is unmasked when the left hemisphere is absent. It was argued that the limited left lesion will not result in right-hemisphere compensation because of this inhibitory effect. If inhibition is exerted by the left hemisphere to prevent the compensation of deficit by the right hemisphere, then it is mysterious how and through what pathways this inhibition is exerted in cases of completed commissurotomies. Brainstem or diencephalic connections have been implicated. In favor of the inhibitory theories is the simple observation that the right hemisphere is indeed capable of producing well-articulated utterances in hemispherectomies and in certain global aphasias with very large left lesions. These utterances are complex phonologically and even syntactically as long as they are automatic, such as: "I don't know," or emotional expletives such as swear words or automatisms. It is likely that the speech capacity of the right hemisphere has developed parallel with that of the left hemisphere but has been suppressed after puberty by hormonal or physiological mechanisms. In this context the phrase "left hemisphere dominance" is justified.

Our evidence from aphasia lesions is against continuing left hemisphere inhibition. Small and limited lesions are more compatible with recovery than larger ones. Indeed, all the converging evidence from dominant hemispherectomies, sodium amobarbital injections in the left internal carotid artery of adults (Milner et al., 1964), and the extensive language testing of the disconnected hemispheres shows the same pattern of relatively good comprehension of single words and poor speech output such as is seen in severe Broca's aphasia. Therefore, evidence from aphasia also supports the model of separation between language encoding and decoding mechanisms of the brain. It is compatible with the fact that children acquire language comprehension before speech, and it is somewhat against the theories of the interdependence of speech perception and speech output (Lieberman, 1974; Mateer, 1983). The left hemisphere seems to specialize in phonological assembly, syntax, and sequencing for propositional speech output that the right hemisphere cannot compensate for in the adult.

REFERENCES

Ades, H. W., and Raab, D. H., 1946, Recovery of motor function after two-stage extirpation of area 4 in monkeys, *J. Neurophysiol.* **9**:55–60.

Astrup, J., Siesjo, B. K., and Symon, L., 1981, Thresholds in cerebral ischemia—the ischemic penumbra, *Stroke* **12**:723–725.

Basser, L. S., 1962, Hemiplegia of early onset and the faculty of speech with special reference to the effects of hemispherectomy, *Brain* **85**:427–460.

Black, S. E., Garnett, E. S., Nicholson, R. L., Carr. T., Nahmias, C., and Kertesz, A., 1984, NMR and PET studies in a crossed dextral aphasic, *Ann. Neurol.* **16**(1):155.

Bryden, M., and Allard, F., 1978, Dichotic listening and the development of linguistic processes, in: *Asymmetrical Function of the Brain* (M. Kinsbourne, ed.), Cambridge University Press, Cambridge, pp. 392–404.

Bucy, P. C. 1934, The relation of the premotor cortex to motor activity, *J. Nerv. Ment. Dis.* **79**:621–630.

Burklund, C. W., and Smith, A., 1977, Language and the cerebral hemispheres, *Neurology (Minneap.)* **27**:627–633.

Cambier, J., Elghozi, D., Signoret, J. L., and Henin, D., 1983, Contribution of the right hemisphere to language in aphasic patients. Disappearance of this language after a right-sided lesion, *Rev. Neurol.* **139**:55–63.

Cummings, J. L., Benson, D. F., Walsh, M. J., and Levine, J. L., 1979, Left-to-right transfer of language dominance: A case study, *Neurology (Minneap.)* **29**:1547–1550.

Czopf, J., 1972, Role of the non-dominant hemisphere in the restitution of speech in aphasia, *Arch. Psychiatr. Nervenkr.* **216**:162–171.

Dax, M., 1865, Lesions de la moitie gauche de l'encephale coincidant avec l'oubli des signes de la pensee (lu a Montpellier en 1836), *Gaz. Hebd. Med. Chir.* **2**:259–262.

Demeurisse, G., Verhas, M., Capon, A., and Paternot, J., 1983, Lack of evolution of the cerebral blood flow during clinical recovery of a stroke, *Stroke* **14**:77–81.

Dennis, M., and Kohn, B., 1975, Comprehension of syntax in infantile hemiplegics after cerebral hemidecortication: Left hemisphere superiority, *Brain Lang.* **2**:472–482.

Franz, S. I., and Oden, R., 1917, On cerebral motor control: The recovery from experimentally produced hemiplegia, *Psychobiology* **1**:3–18.

French, L. A., Johnson, D. R., Brown, I. A., and van Berger, F. B., 1955, Cerebral hemispherectomy for control of intractable convulsive seizures, *J. Neurosurg.* **12**:154–164.

Gardner, W. J., Karnosh, L. J., McClure, C. C., Jr., and Gardner, A. K., 1955, Residual function following hemispherectomy for tumour and for infantile hemiplegia, *Brain* **78**:487–502.

Gazzaniga, M. S., and LeDoux, J., 1978, *The Integrated Mind*, Plenum Press, New York.

Gazzaniga, M. S., and Sperry, R. W., 1967, Language after section of cerebral commissures, *Brain* **90**:131–148.

Geschwind, N., 1974, Late changes in the nervous system: An overview in plasticity and recovery of function in the central nervous system, in: *Plasticity and Recovery of Function in the Central Nervous System* (D. Stein, J. Rosen, and N. Butters, eds.), Academic Press, New York, pp. 467–508.

Geschwind, N., and Levitsky, W., 1968, Human brain, left–right asymmetries in temporal speech regions, *Science* **161**:186–187.

Gloning, I., Gloning, K., Haub, G., and Quartember, R., 1969, Comparison of verbal behavior in right-handed and nonright-handed patients with anatomically verified lesion of one hemisphere, *Cortex* **5**:43–52.

Gloning, K., Trappl, R., Heiss, W. D., and Quartember, R., 1976, *Prognosis and Speech Therapy in Aphasia in Neurolinguistics*, Volume 4, *Recovery in Aphasics*, Swets & Zeitlinger, Amsterdam.

Godfrey, C. M., and Douglass, E., 1959, The recovery process in aphasia, *Can. Med. Assoc. J.* **80**:618–824.

Gordon, H. W., 1980, Right hemisphere comprehension of verbs in patients with complete forebrain commissurotomy. Use of the dichotic method and manual performance, *Brain Lang.* **11**: 76–86.

Gott, P. S., 1973, Language after dominant hemispherectomy, *J. Neurol. Neurosurg. Psychiatry* **36**:1082–1088.

Henschen, S. E., 1920–22, *Klinische und anatomische Beitrage zur Pathologie des Gehirns*, Volumes 5–7, Nordiska Bokhandel, Stockholm.

Hillier, W. F., Jr., 1954, Total left cerebral hemispherectomy for malignant glioma, *Neurology (Minneap.)* **4**:718–721.

Jackson, J. H., 1873, On the anatomical and physiological localization of movements in the brain, *Lancet* **1**:232–234.

Kertesz, A., 1979, *Aphasia and Associated Disorders: Taxonomy, Localization and Recovery*, Grune & Stratton, New York.

Kertesz, A., 1982, *The Western Aphasia Battery*, Grune & Stratton, New York.

Kertesz, A., 1988, What do we learn from recovery from aphasia?, in: *Advances in Neurology*, Vol. 47: *Functional Recovery in Neurological Disease* (S. G. Waxman, ed.) Raven Press, New York, pp. 277–292.

Kertesz, A., and McCabe, P., 1977, Recovery patterns and prognosis in aphasia, *Brain* **100**:1–18.

Kertesz, A., Harlock, W., and Coates, R., 1979, Computer tomographic localization, lesion size and prognosis in aphasia, *Brain Lang.* **8**:34–50.

Kertesz, A., Black, S. E., Polk, M., and Howell, J., 1986, Cerebral asymmetries on magnetic resonance imaging, *Cortex* **22**:117–127.

Kinsbourne, M., 1971, The minor cerebral hemisphere as a source of aphasic speech, *Arch. Neurol.* **25**:302–206.

Kinsbourne, M., 1974, Mechanisms of hemispheric interaction in man, in: *Hemispheric Disconnection and Cerebral Function* (M. Kinsbourne and W. L. Smith, eds.), Charles C. Thomas, Springfield, IL., pp. 260–286.

Knopman, D. S., Selnes, O. A., Niccum, N., and Rubens, A. B., 1983, A longitudinal study of speech fluency in aphasia: CT scan correlates of recovery and persistent nonfluency, *Neurology (Minneap.)* **33**:1170–1178.

Knopman, D. S., Rubens, A. B., Selnes, O. A., Klassen, A. C., and Meyer, M. W., 1984, Mechanisms of recovery from aphasia: Evidence from serial xenon 133 cerebral blood flow studies, *Ann. Neurol.* **15**(6):530–535.

Krashen, S., and Harshman, R., 1972, Lateralization and the critical period, *UCLA Work. Papers Phonet.* **22**:6.

Krynauw, R. A., 1950, Infantile hemiplegia treated by removing one cerebral hemisphere, *J. Neurol. Neurosurg. Psychiatry* **13**:243–267.

Kuhl, D. E., Phelps, M. E., Kowell, A. P., Metter, E. J., Selin, C., and Winter, J., 1980, Effect of stroke on local cerebral metabolism and perfusion: Mapping by emission computed tomography of ^{18}FDG and $^{13}NH_3$, *Ann. Neurol.* **8**:47–60.

Landis, T., Cummings, J. L., and Benson, D. F., 1980, Passage of language dominance to the right hemisphere: Interpretation of delayed recovery after global aphasia, *Rev. Med. Suisse Rom.* **100**:171–177.

Lansdell, H., 1969, Verbal and nonverbal factors in right-hemisphere speech: Relation to early neurological history, *J. Comp. Physiol. Psychol.* **69**:734–738.

Lashley, K. S., 1938, Factors limiting recovery after central nervous lesions, *J. Nerv. Ment. Dis.* **88**:733–755.

Lenneberg, E. H., 1967, *Biological Foundations of Language,* John Wiley & Sons, New York.

Levine, D. M., and Mohr, J. P., 1979, Language after bilateral cerebral infarctions: Role of the minor hemisphere, *Neurology (Minneap.)* **29**:927–938.

Levy, J., and Trevarthen, C., 1976, Metacontrol of hemispheric function in human splitbrain patients, *J. Exp. Psychol. [Hum. Percept.]* **2**:299–312.

Lieberman, A. M., 1974, The specialization of the language hemisphere, in: *The Neurosciences; Third Study Program* (F. O. Schmitt and F. G. Worden, eds.), The MIT Press, Cambridge, pp. 43–56.

Lomas, J., and Kertesz, A., 1978, Patterns of spontaneous recovery in aphasic groups: A study of adult stroke patients, *Brain Lang.* **5**:388–401.

Ludlow, C. L., Rosenberg, J., Fair, C., Buck, D., Schesselman, S., and Salazar, A., 1986, Brain lesions associated with nonfluent aphasia fifteen years following penetrating head injury, *Brain* **109**:55–80.

Martin, W. R. W., and Raichle, M. E., 1983, Cerebellar blood flow and metabolism in cerebral hemisphere infarction, *Ann. Neurol.* **14**:168–176.

Mateer, C., 1983, Motor and perceptual functions of the left hemisphere and their interaction, in: *Language Functions and Brain Organization* (S. J. Segalowitz, ed.), Academic Press, New York, pp. 145–170.

McGlone, J., 1980, Sex differences in human brain asymmetry: a critical survey, *Behav. Brain Sci.* **5**:215–264.

Metter, E. J., Wasterlain, C. G., Kuhl, D. E., Hanson, W. R., and Phelps, M. E., 1981, FDG positron emission computed tomography in a study of aphasia, *Ann. Neurol.* **10**:173–183.

Milner, B., Branch, C., and Rasmussen, T., 1964, Observations on cerebral dominance, in: *Ciba Foundation Symposium on Disorders of Language* (A. V. S. de Reuck and M. O'Connor, eds.), Little, Brown, Boston, pp. 200–214.

Milner, B., Taylor, L., and Sperry, R., 1968, Lateralized suppression of dichotically presented digits after commissural section in man, *Science* **161**:184–185.

Monakow, C., von, 1914, *Die Lokalisation im Grosshirn und der Abbau der Funktionen durch corticale Herde,* Bergmann, Wiesbaden.

Moscovitch, M., 1981, Right hemisphere language, *Top. Lang. Disord.* **1**(3):41–61.

Nagata, K., Yunoki, K., Kabe, S., Suzuki, A., and Araki, G., 1986, Regional cerebral blood flow correlates of aphasia outcome in cerebral hemorrhage and cerebral infarction, *Stroke* **17**(3):417–423.

Nielsen, J. M., 1946, *Agnosia, Apraxia, and Aphasia,* Hoeber, New York.

Perlstein, M. A., and Sugar, O., 1954, Hemispherectomies in infantile hemiplegia, *Arch. Neurol. Psychiatry* **72**:256–257.

Pieniadz, J. M., Naeser, M. A., Koff, E., and Levine, H. L., 1983, CT scan cerebral hemispheric asymmetry measurements in stroke cases with global aphasia: Atypical asymmetries associated with improved recovery, *Cortex* **19**:371–391.

Sands, E., Sarno, M. T., and Shankweiler, D., 1969, Long-term assessment of languge function in aphasia due to stroke, *Arch. Phys. Med. Rehabil.* **50**:202–222.

Sarno, M. T., Silverman, M., and Sands, E., 1970a, Speech therapy and language recovery in severe aphasia, *J. Speech Hear. Res.* **13**:607–623.

Sarno, M. T., Silverman, M., and Levita, E., 1970b, Psychosocial factors and recovery in geriatric patients with severe aphasia, *J. Am. Geriatr. Soc.* **18**:405–409.

Schuell, A., Jenkins, J. J., and Pabon, J., 1964, *Aphasia in Adults,* Harper & Row, New York.

Selnes, O. A., Knopman, D. S., Niccum, N., and Rubens, A. B., 1983, CT scan correlates of auditory comprehension deficits in aphasia: A prospective recovery study, *Ann. Neurol.* **13:**558–566.

Smith, A., 1966, Speech and other functions after left (dominant) hemispherectomy, *J. Neurol. Neurosurg. Psychiatry* **29:**467–471.

Sparks, R., and Geschwind, N., 1968, Dichotic listening in man after section of neocortical commissures, *Cortex* **4:**3–16.

Sperry, R. W., and Gazzaniga, M. S., 1967, Language following surgical disconnection of the hemispheres, in: *Brain Mechanisms Underlying Speech and Language* (C. Millikan and F. Darley, eds.), Grune & Stratton, New York.

Subirana, A., 1969, Handedness and cerebral dominance, in: *Handbook of Clinical Neurology* (P. J. Vinken and G. W. Bruyn, eds.), North Holland, Amsterdam.

Tezner, D., Tzavaras, A., Gruner, J., and Hecaen, H., 1972, L'asymetrie droite–gauche du planum temporale: A propos de l'etude anatomique de 100 cerveaux, *Rev. Neurol.* **126:**444–449.

Wada, J., Clarke, R., and Hamm, A., 1975, Cerebral hemispheric asymmetry in humans, *Arch. Neurol.* **32:**239–246.

Wernicke, C., 1874, *Der aphasische Symptomenkomlex,* Cohn and Weigart, Breslau.

Wernicke, C., 1886, Die neueren Arbeiten uber Aphasie, *Fortschr. Med.* **4:**371–377.

Witelson, S., and Pallie, W., 1973, Left hemispheric specialization for language in the human newborn: Neuroanatomical evidence of asymmetry, *Brain* **96:**641–646.

Woods, B. T., and Teuber, H. L., 1978, Changing pattern of childhood aphasia, *Ann. Neurol.* **3:**273–280.

Zaidel, E., 1976, Auditory vocabulary in the right hemisphere following brain bisection or hemidecortication, *Cortex* **12:**191–211.

Zaidel, E., 1978, Lexical organization in the right hemisphere, in: *Cerebral Correlates of Conscious Experience* (P. A. Buser and A. Rougeul-Buser, eds.), Elsevier–North-Holland, Amsterdam, pp. 177–197.

20

Sensory Substitution and Recovery from "Brain Damage"

PAUL BACH-y-RITA

1. INTRODUCTION

The loss of a major sensory system, such as sight, markedly alters cortical activity. The loss can be considered to produce "brain damage." Our sensory substitution studies were initiated a number of years ago as a model of brain plasticity; congenitally blind persons were considered a Jacksonian model [Hughlings Jackson emphasized the opportunities for discovery offered by the ". . . experiments made on the brain by disease." (excerpts in Clarke and O'Malley, 1968)]. Thus, a thorough study of congenitally blind persons learning to use a vision substitution system offered several unique opportunities:

1. The ability to control and evaluate all aspects of an entirely novel perceptual learning experience, since no relevant visual learning could go on without the use of the substitute receptor matrix (TV camera).
2. The opportunity to evaluate central nervous system mechanisms involved in the perceptual development and sensory substitution process.
3. The opportunity to evaluate the interrelationships of relevant systems, such as the role of motor control (e.g., of camera movement), on spatial localization with a vision substitution system.

Congenitally blind persons trained to use a tactile vision substitution system (consisting of a TV camera, a commutator to convert the output of the TV camera, and a matrix of electrotactile or vibrating stimulators placed against the

PAUL BACH-y-RITA • Department of Rehabilitation Medicine, University of Wisconsin, Madison, Wisconsin 53792.

skin) can identify and correctly locate in space complex forms, objects, figures, and faces. Perspective, parallax, size constancy, including looming and zooming and depth cues, are correctly utilized. The subjective localization of the information obtained through the television camera is not on the skin; it is accurately located in the three-dimensional space in front of the camera, whether the skin stimulation matrix is placed on the back, on the abdomen, on the thigh, or changed from one of these body locations to another (Bach-y-Rita, 1972).

The vision and other sensory substitution studies have demonstrated that the central perceptual mechanisms can adjust to a sensory input originating from an artificial receptor. This capacity can be interpreted in the framework of sensory plasticity, which can be defined as the ability of one sensory system—receptors, afferent pathways, and CNS representation—to assume the functions of another system, with the transducer functions of a set of lost or unavailable receptors (e.g., retina) being mediated by artificial receptors (e.g., a TV camera).

In this chapter, I briefly review vision and other sensory substitution as a basis for a discussion of the relevant physiological and perceptual mechanisms and brain plasticity, primarily in regard to controversial aspects.

2. SENSORY SUBSTITUTION

Tactile vision substitution has been a major focus of study; however, a number of other sensory substitution approaches have been reported, including tactile auditory substitution, tactile substitution for lost hand sensation and for limb prostheses, and auditory vision substitution. Braille and sign language for the deaf are excellent examples of sensory substitution, and electromyographic (EMG) biofeedback, such as to train a hemiplegic to control arm movements, is another example. A broad definition would allow the inclusion of reading and writing in a list of sensory substitution systems.

2.1. Vision Substitution

Our tactile vision substitution studies have been widely reported (e.g., Bach-y-Rita, 1967, 1971, 1972, 1983, 1984; Bach-y-Rita *et al.*, 1969; Bach-y-Rita and Hughes, 1985; Collins and Bach-y-Rita, 1973; Jansson and Brabyn, 1981; White *et al.*, 1970). Pertinent details are discussed in Sections 3 to 5.

In addition to a tactile input from an artificial receptor matrix, two other approaches have been explored. Kay developed a system using projected ultrasonic beams from a head-mounted array and an auditory output. With it, blind persons can perceive the localization of objects in a field. Although some texture information and other object identification are provided, it has been most successfully applied to mobility, allowing blind children to localize objects at a distance, which enables them to plan a clear path (summarized by Warren and

Strelow, 1985). Another approach has been to implant an array of stimulators directly into the human visual cortex and display the image from a television camera by means of the implanted array (cf. Sterling *et al.*, 1971). To date, this approach has not been as successful as the other two.

2.2. Tactile Auditory Substitution

A number of studies have explored the possibility of providing auditory information to deaf persons through arrays of skin stimulators. One such approach has been developed by Saunders (1983) to provide information about environmental sounds, assist lip-reading, augment speech training, and enable a deaf child to monitor his own voice. A bank of third-octave bandpass filters analyzes the frequency components of a signal; the output of each filter represents the energy present in a given band. Each filter controls the intensity of an electrotactile stimulator, with the array of stimulators being mounted on a belt worn around the abdomen. Field tests with profoundly deaf children have shown improvements in speech production, and studies are continuing in regard to its practicality for speech reception.

2.3. Cutaneous Sensory Substitution in Leprosy Patients

In collaboration with C. C. Collins, a sensory substitution system was developed to test the capacity of certain leprosy patients, who had lost hand sensation while retaining motor capacities, to perceive touch and pressure. A glove with strain gauges in the fingertips led to an electrotactile display on the forehead, where cutaneous sensation was present. Preliminary studies with two leprosy patients showed promising results with the ability to perceive touch, pressure, and textures (in Bach-y-Rita, 1976, 1982). Under NASA sponsorship, a comparable approach is being taken in the design of space suit gloves for extravehicular space activities; at present, the thick gloves limit tactile sensitivity, which reduces work capacity.

2.4. Braille and Sign Language

Braille and sign language both were developed as practical solutions to sensory loss and are excellent examples of sensory substitution. In Braille, the information that sighted persons gather by means of the eyes (reading) is translated into a series of raised dots and "read" with the fingertips, thus demonstrating the capacity of the somatosensory system to mediate written information.

American Sign Language (ASL) is a fascinating example of sensory substitution, requiring not only the use of a substitute sensory system but also a major transposition of information delivery mode. In normal use, auditory information is carried by means of a system that can accept signals of more than 10 Hz but has relatively little simultaneous (parallel) information-handling capacity, whereas the visual system operates at the opposite end of the scale, handling relatively low-frequency, high-simultaneous signals. In ASL, the hand signals, which change rather slowly, have a high parallel information content. Thus, conversations can be carried on in real time. This system has been widely studied (Bellugi and Klima, 1979) and has been discussed in the context of sensory substitution elsewhere (Bach-y-Rita, 1981). Bellugi and Klima (1979) consider that ASL differs dramatically from English and other spoken language, with a substitute grammatical pattern, distinct modifications of lexical units, and its own syntax. Thus, ASL demonstrates the brain's capacity to develop a different language system based on hand movements and visual recognition.

2.5. *Electromyographic Sensory Feedback*

Electromyographic sensory feedback, also called EMG biofeedback, is often prescribed for disabilities such as a hemiparetic limb following stroke. In such a case, in the presence of disorganized sensation (proprioception), the accurate positional information reaches the brain through sensory systems (visual or auditory) unaffected by the lesion, thus providing a basis for the brain to reorganize motor function. Other biofeedback systems can also be considered to provide sensory substitution; however, a thorough analysis of this point must evaluate other potential factors of such therapy such as the placebo effect.

3. PHYSIOLOGICAL CONSIDERATIONS

Optical images get no further than the retina. The actual visual information reaches the brain as pulses along nerves, no different, basically, from any other sensory information transmission. The perception of the visual images is dependent on the development of information-handling and decoding mechanisms. In normal development, the mechanisms mature in a genetically programmed sequence when adequate stimulation is present. The ability to process information from an artificial sensor array depends on a number of factors, including the following.

3.1. *Peripheral Factors*

An appropriate way is needed to get the information from the artificial array to the brain. The first, and possibly most difficult step, is the machine–man interface. In most of our studies we have used vibrotactile or electrotactile stimulus arrays applied to various skin areas (e.g., back, forehead, abdomen, finger, waist). Approaches taken by other investigators include auditory displays (e.g., sonic guide: Kay, 1985), direct cortical implantation of the stimulus array (summarized by Sterling *et al.*, 1971), and visual input (e.g., EMG sensory feedback; ASL). The interface is a major information bottleneck. The receptor characteristics and limitations have to be respected. Thus, for a tactile sensory substitution system, skin properties, receptor characteristics, and transduction properties are limitations, and this has dictated stimulator placement, interstimulator separation, and stimulation frequency. Two-point discrimination and other psychophysical measures are of importance in such systems but are not as limiting as might be expected (Section 4).

3.2. Central Nervous System Factors

Sensory cortical representation is neither pure nor immutable. For example, many primary visual cortical neurons receive somatosensory and/or auditory inputs as well as visual inputs (Murata *et al.*, 1965), and there are multiple visual maps in the cortex. The structure and function of the visual cortex and lateral geniculate nucleus respond to functional demand, as demonstrated by a series of studies (e.g., Hubel and Wiesel, 1970) revealing central nervous system effects of lid suture. In a landmark study, Chow and Stewart (1972) replicated Hubel and Weisel's (1970) studies showing that unilateral lid suture in kittens during the months in which vision normally developed led to amblyopia that lasted as long as the cats lived. However, Chow and Stewart showed that an intensive program that included gentling and affective bonding (traditional rewards were ineffective) led to significant recovery of visual function as well as producing morphological changes in the lateral geniculate nucleus and increased numbers of cells in the visual cortex responding to binocular stimulation.

There are many unanswered questions in regard to CNS plasticity. Clinical studies of hemispherectomy cases (Glees, 1980), stroke patients (Bach-y-Rita, 1980, 1981), facial paralysis patients (Balliet *et al.*, 1980), and others suggest that considerable reorganization of brain function is possible following a lesion. Animal studies support this conclusion (cf. Finger and Stein, 1982). Central nervous system reorganization occurs even in the absence of a CNS lesion. In a series of experiments by Wall and his colleagues (summarized by Wall, 1980), the effects of partial destruction of the afferent input to a region were observed. This was done in adult rat thalamus by removal of one of the dorsal column nuclei. The destruction was followed by an expansion over days and weeks of the innervated zone into the denervated zone. Similar types of experiments were carried out in adult cats by sectioning of dorsal roots followed by examination of the denervated zones of dorsal column nuclei and of the spinal cord. In each situation, cells that had lost their input began to respond to intact afferents. In comparable cortical studies, Merzenich *et al.* (1983, 1984) demonstrated cortical reorganization following both median nerve sections and digit amputation in monkeys. But how do CNS mechanisms respond to the functional demands of a sensory substitution system?

The question can certainly extend to more common functional demand changes. For example, what is the CNS response to the intense training of a pianist? Are the cortical representations of the fingers enlarged? In studies of the somatosensory (Feinsod *et al.*, 1973) and auditory (Woods *et al.*, 1985) evoked potentials in blind and sighted persons, significant differences were noted. It remains to be shown whether sensory substitution system use can alter CNS mechanisms, but certainly the behavioral responses of trained subjects suggest that it does.

4. PERCEPTUAL CONSIDERATIONS

A person who becomes blind does not necessarily lose the capacity to see; the central perceptual mechanisms have been shown to be able to adjust to sensory input originating from artificial receptors (Section 2). Although we have generally concluded that trained blind tactile vision-substitution subjects can "see," Morgan (1977) considers that since blind persons are being given similar information to that which causes the sighted to see and are capable of giving similar responses, one is left with little alternative but to admit that they are seeing (and not merely "seeing"). Morgan points out that the structural nature of the perceptual system does not offer any criteria for distinguishing seeing from not seeing (e.g., the horseshoe crab is offered as an example of a biological system with fewer receptors than most mammals but that can nonetheless see). A decision as to whether a person is seeing via a nonvisual perceptual system involves considerations other than solely those of physiological hardware and quantitative capabilities. It ought to be noted, nonetheless, that there are a number of structural similarities between visual and vibrotactile perception: (1) static and continuously transformed images are formed by a lens on a two-dimensional surface; (2) the receptor surface contains discrete elements; (3) the surface can be voluntarily moved to scan the environment; and (4) the source of stimulation is not in direct physical contact with the receptor surface, so that perception can be subject to similar interruptions such as by object occlusion. Morgan (1977) further points out in regard to behavioral equivalence that if blind subjects receive vibrotactile conversions of optical information that would satisfy criteria for seeing in the sighted and respond in an indistinguishable manner, one might concede that the blind are seeing in spite of the quantitative difference in information, since sighted persons still see under impoverished conditions such as in the fog (further discussion in Bach-y-Rita and Hughes, 1985).

In view of the clear demonstrations that sensory substitution is possible, present questions are related to characteristics, mechanisms, and limitations. For example, although we now know that visual information can be delivered to the brain through the skin, to what extent do instrumentation, skin receptors, afferent pathways from the skin, somatosensory cortex, higher perceptual mechanisms, and appropriate behavioral (including training) considerations affect the development of high-resolution tactile vision substitution? What are the major limiting factors? Just how plastic are the central perceptual mechanisms? To what extent does previous visual experience (congenital versus acquired blindness) influence eventual capacity? Our studies to date have used low-resolution systems that may not have challenged the capacity of the system and for the same reason may not have revealed significant differences in learning to use a substitution system between congenital and acquired blind persons (White *et al.*, 1970).

At the highest perceptual levels, modality is ignored, and the information itself is the basis for the percept. Gibson (1966) has pointed out that "fire" is the same whether the information has been obtained by hearing, feeling, looking, or smelling. Similarly, blind vision substitution subjects demonstrate perceptual equivalence even in automatic responses: a telephone across the room, behind and partly occluded by a watering can, is identified as such and in that context. Blind subjects respond behaviorally to the perceptually equivalent (retinal–visual and tactile–visual) information; for example, a blind subject pointing a TV camera at an object in front of him moved backwards when, unknown to him, the experimenter pushed the zoom control, thus producing a sensation of looming. The perception of an object rapidly approaching from in front of him (where the camera under his motor control was pointed) caused the defensive response backwards even though the tactile array was placed on his back. Clearly, the localization of the array was of no perceptual importance during this response, and the visual spatial information was appropriately mediated by the tactile system. In such a case, the perceptual role of the skin is modified, and it becomes a relay, similar in function to a lateral geniculate nucleus. An alternate interpretation is that the skin retains a receptor role, but the subject merely learns to deal with the information in context. At any moment, however, the subject can switch perceptual roles, such as when asked if the skin under the array is warm.

The artificial array (the TV camera) is also linked to the motor system controlling it. The camera can be transferred from hand to hand and to head control with no loss of accurate localization, and the skin stimulation matrix can be moved from one area to another (e.g., forehead to back to abdomen) without loss of orientation or perceptual response. Thus, the artificial receptor matrix and the muscles moving the camera form a perceptual organ (comparable to the eye and the extraocular muscles) but not a fixed organ, since either the area of skin placement or the muscle group controlling camera movement can be changed, simultaneously or separately. This has broad implications for an understanding of sensorimotor interaction and perceptual mechanisms.

The fact that the stimulus array can be moved from one area to another [and from one type of stimulation (electrotactile) to another (vibrotactile), each activating different sensory receptors] suggests that the perceptual learning process is primarily central and at a high level so that the decoding and interpretation, once learned, can be applied to inputs from the entire skin of the body. One physiological substrate for this may be the widespread neural activity induced by cutaneous stimuli: Mountcastle (1961) showed that a large portion of the thalamus is activated by even a localized stimulus, thus suggesting that a great deal of interaction takes place. The distribution of central neural activity from a localized stimulus may be modifiable by training and may be influenced by attention. Wall et al. (1967) showed that the activity produced by localized skin stimulation of a rat varied in accordance to the attention directed by the rat to that area. The fact that a task can be learned using one set of receptors and performed using another is comparable to vision: Lashley (1930) pointed out that entirely different retinal elements could be used in learning and performance of perceptual tasks.

One of the reasons that serious research into tactile sensory substitution has been limited is that extrapolation from psychophysical data led to the belief that a high-resolution system was impossible: Geldard (1960, 1968) produced results demonstrating that it was only feasible to consider using up to ten skin loci (spread over the whole body) for tactile communication. Yet our initial system used 400 stimulators on the skin of the back (which is among the least sensitive areas of the body, with a 6.8-cm two-point discrimination threshold), while our later systems used up to 1032 points on the abdomen, and the Optacon has a 144-point display on a single fingertip (Bach-y-Rita, 1972). We previously noted (White et al., 1970) that we would never have been able to say that it was possible to identify, and lay out in three dimensions, a group of familiar objects if the tactile vision substitution system had been designed to deliver 400 maximally discriminable sensations to the skin.

I noted many years ago (Bach-y-Rita, 1972) that the applicability of tactile sensory substitution systems to complex situations will not be determined by extrapolation from laboratory results. My approach was to begin the studies with devices that could transmit sufficient information to be useful. With such knowledge, a detailed study of the mechanisms is warranted. Our previous results and the modified Optacon (Bach-y-Rita and Hughes, 1985) results demonstrate that the capability of the tactile system is far greater than psychophysical studies (including recent findings, such as Epstein et al., 1986; Schneider et al., 1986) suggest when the subject is presented with patterns of stimuli representing useful tasks and is trained to use the information.

How can an area of skin mediate detailed visual information and send it to the brain? In the past, the sense of touch has been considered an ''inferior sense'' (Wiesner et al., 1949), but in part this has been because of a lack of understanding of the information-handling capabilities of the tactile system. Certainly, less information can be handled simultaneously (parallel input; a principal characteristic of vision), but its frequency capacities are an order of magnitude greater than those of the eye. Information is extracted from patterns of stimuli; although the pattern theory has been seriously challenged by single afferent nerve studies in humans, it remains viable for certain higher-level functions. Thus, although Ochoa and Torebjork (1983) have demonstrated that modality information is carried in single afferent fibers, the extraction of complex optical patterns from the output of an array of stimulators may be the result of complex higher-level elaboration of the patterns of stimulation (Bach-y-Rita, 1972).

Since the skin shows a number of functional similarities to the retina in its capacity to mediate information (see above), and both are capable of mediating displays in two spatial dimensions as well as having the potential for temporal integration, there is no need for complex topological transformation or for temporal coding for the direct presentation of pictorial information onto accessible areas of the skin [however, edge enhancement and temporal display factors are being explored with the goal of transmitting spatial information across the skin more quickly than is possible with present systems (e.g., Kaczmarek et al., 1985)]. Certain types of sensory inhibition, including the Mach band phenomenon and other examples of lateral inhibition originally demonstrated for vision, have been shown to be equally demonstrable in the skin (von Bekesy, 1967). In addition, there is evidence that the skin normally functions as an exteroceptor at least in a limited sense: to some extent both vibration and temperature changes can be felt at a distance [Katz, excerpts translated by Krueger (1970)]. For example, a blind person can "feel" the approach of a warm cylinder at three times the distance required by the sighted individual (Krueger, 1970).

5. PRACTICAL CONSIDERATIONS

In a recent discussion of the long-range feasibility of sensory substitution systems in rehabilitation (Bach-y-Rita, 1983), two key practical points were advanced. The first is that the most successful sensory substitution systems—Braille, sign language, and the long cane—share several common factors. Among these are the absence of complex instrumentation, high reliability, and appropriate transduction of one type of sensory information into another. With these systems, the intact sensory system carries the information to the brain, and information processing occurs in a form and at a rate consistent with its physiological capacities and limitations.

The second point was that each of the three sensory substitution systems also has limitations; in particular, each is capable of transmitting to the brain only a limited range of environmental information. Braille readers, for example, have a limited literature, and long cane users cannot perceive overhangs, nor can they gather information about the space in front of the cane tip. Even with these aids, much environmental information remains totally unavailable. For example, neither Braille nor the long cane can allow a blind child to see a flickering candle flame or the second-story window of a house.

High technology sensory substitution systems, such as the tactile vision substitution system, offer the possibility of harnessing present and future instrumentation and material science advances to the needs of the handicapped. But limitations include (1) cosmesis (including weight, body locus placement, visability, and other subtle factors), (2) cost, (3) technological overload, (4) practicality of low-resolution systems, and (5) reliability. There are psychological factors that limit the acceptability of even the simplest sensory aids, such as glasses and hearing aids, so even the "perfect" sensory substitution system would face obstacles.

Sensory substitution systems have been developed for limited uses, such as for teaching visual spatial concepts to congenitally blind children, to aid deaf persons in lip reading, or for performing certain vocational tasks. In these cases, the complex psychological problems (discussed in part by Bach-y-Rita, 1972) can be avoided, since the use of the substitution system does not alter the subjects' adjustment to the sensory loss.

The adaptation to a sensory substitution device requires a well-planned program. For example, a previously blind person would have to be taught to see, to evaluate and select information, to pool the sensory information from the device with other sensory input, and to relate the sensory information with motor activity. Furthermore the complex self-image and other psychological considerations undoubtably require appropriate counseling during the learning and adaptation process.

6. CONCLUSIONS

Sensory substition studies offer the opportunity to test brain and perceptual plasticity with controlled experiments. Although most of the results have been evaluated in behavioral terms, other possibilities exist: evoked potential and other sophisticated electrophysiological studies have been initiated, and in the future brain metabolism (e.g., positron emission tomography) and possibly morphological (e.g., nuclear magnetic resonance) techniques may be applied to human studies. The development of a good animal model would allow for these as well as invasive and histological techniques to be applied to sensory substitution. The results of Chow and Stewart (1972; Section 3.2) suggest that with proper experimental methods it may be possible to demonstrate morphological and anatomic correlations as well as the behavioral changes.

Under controlled laboratory conditions, complex perceptual and hand–eye coordination tasks are possible with tactile vision substitution, and comparably impressive results have been obtained with other sensory substitution systems. However, for a number of practical reasons, to date only the low-technology sensory substitution approaches have had widespread practical applications. In the present-day scientific climate of interest in brain plasticity, as instrumentation costs decrease while technological sophistication increases, the possibility of developing high-technology sensory substitution devices for everyday use increases.

REFERENCES

Bach-y-Rita, P., 1967, Sensory plasticity: Applications to a vision substitution system, *Acta Neurol. Scand.* **43**:417–426.

Bach-y-Rita, P., 1971, Neural substrates of sensory substitution, in: *Pattern Recognition in the Biological and Technical Systems* (R. Klinke and O. J. Grusser, eds.), Springer-Verlag, Berlin, pp. 130–142.

Bach-y-Rita, P., 1972, *Brain Mechanisms in Sensory Substitution*, Academic Press, New York, p. 181.

Bach-y-Rita, P., 1976/1977, Hirnplastizitat und sensorische Substitution, in: *Mannheimer Formun*, Boehringer Mannheim, Mannheim, pp. 89–144.

Bach-y-Rita, P., 1980, Brain plasticity as a basis for therapeutic procedures, in: *Recovery of Function: Theoretical Considerations for Brain Injury Rehabilitation* (P. Bach-y-Rita, ed.), H. Huber, Bern, Switzerland, University Park Press, Baltimore, pp. 225–263.

Bach-y-Rita, P., 1981, Brain plasticity as a basis of the development of rehabilitation procedures for hemiplegia, *Scand. J. Rehabil. Med.* **13**:73–83.

Bach-y-Rita, P., 1982, Sensory substitution in rehabilitation, in: *Rehabilitation of the Neurological Patient* (L. Illis, M. Sedgwick, and H. Granville, eds.), Blackwell Scientific, Oxford, pp. 361–383.

Bach-y-Rita, P., 1983, Tactile vision substitution: Past and future, *Int. J. Neurosci.* **19**:29–36.

Bach-y-Rita, P., 1984, The relationship between motor processes and cognition in tactile vision substitution, in: *Cognition and Motor Processes* (W. Prinz and A. F. Sanders, eds.), Springer-Verlag, Berlin, Heidelberg, pp. 150–159.

Bach-y-Rita, P., and Hughes, B., 1985, Tactile vision substitution: Some instrumentation and perceptual considerations, in: *Visual Spatial Prosthesis for the Blind* (D. Warren and E. Strelow, eds.), Martinus-Nijhoff, Dordrecht, The Netherlands, pp. 171–186.

Bach-y-Rita, P., Collins, C. C., Saunders, F., White, B., and Scadden, L., 1969, Vision substitution by tactile image projection, *Nature* **221**:963–964.

Balliet, R., Shinn, J. B., and Bach-y-Rita, P., 1982, Facial paralysis rehabilitation: Retraining selective muscle control, *Int. J. Rehabil. Med.* **4**:67–74.

Bellugi, U., and Klima, E., 1979, Language: Perspectives from another modality, *CIBA Found. Symp.* **69**:99–117.

Chow, K. L., and Stewart, D. L., 1972, Reversal of structural and functional effects of long-term visual deprivation in cats, *Exp. Neurol.* **34**:409–433.

Clarke, E., and O'Malley, C. D., 1968, *The Human Brain and Spinal Cord*, University of California Press, Berkeley, pp. 499–505, 507–511.

Collins, C. C., and Bach-y-Rita, P., 1973, Transmission of pictorial information through the skin, *Adv. Biol. Med. Phys.* **14**:285–315.

Collins, C. C., Bach-y-Rita, P., and Loeb, D. R., 1967, Intraocular pressure variation with oculorotary muscle tension, *Am. J. Physiol.* **213**:1039–2043.

Epstein, W., Hughes, B., Schneider, S. L., and Bach-y-Rita, P., 1986, Is anything out there? Distal attribution in response to vibrotactile stimulation, *Perception* **15**:275–284.

Feinsod, M., Bach-y-Rita, P., and Madey, J. M., 1973, Somatosensory evoked responses—latency differences in blind and sighted persons, *Brain Res.* **60**:219–223.

Finger, S., and Stein, D. G., 1982, *Brain Damage and Recovery*, Academic Press, New York.

Geldard, F., 1968, Bod'y Eng'lish, *Psychol. Today* **2**:42–47.

Geldard, F. A., 1960, Some neglected possibilities of communication, *Science* **131**:1583–1588.

Gibson, J. J., 1966, *The Senses Considered as Perceptual Systems*, Houghton Mifflin, Boston.

Glees, P., 1980, Functional reorganization following hemispherectomy in man and after small experimental lesions in primates, in: *Recovery of Function: Theoretical Considerations for Brain Injury Rehabilitation* (Bach-y-Rita, P., ed.), University Park Press, Baltimore, pp. 106–126.

Hubel, D. H., and Wiesel, T. N., 1970, The period of susceptibility to the physiological effects of unilateral eye closure in kittens, *J. Physiol. (Lond.)* **206**:419–436.

Jansson, G., and Brabyn, L., 1981, *Tactually Guided Batting*, Uppsala Psychological Reports No. 304, Uppsala University, Uppsala.

Kaczmarek, K., Bach-y-Rita, P., Tompkins, W., and Webster, J., 1985, A tactile vision substitution system for the blind: Computer-controlled partial image sequencing, *IEEE Trans. Biomed. Eng.* **32**:602–608.

Kay, L., 1985, Sensory aids to spatial perception for blind persons: Their design and evaluation, in: *Electronic Spatial Sensing for the Blind* (D. Warren and E. Strelow, eds.), Martinez-Nijhoff, Dordrecht, The Netherlands, pp. 125–139.

Krueger, L. E., 1970, Der Aufbau der Tastwelt (the world of touch): A synopsis, *Percept. Psychophys.* **7**:337–341.

Lashley, K. S., 1930, Basic neural mechanisms in behavior, *Psychol. Rev.* **13**:1–42.

Merzenich, M. M., Kaas, J. H., Wall, J., Nelson, R. J., Sur, M., and Felleman, D., 1983, Topographic reorganization of somatosensory cortical areas 3B and 1 in adult monkeys following restricted deafferentation, *Neuroscience* **8**:33–55.

Merzenich, M. M., Nelson, R. J., Stryker, J. P., Cynader, M. S., Schoeppman, A., and Zook, J. M., 1984, Somatosensory cortical map changes following digit amputation in adult monkeys, *J. Comp. Neurol.* **224**:591–605.

Morgan, M. J., 1977, *Molyneux's Question*, Cambridge University Press, Cambridge, MA.

Mountcastle, V. B., 1961, Some functional properties of the somatic afferent system, in: *Sensory Communication* (W. A. Rosenblith, ed.), MIT Press, Cambridge, pp. 403–436.

Murata, K., Cramer, H., and Bach-y-Rita, P., 1965, Neuronal convergence of noxious, acoustic and visual stimuli in the visual cortex of the cat, *J. Neurophysiol.* **28**:1223–1239.

Ochoa, J., and Torebjork, E., 1983, Sensations evoked by intraneural microstimulation of single mechanoreceptor units innervating the human hand, *J. Physiol. (Lond.)* **342**:633–654.

Saunders, F. A., 1983, Information transmission across the skin: High-resolution tactile sensory aids for the deaf and the blind, *Int. J. Neurosci.* **19**:21–28.

Schneider, S. L., Hughes, B., Epstein, W., and Bach-y-Rita, P., 1986, The detection of length and orientation changes in dynamic vibrotactile patterns, *Percept. Psychophys.* **40**:293–300.

Sterling, T. D., Bering, E. A., Jr., Pollack, S. V., and Vaughan, H. G., Jr., eds., 1971, *Visual Prosthesis: The Interdiscplinary Dialogue*, Academic Press, New York.

von Bekesy, G., 1967, *Sensory Inhibition*, Princeton University Press, Princeton, NJ.

Wall, P. D., 1980, Mechanisms of plasticity of connection following damage in adult mammalian nervous systems, in: *Recovery of Function: Theoretical Considerations for Brain Injury Rehabilitation* (P. Bach-y-Rita, ed.), University Park Press, Baltimore, pp. 91–105.

Wall, P. D., Freeman, J., and Major, D., 1967, Dorsal horn cells in spinal and in freely moving rats, *Exp. Neurol.* **19**:519–529.

Warren, D., and Strelow, E., eds., 1985, *Electronic Spatial Sensing for the Blind*, Martinez-Nijhoff, Dordrecht, The Netherlands.

White, B. W., Saunders, F. A., Scadden, L., Bach-y-Rita, P., and Collins, C. C., 1970, Seeing with the skin, *Percept. Psychophys.* **7**:23–27.

Wiesner, J., Wiener, N., and Levine, L., 1949, Some problems in sensory prosynthesis, *Science* **110**:512.

Woods, D., Clayworth, C., Bach-y-Rita, P., 1985, Early blindness reorganizes auditory processing in humans, *Neurosci. Abstr.* **1**(134.8):449.

21

Emotion and Motivation in Recovery and Adaptation after Brain Damage

GEORGE P. PRIGATANO

1. INTRODUCTION

As common as affective problems are after brain injury, it is rather surprising that they have not figured prominently in any major theory of recovery of function (see Stein *et al.*, 1974; Finger, 1978). Emotional and motivational variables, however, have been recognized as important for adaptation to the permanent behavioral sequelae associated with various types of brain damage (e.g., Goldstein, 1942; Kotila *et al.*, 1984; Prigatano *et al.*, 1986).

In emphasizing the role of behavioral training to facilitate recovery after brain injury, Gazzaniga (1974) made the following observations and claimed:

> Yet, even if the basic ideas of Von Monokow prove correct, some clinical instances of recovery involving the higher cognitive processes following massive brain damage probably come about through other mechanisms and involve neither disinhibition or actual structural changes. In what follows, I will show instances of recovery of function which can be brought about quickly after a brain lesion or by prelesion prophylactic measures; it will also be demonstrated that recovery can be obtained by the use of proper behavioral training routines long after diaschistic processes are thought to be active (p. 205).

Continuing, Gazzaniga (1974) states:

> Using Premack's (1970) language training system developed for the chimp, we ran a series of tests on global aphasic patients and quickly discovered these patients can learn to perform many language-like operations.
>
> Before beginning language training, a viable social relationship must be established between the patient and the trainer. *The importance of this phase cannot be*

GEORGE P. PRIGATANO • Section of Neuropsychology, Barrow Neurological Institute, Phoenix, Arizona 85013.

*overemphasized, for if the motivational setting is inappropriate, no learning will
occur* (italics added) (p. 208, Gazzaniga, 1974).

The central thesis of the present chapter is that emotional and motivational
factors do play an important role in the recovery of higher cerebral functioning
after brain injury. Moreover, they play a crucial role in adaptation, and unless
they are addressed in the context of rehabilitation programs, the patient can make
less than optimal gains.

The reason for this is that emotional and motivational variables are closely
interconnected with disturbances of arousal, perception, and ultimately self-
awareness. When disorders of self-awareness occur after brain injury, it is difficult, if not impossible, for the patient realistically to recognize altered functional
capacities and to take the steps necessary to improve or compensate for those
capacities. Consequently, unless emotional and motivational disturbances are
actively addressed, the patient can remain childlike and insufficiently aware of
difficulties.

A recent paper by Oddy *et al.* (1985) emphasizes this point. As Table 1
shows, some 7 years after head trauma, 40% of families described their brain-
injured relative as being not only childish but unwilling to admit to any residual
difficulties. It is believed that this lack of awareness or self-recognition is not
purely a function of psychiatric "denial." Rather, it is a disorder of self-
awareness that is organically mediated via disturbances in arousal and perception. Neurological rehabilitation should attempt to facilitate greater recovery of
higher cerebral functioning by dealing with arousal and perceptual disorders so
that disturbances of self-awareness are reduced and do not interfere with the
individual's capacity to evaluate residual strengths and weaknesses. This helps
the patient become more "motivated" in coping with the consequences of the
the brain injury. But, before these concepts are discussed in more detail, the

Table 1. Symptoms Reported by Patients and Relatives 7 Years after Head Injury[a]

Patients	Percent	Relatives	Percent
Trouble remembering things	53	Trouble remembering things	79
Difficulty concentrating	46	Difficulty concentrating	50
Easily affected by alcohol	38	Difficulty speaking	50
Often knocks things over	31	Easily affected by alcohol	43
Often loses temper	31	Difficulty in becoming interested	43
Difficulty becoming interested	28	Becomes tired easily	43
Likes to keep things tidy	28	Often impatient	43
Sometimes loses way	28	Sometimes behaves childishly	40
Eyesight problems	28	Likes to keep things tidy	40
Difficulty following conversation	28	Refuses to admit difficulties	40

[a]From Oddy *et al.* (1985), with permission.

constructs of arousal, emotion, and motivation after brain injury should be considered.

2. AROUSAL, EMOTION, AND MOTIVATION AFTER BRAIN DAMAGE

Both classic neurology and experimental neuropsychology have emphasized the cognitive, linguistic, sensory, and motor deficits associated with localized lesions of the brain. The history of ideas concerning recovery of function is, of course, interconnected with localization versus antilocalization theories of higher cerebral functioning (see Prigatano, 1986a,b). In that debate, it has often been assumed that motivational and emotional disturbances are mainly secondary to cognitive deficits, and virtually no individual has incorporated or studied these affective disturbances over time and related them to actual mechanisms of recovery or adaptation. Yet, in the clinical setting, one is impressed with the importance of these variables in the overall outcome.

The concept of arousal is a familiar one to psychologists. In the early work of Moruzzi and Magoun (1949), the reticular activitating system of the brainstem was associated with alertness and wakefulness. Others, including Luria (1973) and Pribram and McGuinness (1975), have emphasized the importance of arousal level for behavior and have suggested that many factors can contribute to sustained and fluctuating arousal level. Specifically, sensory input, metabolic processes, and cortical inputs (presumably originating in part from perceptions and memories) can have effects on subcortical structures. Malmo (1959) further emphasized that arousal level is associated with behavioral achievement or competency. At extremely low or high arousal levels, behavioral competency is frequently impaired.

Research by Morrow et al. (1981) illustrated that disturbances in arousal level can and do occur after "purely" cerebral insults. The galvanic skin response (GSR), one measure of autonomic nervous system reactivity to emotional and motivational stimuli, was recorded from right and left cerebral vascular accident (CVA) patients (see Fig. 1). The normal response was increased arousal or GSR response to emotional stimuli compared to nonemotional stimuli. Although left hemisphere patients showed this general pattern, their overall arousal level was lower than that of the control group. Right hemispheric lesion patients showed no change in autonomic nervous system reactivity. They had no "arousal response" to either emotional or nonemotional stimuli. Clinically it has been well recognized that right hemisphere patients often seem extremely apathetic and, in the early stages after the injury, even may display anosognosia (or denial or unawareness of illness) (see Weinstein and Kahn, 1955).

The importance of these findings for the present discussion is that brain injury, despite location of lesion, may reduce overall arousal level or activation

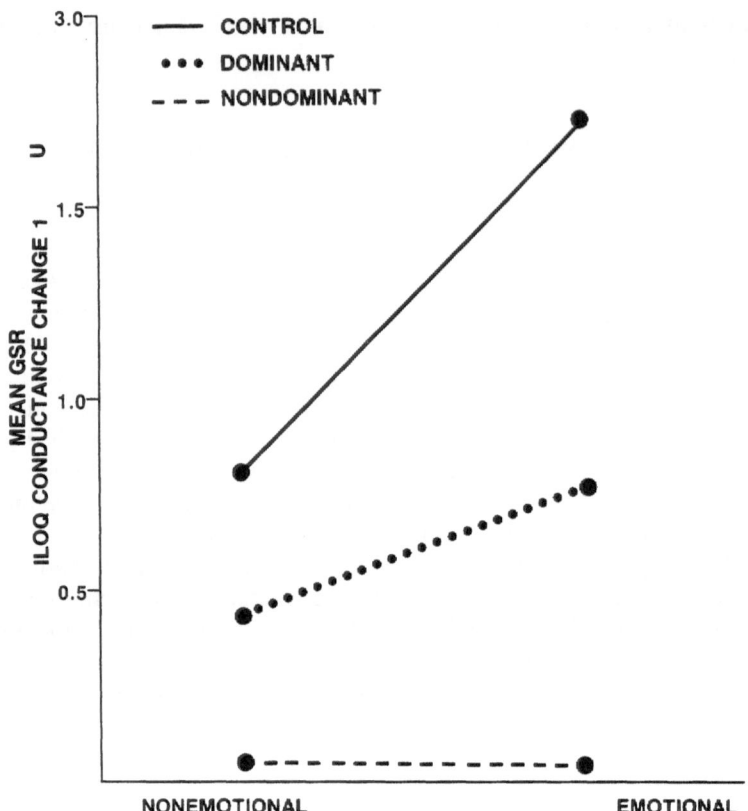

Figure 1. Means of the galvanic skin responses (GSR) to emotional and nonemotional stimuli for three groups, demonstrating source of the significant interaction. (From Morrow *et al.*, 1981. Copyright by Pergamon Press, Ltd. Reprinted by permission.)

responses to typically arousal-generating stimuli. Moreover, some regions of the brain may be more important than others in producing even greater degrees of impaired arousal. Behaviorally, this translates into a patient who is very ineffective in terms of coping with sensory inputs and consequently shows a significant alteration in emotion and motivation.

It is common to hear descriptions of brain-injured patients, especially traumatic brain-injured patients, as being "more emotional" and "less motivated" after their brain insult (see Prigatano, 1987; Prigatano *et al.*, 1986). I believe this is because the basic neurostructures involved in maintaining arousal level and the interpretation (perception) of rudimentary feeling states are more frequently affected after cerebral insults than is commonly recognized. Ultimately this leads to disturbances in self-perception and self-awareness.

Given these concepts, what then are emotion and motivation, and how do they relate to self-awareness? Borrowing from Pribram's (1971) and Simon's (1967) work, I attempt to define these terms commencing with the concept of "feelings."

> Feelings are the most rudimentary, generalized, and differentiated perceptions of internal bodily states. By nature, they have an intensity or an arousal dimension to them (Lindsley, 1970). Core brain receptors involved in the central nervous system's regulation of the organism's metabolic and endocrine functions are probably responsible for the initial or "crude" sensation of feelings, These core brain receptors for temperature, thirst, hunger, pain, and respiration are localized in the brainstem and are near the midline ventricular system involving hypothalamic and midline thalamic nuclei (see Pribram, 1971).
>
> If feelings can be considered the basic representation for homeostatic states of the organism, the terms emotion and motivation can be used to refer to more complex and refined feeling states that incorporated basic homeostasis but also go beyond it. Motivation refers to the complex feeling states that parallel hierarchical goal-seeking behavior (Simon, 1967). As such, it can be described as the arousal component of the behavior which sees to it that a plan of action is developed and executed. In Simon's (1967) information-processing terms, motivation controls attention and thereby influences learning by influencing which program will be followed. A program here is defined simply as a series of steps that are taken in order to achieve a goal (Miller et al., 1960).
>
> Using Simon's (1967) model, emotion, on the other hand, refers to complex feeling states that parallel an interruption of whatever ongoing goal-seeking behaviors or programs are engaged. According to Simon, emotion interrupts an ongoing process, particularly when stimuli enter the perceptual world which must be attended to for survival value. In humans, attention to the activities of other humans is especially important for survival, and thus emotion is crucial and indispensable in social interaction. While there are certainly other models of motivation and emotion, this system has appeal for the neuropsychologically oriented clinician (Prigatano et al., 1986, pp. 30–31).

Given this model, it can be easily seen that emotions and motivation can be affected after brain injury because basic arousal and/or interpretive/perceptive abilities are affected. But how do these variables relate to awareness and particularly self-awareness?

In a theoretical discussion on the relationship among brain, mind, and consciousness, Pribram (1980) distinguished between consciousness and self-consciousness. The phenomenon of consciousness refers to the state of attending to specific internal or external stimuli. For example, individuals are "conscious" or aware of a loud noise or a pretty girl entering the room. By referring to "consciousness," we are referring to the capacity to perceive a stimulus and to give some response or "feedback" acknowledging the relative accuracy of the perception. Consciousness, therefore, is basically a perceptual phenomenon involving attention.

Self-consciousness, on the other hand, refers to the state of subjective awareness. This is the "attentional" perception of the "self" (Pribram notes) that philosophers since James and Brentano have considered to be the essential ingredient in what makes man human.

In further defining self-consciousness, Pribram (1980) suggested that intentions and intentionalities make up self-consciousness. That is, the capacity or the state to intend some action or to prefer to perceive some stimulus (i.e., intentionality) constitutes the "stuff" out of which self-consciousness or self-awareness emerges.

This theoretical distinction is important because it makes an old point and suggests a new one. The old point is that consciousness is not a single phenomenon but seems to have several levels or stages. The second and perhaps new observation is that self-consciousness or awareness has a lot to do with the phenomenon of intentionality. Intentionality and intention are never purely cognitive phenomena. That is, they reflect something about the emotional and motivational characteristics of the organism as well as the perceptual and attentional state. Consciousness and particularly self-consciousness emerge or diminish depending on the intactness of brain structures that allow for attention and perception and the capacity of the organism to interpret feeling states. Thus, arousal level as well as the patient's personal history of feeling states prior to the onset of the brain injury must play an important role in intentionality and ultimately in self-awareness.

Piaget assumed that biological changes in brain function permitted certain cognitive abilities to emerge throughout childhood. Some have interpreted Piaget's work to suggest that "affectivity" or the emotional–motivational characteristics of the organism produce the biological "push" to use these emerging cognitive skills (see Izard, 1980, p. 198). Other theorists who have interpreted Piaget's work, however, see cognitive and affective development emerging simultaneously and being interconnected (see Wadsworth, 1984). In this regard, it is interesting that a recent paper by Lane and Schwartz (1987) specifically attempted to evaluate how awareness of affective functioning may unfold in adults using a Piagetian developmental model. In their work, they tended to use the term "emotional awareness," which I think is a little misleading. In the light of the previous discussion, a better term would have been emotional and motivational awareness. Nevertheless, as Table 2 illustrates, there are five levels that they considered to be important in the development of affective awareness. They related these levels to specific types of cognitive capacities given Piaget's notions. From this table, it can be seen that the first sense of awareness comes through body sensation. The early theorists of anosognosia felt that denial of illness after right hemisphere stroke was truly a disturbance of body image (see Critchley, 1953).

From here, Lane and Schwartz (1987) suggested that awareness next emerges from the general arousal level of the individual. The individual shows a tendency to pursue or avoid a topic based primarily on his hedonic state. From this point, the individual may actually experience extreme emotional and motivational feelings, but the behavioral repertoire and feeling repertoire are limited.

Table 2. Characteristics of Five Levels of Emotional Awareness[a]

Level of emotional awareness	Subjective quality of emotional experience	Differentiation of emotion	Ability to describe emotion
5. Formal operational	Peak differentiation blending	Richer differentiations of quality and intensity	Description of more complex and differentiated states
4. Concrete operational	Differentiated, attenuated emotion	Blends of emotion; concurrence of opposing emotions	Description of differentiated emotions
3. Preoperational	Pervasive emotion	Either/or experience of emotional extremes (limited repertoire)	Description of unidimensional emotion
2. Sensorimotor enactive	Action tendency and/or global arousal	Action tendency or global hedonic state	Description of action tendencies or global hedonic states
1. Sensorimotor reflective	Bodily sensation	Global undifferentiation of arousal	No description or description of bodily sensation

[a]From Lane and Schwartz (1987), with permission.

Eventually, the individual is able to have more complex perceptions of feeling states, and this leads to a richer interpretation of nuances of feelings and thereby enhances the highest degree of "awareness."

I find it fascinating that developmental models concerning cognition and affective functioning are being related to the phenomenon of awareness at this time. Awareness is now seen as emerging from an interconnection of feeling states (complex ones, such as emotion and motivation) and cognitive (perceptual) structures.

It is extremely important, given these theoretical concepts, to appreciate that emotion and motivation can influence the recovery process following brain injury. I would now like to give a few clinical examples of how the reexperiencing of "old" feeling states or the failure to perceive accurately an emotional reaction can either facilitate or retard recovery after brain injury.

3. SOME CLINICAL EXAMPLES OF THE IMPORTANCE OF EMOTION AND MOTIVATION IN RECOVERY AFTER BRAIN DAMAGE

To highlight the importance of emotional and motivational factors in recovery after brain injury, I present specific examples as they emerged with patients who dealt with physical disability, disturbances of voice and language, and disturbances affecting psychological adaptation to the environment.

The first patient was a Chinese woman in her early 30s who had suffered a severe, traumatic brain injury approximately 6 months prior to the clinical episode described. The woman showed major motor deficits in the right arm and leg. Among her deficits was a predictable increase in spasticity with associated elevated muscle tone. During many physical therapy hours, she would attempt to do what the therapist asked her to do with little enthusiasm. During one particular hour, she had the subjective experience of coming into a normal postural alignment, with equal weight bearing through both sides of the body and with pelvis tilted forward. This facilitated a sudden relaxation of her muscle tone. At this precise moment, there was a clear change in her facial expression, which gave the therapist the impression that she was suddenly experiencing and recognizing an old muscular "feeling" that had not been present for a long time. This was an extremely joyful experience for her and seemed to motivate her for more physical therapy. Once she could reexperience this old "feeling" state, she became tremendously motivated to work at her ongoing physical therapy and showed substantial progress a good $2\frac{1}{2}$ years after brain injury. This clinical example highlights how the reexperiencing of an old feeling state produced a level of self-awareness that became crucial for the patient to engage actively in an ongoing rehabilitation activity. This first example highlighted this as it emerged within the context of physical therapy.*

The second example was a patient who worked within the context of speech and language therapy. She was a 17-year-old female who suffered a traumatic brain injury and was approximately 3 months post-trauma.

At the initiation of therapy, her vocal quality was perceived as being abnormally high-pitched and breathy. Prosodic features involved equal stress and monopitch. An indirect laryngoscopy indicated no damage to the vocal folds. The patient was not aware that her voice was different but accepted her parents' judgment that it was not the same.

One therapy task involved speaking in an abnormally low pitch while reading short phrases. The therapist modeled this, and the patient attempted to imitate it. As the patient was reading the phrase "the scum is on the pond," she suddenly stopped with a look of surprise. The therapist recognized that she had just attained her habitual (preinjury) pitch level. The therapist asked, "What is the matter, Kay?" She responded, "I think that's it. That's my voice!" The patient was very excited and said "I can't believe I found my voice on 'the scum is on the pond.'" Although her pitch did not stabilize for another week, the patient went around the rehabilitation unit demonstrating her new voice with the same phrase. With this one major affective experience, she became extremely motivated to work at her voice therapies and subsequently made substantial progress in this area.

A third patient, a young woman in speech and language therapy. was essentially uncommunicative for 1 year following craniocerebral trauma that involved

---

*The author wishes to recognize Joy Dirham, Physical Therapist, for bringing this example to his attention.

both the cerebral hemispheres and the brainstem. All the patient seemed to be able to do was cry out in a loud voice. Her therapists were beginning to believe that she would require a communication board permanently. Finally, a perceptive therapist stated to her, "Jen, are you afraid to talk?" The patient indicated "yes" by nodding her head. The therapist then suggested to her that they start off with some very rudimentary sounds and try to piece together a sequence of sounds that would produce the word "mom." Since Christmas was coming up, the therapist suggested that this would be an excellent Christmas gift for the patient to give her mother. The patient agreed, and after many painstaking sessions, the phonetic combination of M-O-M was slowly put together. Once the patient was able to say this word to her mother as a Christmas gift, a breakthrough occurred in her therapy. She subsequently showed substantial improvement in language function and was very motivated in her speech therapy hours.*

A fourth example comes from the field of psychotherapy and the neuro-psychological examination of brain-injured patients. This patient was a 21-year-old male who had fallen from a bridge while smoking marijuana approximately 1 year prior to entry into a neuropsychological rehabilitation program. The patient was described in some detail in a chapter on cognitive retraining by Prigatano *et al.* (1986). His CT scan showed bilateral atrophy of the frontal lobes. This patient did not recognize that he had any change in his ability to show facial affect or affect in his tone of voice. In fact, he often spoke in a very monotonous voice and showed very little facial (emotional) expression. Many people felt, therefore, that he either did not understand what they were saying to him or was unable or unwilling to show interest in the environment.

When the patient was made aware of how he came across to others, his self-perception was altered. He became extremely motivated to work on a variety of retraining tasks that ultimately aided his ability to show more facial expression and affect. This was paralleled by an increase in his speed of information processing. As reported elsewhere (Prigatano *et al.*, 1986), this did not result in returning to normal levels of speed of information processing or to normal levels of flexibility in demonstrating affective reactions. Yet improvement was achieved to such a point that the individual was able to adapt better than he had previously.

These clinical examples highlight the point that when old feelings are experienced after brain injury, the individual is able to self-perceive in more realistic terms. This can (but will not always) lead to powerfully reinforcing experiences for the patient, which can facilitate recovery because the patient will practice the rehabilitative tasks in a more motivated manner. It is important to recognize that feelings, emotion, and motivation all contribute to self-awareness, and self-awareness, in turn, contributes to enhanced motivation. This is simply a feedback system. Clearly, there is nothing automatically therapeutic about motivation or the patient becoming more aware. In fact, in some instances, as patients become more aware, they become more depressed. For a period of time they may

*The author wishes to thank Karen Tripp, Speech and Language Therapist, for sharing examples 2 and 3 with him and allowing them to be used in this chapter.

not be able to work as aggressively in rehabilitation. This does not mean, however, that one should not attempt to help the patient become more self aware. Ultimately, self-awareness is necessary for maximum adaptation. There is also nothing magical about motivation. Certainly, there are limits as to how much an individual can recover, but the problem of motivation has to be seriously considered in neurological rehabilitation.

Before considering the problem of motivation in neurological rehabilitation, the question of whether or not changes in motivation and self-awareness are the cause or the effect of changed brain functions should be considered. Stated in other terms, is this phenomenon really an epiphenomenon? I think not. First, these changes in self-awareness can occur several months or even years after the period of so-called spontaneous recovery. Second, as one of the examples illustrated, awareness of old "feeling states" or one's "old voice" is not accompanied by immediate improvement. It only provides the basis for motivation to improve. Third, in a large number of patients, this enhanced self-awareness may not occur spontaneously. As the study by Oddy et al. (1985) suggested, a large portion of young adult traumatic brain-injured patients may remain unaware of the severity of residual deficits at least as long as 7 years posttrauma, according to relatives' reports.

It appears that when the brain is injured, it may not, like other bodily organs, "tell" the person (i.e., the brain) that something is "wrong." In fact, the opposite may often be true. In the practice of clinical neuropsychology, even subtle disorders of brain function may produce disturbances in self-consciousness, and these disturbances are not necessarily registered by intact brain structures (see Bisiach et al., 1986). Although this phenomenon may be, in part, a disturbance in the abstract attitude (as suggested by Goldstein as early as 1942), it appears to be directly related to alterations in self-perception and possibly attention or arousal level.

4. THE PROBLEM OF MOTIVATION IN NEUROLOGICAL REHABILITATION AND THE LIMITS OF THE DAMAGED NEUROLOGICAL SYSTEM

Earlier researchers of recovery of function mentioned the importance of motivation. In his work on hemiplegia and motor functioning, Denny-Brown (1950) noted that under intense excitement or emergency situations, patients might use a hemiparetic limb when they could not do so otherwise. He emphasized that with intense motivation, control of motor function was, perhaps, greater. Lashley (1938) introduced a patient who failed to learn the alphabet after several trials. Lashley bet the patient 100 cigarettes that he could not learn it within a week. With the appropriate reinforcement, the patient accomplished the task within 1 to 2 days. As an experimentally oriented psychologist, Lashley emphasized that we have yet to understand what motivates certain brain-injured people. If we could, functional outcomes might improve. In his review of the literature on the effectiveness of speech and language therapy, Darley (1972) discussed the importance of motivation specifically for aphasic patients. Also, as noted in Section 1, Gazzaniga (1974) argued for the effects of motivation on animal learning after brain lesions. He demonstrated that with appropriate reinforcement contingencies, brain-injured rats would perform a variety of tasks that they normally would not do.

Not all studies on motivation and recovery of function, however, have been positive. For example, Schwartz (1969) produced experimental lesions in monkeys and assessed their abilities to learn a complex motor act under different levels of reinforcement (mild versus strong). His results did not indicate that level of reinforcement affected motor recovery. However, it should be noted that the motor tasks required of these monkeys may have been outside the "biological constraints" imposed by the lesion. A careful reading of this paper reveals that monkeys became quite excited under conditions of strong reinforcement. Apparently, they attempted to perform the motor act but lacked the basic motor skills.

These results suggest an important issue. Motivation does seem to influence the performance of actions and the acquisition of new information if the actions and information fall within the limits of the system involved. Beyond those limits, only hyperactive, agitated subjects may be produced under conditions of very strong reinforcement. Such data emphasize that one of the primary goals of researchers and clinicians is to balance the patient's actual capacity with the salient, motivating stimuli. This is not an easy therapeutic task and may well require a period of "give-and-take" to define the limits. Although motivation produces nothing magical, it does establish the necessary internal milieu (i.e., arousal) to facilitate or augment new learning in an individual.

5. THE RELATIVE IMPORTANCE OF FRONTAL LOBE INJURY VERSUS TEMPORAL LOBE INJURY FOR RECOVERY OF EMOTIONAL AND MOTIVATIONAL DEFICITS

As noted earlier, emotional and motivational disturbances can emerge after injury to almost any part of the brain (see Prigatano et al., 1986). However, the patient may not cognitively "grasp" a situation properly and may therefore show an inappropriate emotional–motivational response when the frontal lobes are damaged. Goldstein (1958) described this clinical phenomenon in some detail years ago.

In my opinion, teaching some "frontal" patients that they do not "see the big picture" can be achieved. The therapist has to walk patiently through various facts in a slow, systematic way. When these frontal lobe patients are able to "see the big picture" or develop an abstract idea with the help of a therapist, they frequently become cooperative and more "motivated." Awareness of the situation is improved, and consequently the patient appears eager to engage in rehabilitation and various associated life events.

When the temporal lobe is badly damaged, however, the clinical picture is often different, and the outcome is frequently much worse. Disorders of the temporal lobe can lead to faulty information processing, and cognitive restructuring by the therapist often is not achieved. The patient may literally misperceive visual and auditory information in such a way that it fosters suspiciousness or paranoid thinking. For example, in looking at a therapist's face, a patient with a badly contused temporal lobe may not know if the therapist is smiling or sneering. As they perceive only fragmented information and have associated memory difficulties, they can become quite confused in interpreting the events surrounding them. Trying to work with these patients by improving an external logic does not overcome their basic information-processing deficits. Good verbal skills may be useless in helping them cope because of their basic underlying perceptual and memory disorders.

> A young woman, S.D., provides an example of this. She was able to communicate reasonably well but persisted in believing that others were mistreating her. This resulted in her becoming emotionally out of control and extremely paranoid at times. It was assumed that deep brain structures involving the mesial portion of the temporal lobe were damaged, especially the amygdala and hippocampus. This patient was unable to control her emotional outbursts. Her poor memory, coupled with perceptual disorders, made it impossible for S.D. to recognize what was occurring in an objective way.

Information-processing deficits associated with damage to the inferotemporal cortex and the convexity of the temporal lobe have been implicated in producing not only visual discrimination difficulties but certain psychiatric problems. The literature on the presence of psychosis after traumatic brain injury frequently implies involvement of the temporal lobe and portions of the hypothalamus (see Davison and Bagley, 1969). Although there is growing evidence that the left temporal lobe especially may be at high risk for psychiatric disturbance (Lishman, 1968), clinicians working in the field have reported similar problems with both right and left temporal lobe damage (Leftoff, 1985; Prigatano, 1987). In my clinical experience, patients with significant temporal lobe disorders are harder to treat than patients with frontal lobe problems.

6. A NOTE ABOUT AWARENESS AND ITS IMPORTANCE IN PSYCHIATRIC AND NEUROLOGICALLY ORIENTED THERAPIES

The process of being aware of one's self requires certain functional states and some internal "push" to attend or "desire" to perceive or act. At the turn of the century, Freud suggested that feeling states of the organism also can serve the converse role. That is, emotional and motivational forces can block self-awareness rather than facilitate it; he suggested that in some cases this can lead to pathological behavior (Freud, 1924).

In neurology, psychiatry, and psychology, the debate continues on the relative importance of understanding the phenomena of consciousness, self-consciousness (as Pribram distinguishes them), and the unconscious (as described by Freud and Jung). Yet in daily clinical practice, working with young adult brain-injured patients, I am impressed with how frequently disturbances of self-awareness occur and how important they are for recovery of function and adaptation to permanent behavioral changes. It is somewhat surprising that these apparent disturbances in self-awareness have not been adequately discussed in the literature on recovery and rehabilitation following a brain injury. This may reflect how we have conceptualized the problems and our medical tradition of diagnosing brain disturbances without attempting to rehabilitate such patients over time. That is, when we attempt to help a patient improve higher cerebral functioning and "live with them" during their course of rehabilitation and struggle over what they are able to recognize about their problems, the importance of the disturbance of self-awareness emerges. These problems may not emerge in a brief neurological examination or while the patient is medicated and in a hospital setting.

It is, of course, fairly obvious that one reason that scientists have not addressed this variable in experimental studies is that altered self-awareness is extremely difficult to conceptualize and measure. This variable is not like giving a patient or animal "a drug" or some experimental treatment so that one can study group effects in manipulating the variable. By its very nature, this phenomenon does not allow itself to be easily investigated in parametric studies.

The early work of Franz and the more recent studies of LaVere should be considered here. As noted by Finger and Stein (1982), Franz, as early as 1923, emphasized that "a will to get well must be created if it does not exist" in the neurological patient. Without using the words "emotion" and "motivation" or "self-awareness," Franz emphasized that the discouraged patient who was not hopeful would not take his rehabilitation seriously, and this, undoubtedly, would result in fewer gains.

The recent experimental work of LeVere and his colleagues (e.g., LeVere and Davis, 1977; LeVere et al., 1979; LeVere, 1984) demonstrated that relearning a brightness discrimination in the rat following primary visual neocortex lesions is related to the reinforcement history of the animal. In a series of provocative and interesting papers, LeVere and Davis (1977) showed that if the "preoperative and postoperative motivational states are not the same, postoperative recovery of a brightness discrimination is poor. If they are the same, postoperative recovery can be much better."

Summarizing this work and considering its theoretical implications, LeVere

(1984) suggested that motivation is crucial for recovery following brain lesions. He argued that what usually is ''lost'' is not the engram nor the memory to do something. Rather, retained memories are not easily accessed if the motivational state during the recovery process differs from the motivational state when the initial learning occurred. What is done in rehabilitation should take into consideration how the individual learned a given skill in the first place.

7. SUMMARY AND CONCLUSIONS

In summary, emotional and motivational disturbances are common after traumatic brain injury and may be crucial determinants for functional recovery. These dimensions influence how an individual will engage in rehabilitation programs. They also impact on how the patient adapts to higher cerebral deficits. Emotional and motivational problems have not figured prominently in any major theory of recovery of function. This appears to be in part because these two concepts have not been adequately defined and are not easily investigated in highly controlled ways. Both include an arousal component that can have reinforcing qualities. Future clinical work, as well as basic neuroscience research, should explore how these two dimensions can enhance recovery of function and be tied into a general theory of brain organization. In this regard, Stuss and

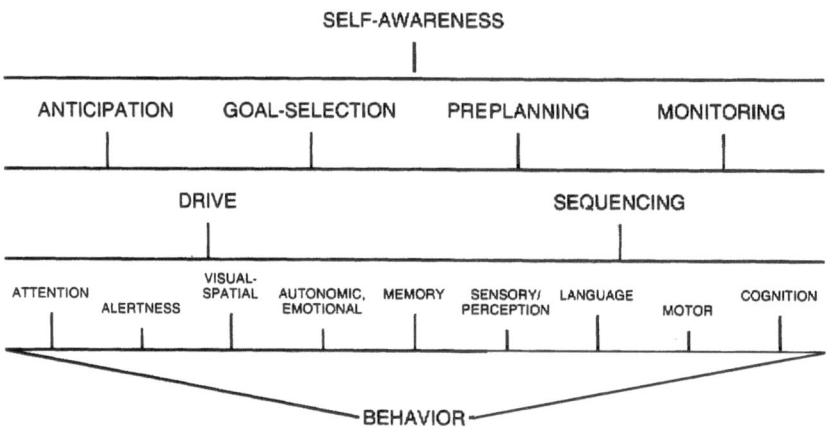

Figure 2. This figure illustrates the proposed model of brain functioning. External behavior is dependent on various organized integrated fixed functional systems such as attention, alertness, and so on, hypothesized to be based, in relation to the frontal lobes, in more posterior/basal brain regions. The next three levels are all more intimately associated with the frontal lobes, with self-awareness (self-consciousness, self-reflectiveness) hypothesized to be the highest attribute of the frontal lobes. (From Stuss and Benson, 1986, with permission.)

Benson's (1986) recent book on the frontal lobes emphasizes how important drive and arousal are for the emergence of self-awareness, which they view as perhaps the highest of all cerebral functions given their theoretical model (see Fig. 2).

From a clinician's point of view, it certainly appears that the self-awareness dimension is easily affected after brain injury and very much guides what is accomplished in the rehabilitative setting. Thus, from both theoretical and practical clinical efforts, there seems to be an emergence of interest in considering emotional and motivational variables and how they interact with consciousness and self-awareness to facilitate rate and possibly level of recovery following brain injury.

REFERENCES

Bisiach, E., Vallar, G., Perani, D., Papagno, C., and Berti, A., 1986, Unawareness of disease following lesions of the right hemisphere: Anosognosia for hemiplegia and anosognosia for hemianopia, *Neuropsychologia* **24**(4):471–482.

Critchley, M., 1953, *The Parietal Lobes*, Hafner Press, New York.

Darley, F. L., 1972, The efficacy of language rehabilitation in aphasia, *J. Speech Hear. Disord.* **30**:3–22.

Davison, K., and Bagley, C. R., 1969, Schizophrenia-like psychosis associated with organic disorders of the central nervous system: A review of the literature, *Br. J. Psychiatry* **4**:113–184.

Denny-Brown, D., 1950, Disintegration of motor function resulting from cerebral lesion, *J. Nerv. Ment. Dis.* **112**:1–45.

Finger, S., ed., 1978, *Recovery from Brain Damage*, Plenum Press, New York.

Finger, S., and Stein, D. G., eds., 1982, *Brain Damage and Recovery*, Academic Press, New York.

Freud, S., 1924, *A General Introduction to Psychoanalysis*, Pocket Books, New York.

Gazzaniga, M. S., 1974, Determinants of cerebral recovery, in: *Plasticity and Recovery of Function in the Central Nervous System* (D. G. Stein, J. J. Rosen, and N. Butters, eds.), Academic Press, New York, pp. 203–215.

Goldstein, K., 1942, *Aftereffects of Brain Injury in War*, Grune & Stratton, New York.

Goldstein, K., 1958, The effect of brain damage on the personality, *Psychiatry* **15**:245–260.

Izard, C. E., 1980, The emergence of emotions and the development of consciousness in infancy, in: *The Psychobiology of Consciousness* (J. Davidson and R. Davidson, eds.), Plenum Press, New York, pp. 193–214.

Klove, H., 1979, The hyperkinetic syndrome: Criteria for diagnosis, in: *Hyperactivity in Children* (R. L. Trites, ed.), University Park Press, Baltimore, pp. 1–26.

Kotila, M., Waltimo, O. L., Marja, L. N., Laaksonen, R., and Lempinen, M., 1984, The profile of recovery from stroke and factors influencing outcome, *Stroke* **15**:1039–1044.

Lane, R. D., and Schwartz, G. E., 1987, Levels of emotional awareness: A cognitive–developmental theory and its application to psychopathology, *Am. J. Psychiatry* **144**:138–143.

Lashley, K. D., 1938, Factors limiting recovery after central nervous lesions, *J. Nerv. Ment. Dis.* **88**:733–755.

Leftoff, S., 1983, Psychopathology in light of brain injury: A case study, *J. Clin. Neuropsychol.* **5**:51–63.

LeVere, T. E., 1984, Recoveries of function after brain damage: Variables influencing retrieval of latent memories, *Physiol. Psychol.* **12**:73–80.

LeVere, T. E., and Davis, N., 1977, Recovery of function after brain damage: The motivational specificity of spared neural traces, *Exp. Neurol.* **57**:883–899.

LeVere, T. E., Davis, N., and Gonder, L., 1979, Recovery of function after brain damage: Toward understanding the deficit, *Physiol. Psychol.* **7**:317–326.

Lindsley, D. B., 1970, The role of nonspecific reticulothalamocortical systems in emotion, in: *Physiological Correlates of Emotion* (P. Black, ed.), Academic Press, New York.

Lishman, W. A., 1968, Brain damage in relation to psychiatric disability after head injury, *Br. J. Psychiatry* **114**:373–410.

Luria, A., 1973, *The Working Brain*, Penguin Press, London.

Malmo, R. B., 1959, Activation: A neuropsychological dimension, *Psychol. Rev.* **66**:367–386.

Miller, G. A., Galanter, E., and Pribram, K. H., 1960, *Plans and the Structure of Behavior*, Holt, Rinehart, and Winston, New York.

Morrow, L., Vrtunski, K., Kim, Y., and Boller, F., 1981, Arousal responses to emotional stimuli and laterality of lesion, *Neuropsychologia* **19**:65–71.

Moruzzi, G., and Magoun, H. W., 1949, Brainstem reticular formation and activation of the EEG, *Clin. Neurophysiol.* **1**:455–473.

Oddy, M., Coughlan, T., and Tyerman, A., 1985, Social adjustment after closed head injury: A further follow-up seven years after injury, *J. Neurol. Neurosurg. Psychiatry* **48**:564–568.

Pribram, K. H., 1971, *Languages of the Brain: Experimental Paradoxes and Principles in Neuropsychology*, Prentice Hall, Englewood Cliffs, NJ, pp. 47–61.

Pribram, K. H., 1980, Mind, brain, and consciousness: The organization of competence and control, in: *The Psychobiology of Consciousness* (J. Davidson and R. Davidson, eds.), Plenum Press, New York, pp. 47–61.

Pribram, K. H., and McGuinness, D., 1975, Arousal, activation, and effect in the control of attention, *Psychol. Rev.* **82**:116–149.

Prigatano, G. P., 1986a, Higher cerebral deficits: History of methods of assessment and approaches to rehabilitation: Part II, *BNI Q.* **2**(4):9–17.

Prigatano, G. P., 1986b, Higher cerebral deficits: The history of methods of assessment and approaches to rehabilitation: Part I, *BNI Q.* **2**(3):15–26.

Prigatano, G. P., 1987, Psychiatric aspects of head injury: Problem areas and suggested guidelines for research, *BNI Q.* **3**:2–9.

Prigatano, G. P., and Others, 1986, *Neuropsychological Rehabilitation after Brain Injury*, Johns Hopkins University Press, Baltimore.

Schwartz, A. S., 1969, Recovery from motor deficit under different motivational conditions, *Physiol. Behav.* **4**:57–60.

Simon, H. A., 1967, Motivation and emotional controls of cognition, *Psychol. Rev.* **74**:29–39.

Stein, D. G., Rosen, J. J., Butters, N., eds., 1974, *Plasticity and Recovery of Function in the Central Nervous System*, Academic Press, New York.

Stuss, D. T., and Benson, D. F., 1986, *The Frontal Lobes*, Raven Press, New York.

Wadsworth, B. J., 1984, *Piaget's Theory of Cognitive and Effective Development*, Longmans, New York.

Weinstein, E., and Kahn, R. L., 1955, *Denial of Illness: Symbolical and Physiological Aspects*, Charles C. Thomas, Springfield, IL.

22

Recovery of Function
Sources of Controversy

STANLEY FINGER, T. E. LeVERE,
C. ROBERT ALMLI, and DONALD G. STEIN

1. INTRODUCTION

Individuals engaged in the study of recovery from brain damage and those looking at the literature in this area from a distance would agree that at present the subject is controversial. In fact, historically minded individuals would point out that this has always been the case. For example, Flourens (1842) believed that recovery could occur, but when one function returned all functions returned. In Broca's 1865 paper, he argued that one hemisphere could function for the other under certain pathological conditions, and Munk (1877) generated disagreement when he proposed that neighboring zones may take over for injured cortical areas through a process that might best be described as "reeducation."

An analysis of hypotheses like these from the 19th century, as well as proposals from more contemporary writers, reveals that most theories and controversies about recovery raise two basic, related questions, one behavioral and the other biological. The fundamental behavioral question is whether recovery occurs at all. If one is willing to grant that it does, then closely allied behavioral questions that might follow are "Why is recovery not always seen?" and "What are the conditions that may maximize or minimize the chances of promoting

STANLEY FINGER • Department of Psychology and Neural Sciences Program, Washington University, St. Louis, Missouri 63130.   T. E. LeVERE • Behavioral Neuropsychology Laboratory, Department of Psychology, North Carolina State University, Raleigh, North Carolina 27695.   C. ROBERT ALMLI • Programs in Occupational Therapy and Neural Sciences and Departments of Anatomy and Neurobiology, Preventive Medicine, and Psychology, Washington University School of Medicine, St. Louis, Missouri 63110.   DONALD G. STEIN • Dean of the Graduate School and Associate Provost for Research, Rutgers University, Newark, New Jersey 07102.

recovery?'' Many experiments concerned with age at the time of brain damage, environmental interaction effects, and lesion momentum differences have been conducted in this context (Finger and Stein, 1982). Additional questions are being raised about whether the behavioral recovery process proceeds in discrete steps or stages that emulate those seen in normal development and whether seemingly recovered subjects respond to challenges such as stressors, drugs, and additional lesions in ways that are comparable to those of neurally intact organisms.

The fundamental biological question asks about the brain changes that might account for recovery. Experimenters first asked, ''Which part of the brain is responsible for the recovery process?'' Answers usually centered on the role of tissue near the lesion site or of homologous areas on the opposite side of the brain (see Slavin, Laurence, and Stein, Chapter 11, this volume). The position that more primitive structures, which guide behavior earlier in development, may also contribute to recovery of function never seemed to have as strong a following. However, the latter position is consistent with the thinking of John Hughlings Jackson (see Greenblatt, Chapter 12, this volume).

The biological approach has become much more molecular in the past few decades. This is because earlier scientists did not always have the means, materials, or philosophical frame of reference needed to see and evaluate some of the events that we now know take place after brain lesions. Indeed, one can say that in initial studies of recovery, the whole issue of neuronal adaptation was skillfully avoided. This was followed by a period of tautological statements and poorly defined terms, such as neuronal ''reorganization of function'' and ''plasticity.'' Here, description and explanation, and observation and theory, were not always distinguished. At the present time, advances in neuroanatomy allow scientists to refer specifically to events such as collateral sprouting, axons failing to retract on schedule, and fibers going to unusual loci after injuries. Nevertheless, even with this new knowledge, there still is controversy. Two of the most important questions that now are being asked are whether the observed biological events are causally related to recovery or only correlated with it, and whether one can even isolate the critical events from all of the other processes that may be occurring at the same time (Finger and Almli, 1985).

It is in this context of perplexing and challenging issues that it seems useful to ask why CNS recovery of function has always been a controversial subject and whether the controversies are likely to continue. We recognize that there are a number of probable causes for disagreements in the field, and that some are more fundamental then others. Among the most important considerations are (1) how little we know about the ''functional organization'' of the healthy brain, not to mention the damaged brain, (2) the observation that there often is great variability in response to brain damage, (3) the difficulties involved in showing that specific neuronal changes are causally related to recovery phenomena, and (4) the fact that we really cannot ''prove'' that recovery has occurred. A related fifth issue is the general lack of agreement on how the term "recovery" should be defined. This important topic, however, has already been addressed in detail in this book (Almli and Finger, Chapter 1).

2. ASSESSING THE FUNCTIONAL ORGANIZATION OF THE BRAIN

One primary problem stems from the lack of agreement about the functions of the different "parts" of the brain and how the various components work together. Here it can be argued that one cannot adequately discuss recovery of function in the damaged brain if the functional organization of the healthy brain is not well understood (Finger and Stein, 1982).

Much of the difficulty lies in the inadequacy of the methods used to assess the behavioral functions of different brain areas, especially those areas that appear to be neither sensory nor motor in the classical sense. Although clinical findings from brain damage cases traditionally have been used to define the functions of many brain areas, these observations can be interpreted in widely divergent ways. Lesion studies allow the investigator only to ask what the rest of the brain can do in the absence of one or more of its parts. What these areas might be doing in the healthy brain is a matter of supposition.

Because there are many secondary effects of brain lesions, such as primary and secondary degeneration, edema, or changes in vascular patterns (see Section 4), it is a certainty that a lesion anywhere in the brain will also have effects on other, "distant," areas. From this it follows that the presence of a "lesion effect" cannot be regarded as proof that the area damaged was directly responsible for the behavioral deficit. Although deficits could reflect the loss of an area, they could also reflect changes remote from the lesion site or an interaction between the changes in the damaged area and those in secondary loci.

Further, the absence of a "lesion effect" does not constitute proof that the damaged area had nothing to do with the function in question in the intact brain. If the damaged system were diffuse or redundant, or if the test permitted multiple solutions to a problem, the lesion effect could be negligible and fail to provide an adequate portrayal of the possible functional contribution of the now-damaged neural area.

A perception of static, discrete, well-defined functions housed in limited parcels of tissue may be just one way of synthesizing and treating data beset with ambiguity. Even when divergent operations (anatomy, electrophysiology, metabolic scans, lesions) are used to define behavioral "functions," the results will still be theoretical and lack the precision that one strives for in the sciences. This is because none of the methods may be capable of directly and unambiguously answering the question being asked.

Although textbooks of anatomy, physiology, neuroscience, and psychobiology may make it sound as though the different parts of the brain can be considered as discrete functional entities, one must question whether any part of the brain functions on its own, insulated from the activities of the other parts of the overall system. Much as the structural analysis of a transistor in isolation may shed little light on its precise function in a complex piece of electrical equipment, it may be that only when we are eventually capable of observing dynamic interactions between brain areas that we may begin to understand the complex and perhaps subtle roles played by particular brain parts, if, in fact, the concept of separable components remains viable.

Thus, some of the confusion regarding recovery may not diminish until it is recognized that the arbitrary separation of the brain into anatomically discrete functional areas capable only of sequential processing merely represents a model of brain function rather than an established fact. The model may be parsimonious, but it may be misrepresenting reality. As noted by Kurt Goldstein (Frommer and Smith, Chapter 5, this volume), although the brain is obviously not equipotential, attempts to localize psychological functions into smaller and smaller parcels of tissue may be exaggerating the error of conferring on these areas certain static properties and artificial limits that may deviate markedly from the way in which they function in everyday life. With so little known about structure–function relationships in the CNS, and the growing belief among some scientists that the brain may be even more dynamic, interactive, and adaptive than previously believed, it should be no surprise that research on recovery of function continues to stimulate debate.

3. VARIABILITY AND THE CONCEPT OF "NORMATIVE" PERFORMANCE

Extensive variability among individuals typically characterizes post-traumatic performance. This variability is not always well understood, raising questions about scientific rigor and reliability that serve to fuel additional controversy about recovery phenomena. In this context it is often surprising to learn that many people with no noticeable signs of disease, or perhaps only minor symptoms, exhibit considerable neuropathology on autopsy. Further, almost every investigator who has worked with laboratory animals has had some animals with lesions performing more like control subjects than other members of the lesion group. Although lesion characteristics might be expected to account for performance differences, relationships between lesion features and behavior are not always apparent. Sometimes the outlying cases are treated as if they were curious anomalies. But more often they are simply lost or obscured when the data are analyzed with ''group'' statistics. Rarely are they resurrected for further analysis or discussion.

The fact that one may see sparing of function, improvement, or no recovery in different cases after damage to the same neural area requires that one attend not only to the characteristics of the lesion but also to factors that could differentially affect the condition of the remaining brain (Finger and Stein, 1982). In this context, some variables that might account for individual differences have been identified, although this is rarely discussed in most textbooks of neurology and neuroscience.

For example, slow-growing lesions may not result in the same impairments as lesions of rapid onset, a phenomenon referred to as the "law of momentum of lesions" in the human clinical literature and as the "serial lesion effect" in laboratory studies involving staged or sequential lesions (Finger and Stein, 1982). As noted by Riese (1948, p. 75):

> Sudden lesions, i.e., those of greatest momentum, are most likely to produce aphasia, although only transitory in uncomplicated cases. In lesions of slow momentum, speech may be preserved either throughout the whole history or, at least, for a long time.

Different environments (e.g., complex, impoverished) either before or after trauma can also affect some measures of behavior (Will and Eclancher, 1984), as can early nutritional history (Mangold et al., 1981). It is now known that even hormonal status can affect the response to a brain injury (Nance, 1984). For example, in one recent study involving frontal association cortex lesions in rats (Attella et al., 1987), estrogen and progesterone levels were associated with differences in performance on a spatial learning task. Factors such as ovarian hormone levels typically are not considered when planning for surgical interventions in patients with brain disease or when drawing conclusions about structure–function relationships from lesion data. Yet findings such as these show that the inferences generated by lesion material depend on more than just the site of the damage itself.

The response to brain damage may also change over time. In fact, we now know that behavioral improvement after left hemispheric injury can continue over many years, although, typically, clinical evaluation is conducted within weeks or months after trauma (see Geschwind, 1985, for a discussion of this issue). The importance of the time factor is also brought out by Goldman (1974), who noted that lesions of frontal association cortex in neonatal monkeys led to sparing of function on spatial learning tasks when the animals were tested early in life. As the young monkeys reached maturity, however, they "grew into" spatial learning deficits. Comparable findings were noted after motor cortex damage by Kennard in the 1930s (see Finger and Almli, Chapter 8, this volume). It is also interesting to note that Stein and Firl (1976) observed that "old" rats with frontal cortex lesions were no worse than their sham-operated counterparts in spatial performance, whereas "young" animals with these lesions showed severe deficits. In this case, it appeared as if the frontal cortex forfeited its ability to mediate spatial performance as its cells diminished in old age.

These findings show that the outcome of a brain injury can be influenced by the duration of the recovery period and by the developmental status of the subject. One implication of the data is that the sequence of events that follows a brain injury should be viewed within a broadly defined developmental context. Yet, as we have seen, recovery time, age at the time of insult, and maturational status at the time of testing are only some of the variables that can affect performance after brain lesions. Speed of lesion growth, nutritional and environmental factors, genetics, and perhaps even the time of day when an injury was sustained may also account for the fact that so-called "textbook" responses to specific brain injuries are not common on an individual level.

Although laboratory scientists have made significant progress in understanding some of the factors that can affect the response to a brain injury and how they might be differentially affecting the condition of the remaining brain, we still know very little about how these factors should be mathematically weighted and how they might interact with each other. As a result, especially when dealing with material as complex as human head injury cases, the predictive value of this new knowledge may not always seem very impressive. Moreover, our new appreciation of variables such as environment and nutrition comes primarily from tests on relatively extreme cases, and one can only wonder how smaller variations in some of these factors (e.g., "subclinical undernutrition") might affect brain and behavior after insult.

Until we achieve a better understanding of the nature and potential significance of factors such as these and how they may interact with each other, cases of sparing and recovery, especially in the clinical literature, may continue to be treated as curious anomalies or unwanted sources of statistical noise. The contention that data showing recovery are unreliable and frought with so much ambiguity that one can draw almost any conclusion from them must be met with improved appreciation of the conditions that can account for the variablity in these studies. Efforts in this direction are just beginning.

4. MULTIPLE BRAIN CHANGES AND CAUSALITY

Damage to the brain does not merely produce a circumscribed, static "hole," even in the case of a presumptive "focal lesion" (Isaacson, 1965; Raichle, 1982; Schoenfeld and Hamilton, 1977). Brain damage results in the death of nerve, glial, and ependymal cells at the site of injury and may also be associated with transsynaptic and transneural degeneration of distant nerve cells (retrograde and anterograde). Neuronal death can then result in disruption of neural activity in injured (but not dying) and noninjured nerve cells adjacent to, and far removed from, the initial site of damage. This disruption of neural activity could be related to "irritative reactions" at the edge of the lesion, deafferentation, and altered neurotransmitter pools and synaptic drive.

Glial reactions also occur in response to neural injury, including astrocytic responses at the damage site and beyond, as well as phagocytosis and invasion by microglia. Disruption of blood vessels at the site of damage and in nearby regions can influence subsequent neuronal viability and affect neural activity, as can the subsequent proliferation of blood vessels into the perilesion area.

Brain damage is also typically associated with regional or widespread edema, bleeding, and changes in cerebrospinal fluid composition and pressure. These changes are associated with compression and pressure on neural tissue and blood vessels and can further alter the extracellular ionic environment.

Synaptic changes such as supersensitivity or hyposensitivity can occur and may exert permanent or transient effects on activity in circumscribed or divergent pathways. Axonal sprouting and the formation of "aberrant" neuronal pathways, as well as widespread and general changes in brain size and composition, have also been shown to occur in conjunction with brain damage.

Thus, a wide variety of degenerative and regenerative neural changes, as well as other biochemical effects, can occur in response to brain damage. Nevertheless, a multivariate approach to the events following brain damage has been virtually nonexistent in the recovery literature. Most studies of changes after brain damage have focused on only one variable at a time, and, as a result, the conclusions drawn in these studies about "cause-and-effect relationships" between a specific brain change and behavior must be questioned.

For example, individual studies demonstrating axonal sprouting after brain damage are obviously important (Cotman and Nadler, 1978). Yet, the assumption that sprouting underlies behavioral recovery must be considered premature. Even when functional axonal sprouting can be correlated with recovery, it should be recognized that sprouting is only one of a multitude of regenerative and degenerative neural phenomena that may share a relatively common time course following neural injury.

Because injury-induced primary and secondary brain changes are widespread and complex, they are difficult to deal with in contemporary research designs. However, a better understanding of brain damage and recovery of function is not going to be facilitated by ignoring concurrent changes or merely classifying them under a general umbrella, such as "shock" effects. Ambiguity and confusion will be present until the relative contributions of these transient and permanent changes can be better understood, not only in relation to behavioral outcome but in relation to each other. The task is easier to state than to accomplish, but until a better appreciation of these multiple events can be achieved, we should not be impressed that some of these brain changes appear to be temporally correlated with recovery. The surprise would be if they were not.

5. RECOVERY AND THE NULL HYPOTHESIS

Another problem in the recovery field stems from the fact that "recovery," from a scientific orientation, means proving that there are no deficits. By definition, an individual has recovered from the consequences of neural injury if, and when, his or her behavior does not differ from his or her behavior prior to the injury or from the behavior of a comparable individual who has not suffered brain injury (Almli and Finger, Chapter 1, this volume). To establish this requires accepting the hypothesis of no differences.

The argument stems from the formal logic of experimental design, where the traditional procedure is to establish and then evaluate two mutually exclusive hypotheses. These hypotheses are called the "null hypothesis," which states that there are no differences between the groups, and the alternate hypothesis, which holds that there are differences that are unlikely to be accounted for by chance alone. The problem is that one cannot prove that the null hypothesis is true, i.e., that there are no deficits.

This relates to the issues of when to stop testing and "insensitive tests." Regarding the former, it can always be claimed that with additional observations a deficit might appear. As for the latter, no matter how sensitive an investigator may feel a test is, one could always argue that with a more sensitive instrument, group differences would emerge. Although some experiments on recovery allow one to argue that the tests are obviously sensitive enough to reveal deficits in the group not given the special treatment (e.g., one-stage versus staged lesions, vehicle alone versus drug), this counterargument still falls short when comparisons are made between the lesion-treatment group and healthy, unoperated control groups. No matter how closely the control and lesion groups resemble each other, the argument that the groups are using different strategies or cues or are showing cognitive changes that are too subtle to be picked up with present testing procedures cannot be refuted.

In some cases, the issue of insensitive tests is clearly worth emphasizing. Perhaps the best example of this deals with commissural sectioning, an historically accepted treatment for severe cases of epilepsy (van Wagenen and Herren, 1940). This operation was performed because it had beneficial medical consequences and seemingly no effects on the patient's general behavior. In 15 cases of partial and nine cases of complete division of the corpus callosum, Akelaitis and his colleagues were unable to detect any behavioral effects that could be attributed to the commissurotomy (Akelaitis, 1940, 1941a–c, 1942a,b; Akelaitis *et al.*, 1941, 1942, 1943).

This view of no behavioral difference between individuals with and without a corpus callosum persisted for some 10 years until Myers reported the results of his animal studies (Myers, 1955, 1956, 1959, 1961, 1965). These experiments demonstrated profound behavioral effects of sectioning the corpus callosum if information were presented to one hemisphere and a test required the use of the opposite hemisphere. Sperry and his group (Sperry, 1961, 1964; Gazzaniga *et al.*, 1962) later extended these results to humans who had undergone commissurotomy. In fact, Goldstein and Joynt (1969) demonstrated identical findings in one of Akelaitis' original patients. Clearly, accepting the hypothesis of no differences, the null hypothesis, was incorrect and was far from the mark of establishing normal functioning.

The message provided by this example is not that recovery-of-function research should be abandoned. Rather, it is that the notion of no deficits may only be an ideal, in part because we can never prove that the organism has fully recovered.

6. CONCLUSIONS

In this chapter we have examined four sources of controversy that seem unlikely to disappear completely soon. However, if we can (1) learn more about structure–function relationships in the healthy brain, (2) understand more about the many interactive factors that might account for behavioral variability after specific brain lesions, (3) recognize that a myriad of concurrent changes will inevitably take place after CNS injury, and (4) realize that even our best experiments may fall short of actually "proving" hypotheses about recovery, at least some of this controversy might be expected to diminish.

The more pessimistic among us might suggest that these issues and controversies are so serious at this time that it is impossible to do good research on recovery of function. Yet, in the scientific arena, controversy need not have negative connotations, and debate and disagreement may correlate more highly with advancement and new ideas than would uncritical acceptance of one view. For example, for hundreds of years there was a consensus among physicians that Galen's conclusions about the roles of the cerebral ventricles and the rete mirable could not be challenged (Finger and Stein, 1982). It is generally accepted that this resulted in a virtual standstill in anatomic and medical knowledge about the brain.

This is not to say that establishing agreed-on facts and striving for rules and laws are not the business of science. It most certainly is. But it is also true that queries such as "Is that really true?" "If it is true, how can this be?" and "Isn't there an alternative explanation?" are never out of order in science. Thus, controversy can be viewed as grist for the scientific mill that serves to stimulate new research and theories and, in the long run, should increase knowledge and understanding. On an individual level, healthy controversy forces us to consider our speculations more carefully, to evaluate our conclusions more critically, and, perhaps most importantly, to temper our impulses to reject too rapidly those positions, models, and hypotheses that differ from our own.

In this context, it is important to emphasize that the field of controlled recovery research is relatively new and that much remains to be discovered. For example, 20 years ago, who would have thought that the injured brain could produce substances that might even promote its own healing? Today, the question is not whether such substances exist but, rather, how many different types of "endogenous" neurotrophic factors can be identified and whether they are specific to different CNS regions (Cotman and Nieto-Sampedro, 1985). Likewise, there is now serious discussion and a growing literature on the possibility of neurogenesis in the CNS of even mature mammals (Bayer, 1985).

For the patient, it will probably matter little if one or many biological events are responsible for recovery of function or whether one can or cannot "prove" a particular hypothesis. The victims of brain or spinal cord injury only want to know if the quality of their lives will improve and if a return to a productive and independent existence will be possible. Over the last few decades the rapid advance of research in the neuropharmacology, anatomy, and behavior of brain-damaged subjects has moved us to the stage where we can begin to intervene effectively to modify some of the effects of CNS injuries. Thus, although there are sources of controversy that may persist, there also are reasons for optimism, especially when broader and perhaps more important questions are asked about brain damage and the potential for "meaningful" improvement.

REFERENCES

Akelaitis, A. J., 1940, A study of gnosis, praxis and language following partial and complete section of the corpus callosum, *Trans. Am. Neurol. Assoc.* **66**:182–185.

Akelaitis, A. J., 1941a, Psychobiological studies following section of the corpus callosum, *Am. J. Psychiatry* **97**:1147–1157.

Akelaitis, A. J., 1941b, Studies on the corpus callosum. II, *Arch. Neurol. Psychiatry* **45**:788.

Akelaitis, A. J., 1941c, Studies on the corpus callosum. VIII, *Am. J. Psychiatry* **98**:409–414.

Akelaitis, A. J., 1942a, Studies on the corpus callosum. V., *Arch. Neurol. Psychiatry* **47**:971–1008.

Akelaitis, A. J., 1942b, Studies on the corpus callosum. VI. Orientation (temporal–spatial gnosis) following section of the corpus callosum, *Arch. Neurol. Psychiatry* **48**:914–937.

Akelaitis, A. J., Risteen, W. A., and van Wagenen, W. P., 1941, A contribution to the study of dyspraxia and apraxia following partial and complete section of the corpus callosum, *Trans. Am. Neurol. Assoc.* **67**:75–78.

Akelaitis, A. J., Risteen, W. A., and van Wagenen, W. P., 1942, Studies on the corpus callosum. III. A contribution to the study of dyspraxia following partial and complete section of the corpus callosum, *Arch. Neurol. Psychiatry* **48**:914–937.

Akelaitis, A. J., Risteen, W. A., and van Wagenen, W. P., 1943, Studies on the corpus callosum. IX. Relation of the grasp reflex to section of the corpus callosum, *Arch. Neurol. Psychiatry* **49**:820–825.

Attella, M., Nattinville, A., and Stein, D. G., 1987, Hormonal state affects recovery from frontal cortex lesions in adult female rats, *Behav. Neural. Biol.* **48**:352–367.

Bayer, S. A., 1985, Neuron production in the hippocampus and olfactory bulb of the adult rat brain: Addition or replacement? *Ann. N.Y. Acad. Sci.* **457**:163–172.

Broca, P., 1865, Sur la siege de la faculte de language articule, *Bull. Soc. Anthropol.* **6**:337–393.

Cotman, C. W., and Nadler, J. V., 1978, Reactive synaptogenesis in the hippocampus, in: *Neuronal Plasticity* (C. W. Cotman, ed.), Raven Press, New York, pp. 227–271.

Cotman, C., and Nieto-Sampedro, M., 1985, Progress in facilitating the recovery of function after central nervous system trauma, *Ann. N.Y. Acad. Sci.* **457**:83–104.

Finger, S., and Almli, C. R., 1985, Brain damage and neuroplasticity: Mechanisms of recovery or development? *Brain Res. Rev.* **10**:177–186.

Finger, S., and Stein, D. G., 1982, *Brain Damage and Recovery*, Academic Press, New York.

Flourens, J.-P.-M., 1842, Recherches experimentales sur les proprietes et les fonctions du systeme nerveux dans les animaux vertebres, 2nd ed., Balliere, Paris.

Gazzaniga, M. S., Bogen, J. E., and Sperry, R. W., 1965, Some functional effects of sectioning the cerebral commissures in man. *Proc. Natl. Acad. Sci. U.S.A.* **48**:1765–1796.

Geschwind, N., 1985, Mechanisms of change after brain lesions, *Ann. N.Y. Acad. Sci.* **457**:1–12.

Goldman, P. S., 1974, An alternative to developmental plasticity: Heterology of CNS structures in infants and adults, in: *Plasticity and Recovery of Function in the Central Nervous System* (D. G. Stein, J. J. Rosen, and N. Butters, eds.), Academic Press, New York, pp. 149–174.

Goldstein, M. N., and Joynt, R. J., 1969, Long-term follow-up of a callosal-sectioned patient, *Arch. Neurol.* **10**:96–102.

Isaacson, R. L., 1975, The myth of recovery from early brain damage, in: *Aberrant Development in Infancy,* (N. R. Ellis, ed.), Lawrence Erlbaum, New York, pp. 1–25.

Mangold, R. M., Bell, J., Gruenthal, M., and Finger, S., 1981, Undernutrition and recovery from brain damage: A preliminary investigation, *Brain Res.* **230**:406–411.

Munk, H., 1877, Zur Physiologie der Grossirnrinde, *Berl. Klin. Wochenschr.* **14**:505–506.

Myers, R. E., 1955, Interocular transfer of pattern discrimination in cats following section of crossed optic fibers, *J. Comp. Physiol. Psychol.* **48**:470–473.

Myers, R. E., 1956, Functions of corpus callosum in interocular transfer, *Brain* **79**:358–363.

Myers, R. E., 1959, Interhemispheric communication through the corpus callsoum: Limitations under conditions of conflict, *J. Comp. Physiol. Psychol.* **52**:6–9.

Myers, R. E., 1961, Corpus callosum and visual gnosis, in: *Brain Mechanisms and Learning* (A. Fessard, R. W. Gerard, J. Konorski, and J. F. Delafresnaye, eds.), Blackwell Scientific, Oxford, pp. 481–505.

Myers, R. E., 1965, The neocortical commissures and interhemispheric transmission of information, in: *Functions of the Corpus Callosum* (E. G. Ettlinger, ed.), Churchill, London, pp. 1–17.

Nance, D. W., 1984, Sex-steroid-induced alterations in the behavioral effects of brain damage, in: *Early Brain Damage,* Volume 2, *Neurobiology and Behavior* (S. Finger and C. R. Almli, eds.), Academic Press, Orlando, FL, pp. 313–325.

Raichle, M. E., 1982, Restoration of function, in: *Repair and Regeneration of the Nervous System* (J. G. Nicholls, ed.), Springer-Verlag, New York, pp. 383–397.

Riese, W., 1948, Aphasia in brain tumors, *Confin. Neurol.* **9**:64–79.

Schoenfeld, T. A., and Hamilton, L. W., 1977, Secondary brain changes following lesions: A new paradigm for lesion experimentation, *Physiol. Behav.* **18**:951–967.

Sperry, R. W., 1961, Cerebral organization and behavior, *Science* **133**:1749–1757.

Sperry, R. W., 1964, The great cerebral commissure, *Sci. Am.* **210**:42–52.

Stein, D. G., and Firl, A. C., 1976, Brain damage and reorganization of function in old age, *Exp. Neurol.* **52**:157–167.

van Wagenen, W. P., and Herren, R. Y., 1940, Surgical division of commissural pathways in the corpus callosum: Relation to spread of an epileptic attack, *Arch. Neurol. Psychiatry* **44**:740–759.

Will, B., and Eclancher, F., 1984, Early brain damage and early environment, in: *Early Brain Damage,* Volume 2, *Neurobiology and Behavior* (S. Finger and C. R. Almli, eds.), Academic Press, Orlando, FL, pp. 349–367.

Index